Clean-In-Place for Biopharmaceutical Processes

DRUGS AND THE PHARMACEUTICAL SCIENCES
A Series of Textbooks and Monographs

1. Pharmacokinetics, *Milo Gibaldi and Donald Perrier*
2. Good Manufacturing Practices for Pharmaceuticals: A Plan for Total Quality Control, *Sidney H. Willig, Murray M. Tuckerman, and William S. Hitchings IV*
3. Microencapsulation, *edited by J. R. Nixon*
4. Drug Metabolism: Chemical and Biochemical Aspects, *Bernard Testa and Peter Jenner*
5. New Drugs: Discovery and Development, *edited by Alan A. Rubin*
6. Sustained and Controlled Release Drug Delivery Systems, *edited by Joseph R. Robinson*
7. Modern Pharmaceutics, *edited by Gilbert S. Banker and Christopher T. Rhodes*
8. Prescription Drugs in Short Supply: Case Histories, *Michael A. Schwartz*
9. Activated Charcoal: Antidotal and Other Medical Uses, *David O. Cooney*
10. Concepts in Drug Metabolism (in two parts), *edited by Peter Jenner and Bernard Testa*
11. Pharmaceutical Analysis: Modern Methods (in two parts), *edited by James W. Munson*
12. Techniques of Solubilization of Drugs, *edited by Samuel H. Yalkowsky*
13. Orphan Drugs, *edited by Fred E. Karch*
14. Novel Drug Delivery Systems: Fundamentals, Developmental Concepts, Biomedical Assessments, *Yie W. Chien*
15. Pharmacokinetics: Second Edition, Revised and Expanded, *Milo Gibaldi and Donald Perrier*
16. Good Manufacturing Practices for Pharmaceuticals: A Plan for Total Quality Control, Second Edition, Revised and Expanded, *Sidney H. Willig, Murray M. Tuckerman, and William S. Hitchings IV*
17. Formulation of Veterinary Dosage Forms, *edited by Jack Blodinger*
18. Dermatological Formulations: Percutaneous Absorption, *Brian W. Barry*
19. The Clinical Research Process in the Pharmaceutical Industry, *edited by Gary M. Matoren*
20. Microencapsulation and Related Drug Processes, *Patrick B. Deasy*
21. Drugs and Nutrients: The Interactive Effects, *edited by Daphne A. Roe and T. Colin Campbell*
22. Biotechnology of Industrial Antibiotics, *Erick J. Vandamme*
23. Pharmaceutical Process Validation, *edited by Bernard T. Loftus and Robert A. Nash*
24. Anticancer and Interferon Agents: Synthesis and Properties, *edited by Raphael M. Ottenbrite and George B. Butler*
25. Pharmaceutical Statistics: Practical and Clinical Applications, *Sanford Bolton*

26. Drug Dynamics for Analytical, Clinical, and Biological Chemists, *Benjamin J. Gudzinowicz, Burrows T. Younkin, Jr., and Michael J. Gudzinowicz*
27. Modern Analysis of Antibiotics, *edited by Adjoran Aszalos*
28. Solubility and Related Properties, *Kenneth C. James*
29. Controlled Drug Delivery: Fundamentals and Applications, Second Edition, Revised and Expanded, *edited by Joseph R. Robinson and Vincent H. Lee*
30. New Drug Approval Process: Clinical and Regulatory Management, *edited by Richard A. Guarino*
31. Transdermal Controlled Systemic Medications, *edited by Yie W. Chien*
32. Drug Delivery Devices: Fundamentals and Applications, *edited by Praveen Tyle*
33. Pharmacokinetics: Regulatory • Industrial • Academic Perspectives, *edited by Peter G. Welling and Francis L. S. Tse*
34. Clinical Drug Trials and Tribulations, *edited by Allen E. Cato*
35. Transdermal Drug Delivery: Developmental Issues and Research Initiatives, *edited by Jonathan Hadgraft and Richard H. Guy*
36. Aqueous Polymeric Coatings for Pharmaceutical Dosage Forms, *edited by James W. McGinity*
37. Pharmaceutical Pelletization Technology, *edited by Isaac Ghebre-Sellassie*
38. Good Laboratory Practice Regulations, *edited by Allen F. Hirsch*
39. Nasal Systemic Drug Delivery, *Yie W. Chien, Kenneth S. E. Su, and Shyi-Feu Chang*
40. Modern Pharmaceutics: Second Edition, Revised and Expanded, *edited by Gilbert S. Banker and Christopher T. Rhodes*
41. Specialized Drug Delivery Systems: Manufacturing and Production Technology, *edited by Praveen Tyle*
42. Topical Drug Delivery Formulations, *edited by David W. Osborne and Anton H. Amann*
43. Drug Stability: Principles and Practices, *Jens T. Carstensen*
44. Pharmaceutical Statistics: Practical and Clinical Applications, Second Edition, Revised and Expanded, *Sanford Bolton*
45. Biodegradable Polymers as Drug Delivery Systems, *edited by Mark Chasin and Robert Langer*
46. Preclinical Drug Disposition: A Laboratory Handbook, *Francis L. S. Tse and James J. Jaffe*
47. HPLC in the Pharmaceutical Industry, *edited by Godwin W. Fong and Stanley K. Lam*
48. Pharmaceutical Bioequivalence, *edited by Peter G. Welling, Francis L. S. Tse, and Shrikant V. Dinghe*
49. Pharmaceutical Dissolution Testing, *Umesh V. Banakar*
50. Novel Drug Delivery Systems: Second Edition, Revised and Expanded, *Yie W. Chien*

51. Managing the Clinical Drug Development Process, *David M. Cocchetto and Ronald V. Nardi*
52. Good Manufacturing Practices for Pharmaceuticals: A Plan for Total Quality Control, Third Edition, *edited by Sidney H. Willig and James R. Stoker*
53. Prodrugs: Topical and Ocular Drug Delivery, *edited by Kenneth B. Sloan*
54. Pharmaceutical Inhalation Aerosol Technology, *edited by Anthony J. Hickey*
55. Radiopharmaceuticals: Chemistry and Pharmacology, *edited by Adrian D. Nunn*
56. New Drug Approval Process: Second Edition, Revised and Expanded, *edited by Richard A. Guarino*
57. Pharmaceutical Process Validation: Second Edition, Revised and Expanded, *edited by Ira R. Berry and Robert A. Nash*
58. Ophthalmic Drug Delivery Systems, *edited by Ashim K. Mitra*
59. Pharmaceutical Skin Penetration Enhancement, *edited by Kenneth A. Walters and Jonathan Hadgraft*
60. Colonic Drug Absorption and Metabolism, *edited by Peter R. Bieck*
61. Pharmaceutical Particulate Carriers: Therapeutic Applications, *edited by Alain Rolland*
62. Drug Permeation Enhancement: Theory and Applications, *edited by Dean S. Hsieh*
63. Glycopeptide Antibiotics, *edited by Ramakrishnan Nagarajan*
64. Achieving Sterility in Medical and Pharmaceutical Products, *Nigel A. Halls*
65. Multiparticulate Oral Drug Delivery, *edited by Isaac Ghebre-Sellassie*
66. Colloidal Drug Delivery Systems, *edited by Jörg Kreuter*
67. Pharmacokinetics: Regulatory • Industrial • Academic Perspectives, Second Edition, *edited by Peter G. Welling and Francis L. S. Tse*
68. Drug Stability: Principles and Practices, Second Edition, Revised and Expanded, *Jens T. Carstensen*
69. Good Laboratory Practice Regulations: Second Edition, Revised and Expanded, *edited by Sandy Weinberg*
70. Physical Characterization of Pharmaceutical Solids, *edited by Harry G. Brittain*
71. Pharmaceutical Powder Compaction Technology, *edited by Göran Alderborn and Christer Nyström*
72. Modern Pharmaceutics: Third Edition, Revised and Expanded, *edited by Gilbert S. Banker and Christopher T. Rhodes*
73. Microencapsulation: Methods and Industrial Applications, *edited by Simon Benita*
74. Oral Mucosal Drug Delivery, *edited by Michael J. Rathbone*
75. Clinical Research in Pharmaceutical Development, *edited by Barry Bleidt and Michael Montagne*

76. The Drug Development Process: Increasing Efficiency and Cost Effectiveness, *edited by Peter G. Welling, Louis Lasagna, and Umesh V. Banakar*

77. Microparticulate Systems for the Delivery of Proteins and Vaccines, *edited by Smadar Cohen and Howard Bernstein*

78. Good Manufacturing Practices for Pharmaceuticals: A Plan for Total Quality Control, Fourth Edition, Revised and Expanded, *Sidney H. Willig and James R. Stoker*

79. Aqueous Polymeric Coatings for Pharmaceutical Dosage Forms: Second Edition, Revised and Expanded, *edited by James W. McGinity*

80. Pharmaceutical Statistics: Practical and Clinical Applications, Third Edition, *Sanford Bolton*

81. Handbook of Pharmaceutical Granulation Technology, *edited by Dilip M. Parikh*

82. Biotechnology of Antibiotics: Second Edition, Revised and Expanded, *edited by William R. Strohl*

83. Mechanisms of Transdermal Drug Delivery, *edited by Russell O. Potts and Richard H. Guy*

84. Pharmaceutical Enzymes, *edited by Albert Lauwers and Simon Scharpé*

85. Development of Biopharmaceutical Parenteral Dosage Forms, *edited by John A. Bontempo*

86. Pharmaceutical Project Management, *edited by Tony Kennedy*

87. Drug Products for Clinical Trials: An International Guide to Formulation • Production • Quality Control, *edited by Donald C. Monkhouse and Christopher T. Rhodes*

88. Development and Formulation of Veterinary Dosage Forms: Second Edition, Revised and Expanded, *edited by Gregory E. Hardee and J. Desmond Baggot*

89. Receptor-Based Drug Design, *edited by Paul Leff*

90. Automation and Validation of Information in Pharmaceutical Processing, *edited by Joseph F. deSpautz*

91. Dermal Absorption and Toxicity Assessment, *edited by Michael S. Roberts and Kenneth A. Walters*

92. Pharmaceutical Experimental Design, *Gareth A. Lewis, Didier Mathieu, and Roger Phan-Tan-Luu*

93. Preparing for FDA Pre-Approval Inspections, *edited by Martin D. Hynes III*

94. Pharmaceutical Excipients: Characterization by IR, Raman, and NMR Spectroscopy, *David E. Bugay and W. Paul Findlay*

95. Polymorphism in Pharmaceutical Solids, *edited by Harry G. Brittain*

96. Freeze-Drying/Lyophilization of Pharmaceutical and Biological Products, *edited by Louis Rey and Joan C. May*

97. Percutaneous Absorption: Drugs–Cosmetics–Mechanisms–Methodology, Third Edition, Revised and Expanded, *edited by Robert L. Bronaugh and Howard I. Maibach*

98. Bioadhesive Drug Delivery Systems: Fundamentals, Novel Approaches, and Development, *edited by Edith Mathiowitz, Donald E. Chickering III, and Claus-Michael Lehr*

99. Protein Formulation and Delivery, *edited by Eugene J. McNally*

100. New Drug Approval Process: Third Edition, The Global Challenge, *edited by Richard A. Guarino*

101. Peptide and Protein Drug Analysis, *edited by Ronald E. Reid*

102. Transport Processes in Pharmaceutical Systems, *edited by Gordon L. Amidon, Ping I. Lee, and Elizabeth M. Topp*

103. Excipient Toxicity and Safety, *edited by Myra L. Weiner and Lois A. Kotkoskie*

104. The Clinical Audit in Pharmaceutical Development, *edited by Michael R. Hamrell*

105. Pharmaceutical Emulsions and Suspensions, *edited by Francoise Nielloud and Gilberte Marti-Mestres*

106. Oral Drug Absorption: Prediction and Assessment, *edited by Jennifer B. Dressman and Hans Lennernäs*

107. Drug Stability: Principles and Practices, Third Edition, Revised and Expanded, *edited by Jens T. Carstensen and C. T. Rhodes*

108. Containment in the Pharmaceutical Industry, *edited by James P. Wood*

109. Good Manufacturing Practices for Pharmaceuticals: A Plan for Total Quality Control from Manufacturer to Consumer, Fifth Edition, Revised and Expanded, *Sidney H. Willig*

110. Advanced Pharmaceutical Solids, *Jens T. Carstensen*

111. Endotoxins: Pyrogens, LAL Testing, and Depyrogenation, Second Edition, Revised and Expanded, *Kevin L. Williams*

112. Pharmaceutical Process Engineering, *Anthony J. Hickey and David Ganderton*

113. Pharmacogenomics, *edited by Werner Kalow, Urs A. Meyer and Rachel F. Tyndale*

114. Handbook of Drug Screening, *edited by Ramakrishna Seethala and Prabhavathi B. Fernandes*

115. Drug Targeting Technology: Physical • Chemical • Biological Methods, *edited by Hans Schreier*

116. Drug–Drug Interactions, *edited by A. David Rodrigues*

117. Handbook of Pharmaceutical Analysis, *edited by Lena Ohannesian and Anthony J. Streeter*

118. Pharmaceutical Process Scale-Up, *edited by Michael Levin*

119. Dermatological and Transdermal Formulations, *edited by Kenneth A. Walters*

120. Clinical Drug Trials and Tribulations: Second Edition, Revised and Expanded, *edited by Allen Cato, Lynda Sutton, and Allen Cato III*

121. Modern Pharmaceutics: Fourth Edition, Revised and Expanded, *edited by Gilbert S. Banker and Christopher T. Rhodes*

122. Surfactants and Polymers in Drug Delivery, *Martin Malmsten*

123. Transdermal Drug Delivery: Second Edition, Revised and Expanded, *edited by Richard H. Guy and Jonathan Hadgraft*

124. Good Laboratory Practice Regulations: Second Edition, Revised and Expanded, *edited by Sandy Weinberg*

125. Parenteral Quality Control: Sterility, Pyrogen, Particulate, and Package Integrity Testing: Third Edition, Revised and Expanded, *Michael J. Akers, Daniel S. Larrimore, and Dana Morton Guazzo*

126. Modified-Release Drug Delivery Technology, *edited by Michael J. Rathbone, Jonathan Hadgraft, and Michael S. Roberts*

127. Simulation for Designing Clinical Trials: A Pharmacokinetic-Pharmacodynamic Modeling Perspective, *edited by Hui C. Kimko and Stephen B. Duffull*

128. Affinity Capillary Electrophoresis in Pharmaceutics and Biopharmaceutics, *edited by Reinhard H. H. Neubert and Hans-Hermann Rüttinger*

129. Pharmaceutical Process Validation: An International Third Edition, Revised and Expanded, *edited by Robert A. Nash and Alfred H. Wachter*

130. Ophthalmic Drug Delivery Systems: Second Edition, Revised and Expanded, *edited by Ashim K. Mitra*

131. Pharmaceutical Gene Delivery Systems, *edited by Alain Rolland and Sean M. Sullivan*

132. Biomarkers in Clinical Drug Development, *edited by John C. Bloom and Robert A. Dean*

133. Pharmaceutical Extrusion Technology, *edited by Isaac Ghebre-Sellassie and Charles Martin*

134. Pharmaceutical Inhalation Aerosol Technology: Second Edition, Revised and Expanded, *edited by Anthony J. Hickey*

135. Pharmaceutical Statistics: Practical and Clinical Applications, Fourth Edition, *Sanford Bolton and Charles Bon*

136. Compliance Handbook for Pharmaceuticals, Medical Devices, and Biologics, *edited by Carmen Medina*

137. Freeze-Drying/Lyophilization of Pharmaceutical and Biological Products: Second Edition, Revised and Expanded, *edited by Louis Rey and Joan C. May*

138. Supercritical Fluid Technology for Drug Product Development, *edited by Peter York, Uday B. Kompella, and Boris Y. Shekunov*

139. New Drug Approval Process: Fourth Edition, Accelerating Global Registrations, *edited by Richard A. Guarino*

140. Microbial Contamination Control in Parenteral Manufacturing, *edited by Kevin L. Williams*

141. New Drug Development: Regulatory Paradigms for Clinical Pharmacology and Biopharmaceutics, *edited by Chandrahas G. Sahajwalla*

142. Microbial Contamination Control in the Pharmaceutical Industry, *edited by Luis Jimenez*
143. Generic Drug Product Development: Solid Oral Dosage Forms, *edited by Leon Shargel and Izzy Kanfer*
144. Introduction to the Pharmaceutical Regulatory Process, *edited by Ira R. Berry*
145. Drug Delivery to the Oral Cavity: Molecules to Market, *edited by Tapash K. Ghosh and William R. Pfister*
146. Good Design Practices for GMP Pharmaceutical Facilities, *edited by Andrew Signore and Terry Jacobs*
147. Drug Products for Clinical Trials, Second Edition, *edited by Donald Monkhouse, Charles Carney, and Jim Clark*
148. Polymeric Drug Delivery Systems, *edited by Glen S. Kwon*
149. Injectable Dispersed Systems: Formulation, Processing, and Performance, *edited by Diane J. Burgess*
150. Laboratory Auditing for Quality and Regulatory Compliance, *Donald Singer, Raluca-Ioana Stefan, and Jacobus van Staden*
151. Active Pharmaceutical Ingredients: Development, Manufacturing, and Regulation, *edited by Stanley Nusim*
152. Preclinical Drug Development, *edited by Mark C. Rogge and David R. Taft*
153. Pharmaceutical Stress Testing: Predicting Drug Degradation, *edited by Steven W. Baertschi*
154. Handbook of Pharmaceutical Granulation Technology: Second Edition, *edited by Dilip M. Parikh*
155. Percutaneous Absorption: Drugs–Cosmetics–Mechanisms–Methodology, Fourth Edition, *edited by Robert L. Bronaugh and Howard I. Maibach*
156. Pharmacogenomics: Second Edition, *edited by Werner Kalow, Urs A. Meyer and Rachel F. Tyndale*
157. Pharmaceutical Process Scale-Up, Second Edition, *edited by Michael Levin*
158. Microencapsulation: Methods and Industrial Applications, Second Edition, *edited by Simon Benita*
159. Nanoparticle Technology for Drug Delivery, *edited by Ram B. Gupta and Uday B. Kompella*
160. Spectroscopy of Pharmaceutical Solids, *edited by Harry G. Brittain*
161. Dose Optimization in Drug Development, *edited by Rajesh Krishna*
162. Herbal Supplements-Drug Interactions: Scientific and Regulatory Perspectives, *edited by Y. W. Francis Lam, Shiew-Mei Huang, and Stephen D. Hall*
163. Pharmaceutical Photostability and Stabilization Technology, *edited by Joseph T. Piechocki and Karl Thoma*
164. Environmental Monitoring for Cleanrooms and Controlled Environments, *edited by Anne Marie Dixon*

165. Pharmaceutical Product Development: In Vitro-In Vivo Correlation, *edited by Dakshina Murthy Chilukuri, Gangadhar Sunkara, and David Young*

166. Nanoparticulate Drug Delivery Systems, *edited by Deepak Thassu, Michel Deleers, and Yashwant Pathak*

167. Endotoxins: Pyrogens, LAL Testing and Depyrogenation, Third Edition, *edited by Kevin L. Williams*

168. Good Laboratory Practice Regulations, Fourth Edition, *edited by Anne Sandy Weinberg*

169. Good Manufacturing Practices for Pharmaceuticals, Sixth Edition, *edited by Joseph D. Nally*

170. Oral Lipid-based Formulations: Enhancing the Bioavailability of Poorly Water-Soluble Drugs, *edited by David J. Hauss*

171. Handbook of Bioequivalence Testing, *edited by Sarfaraz K. Niazi*

172. Advanced Drug Formulation Design to Optimize Therapeutic Outcomes, *edited by Robert O. Williams III, David R. Taft, and Jason T. McConville*

173. Clean-in-Place for Biopharmaceutical Processes, *edited by Dale A. Seiberling*

Clean-In-Place for Biopharmaceutical Processes

edited by

Dale A. Seiberling

Vice President - Electrol Specialties Company
South Beloit, Illinois, USA

informa
healthcare

New York London

Informa Healthcare USA, Inc.
52 Vanderbilt Avenue
New York, NY 10017

Printed in the United States of America on acid-free paper
10 9 8 7 6 5 4 3 2 1

International Standard Book Number-10: 0-8493-4069-1 (Hardcover)
International Standard Book Number-13: 978-0-8493-4069-7 (Hardcover)

Library of Congress Cataloging-in-Publication Data

Clean-in-place for biopharmaceutical processes / edited by Dale A. Seiberling.
p. ; cm. – (Drugs and the pharmaceutical sciences ; v. 173)
Includes bibliographical references.
ISBN-13: 978-0-8493-4069-7 (hardcover : alk. paper)
ISBN-10: 0-8493-4069-1 (hardcover : alk. paper)
1. Drugs–Sterilization.
2. Clean rooms.
3. Pharmaceutical technology–Equipment and supplies–Cleaning.
4. Pharmaceutical technology–Quality control.
5. Pharmaceutical technology–Standards.
I. Seiberling, Dale A. II. Series.
[DNLM: 1. Technology, Pharmaceutical–methods. 2. Biotechnology.
3. Disinfection–methods. 4. Drug Contamination–prevention & control.
5. Equipment Contamination–prevention & control. 6. Technology, Pharmaceutical–standards. W1 DR893B v.173 2007 / QV 778 C623 2007]

RS199.S73C54 2007
615′.7–dc22 2007023339

Visit the Informa Web site at
www.informa.com

and the Informa Healthcare Web site at
www.informahealthcare.com

Preface

Clean-in-Place for Biopharmaceutical Processes is intended to be a source of information for personnel from all disciplines who are involved in the manufacture of a pharmaceutical product—from the scientist who creates the drug to those responsible for the design, construction, validation, and operation and maintenance of a facility, as well as for those responsible for regulation.

The objective of this book is to combine the experience and knowledge of experts familiar with the many items of equipment required for the pharmaceutical process with the knowledge of people who have had long experience in the successful application of clean-in-place technology to a variety of non-biopharmaceutical and biopharmaceutical processes. The unit operations of the process are analyzed with respect to whether or not clean-in-place is a possible or preferred method of cleaning, and examples of successful applications are included. While each new user of clean-in-place has a tendency to "reinvent the wheel," this book is an attempt to show that, for the most part, there are no new problems; rather there are problems solved previously, in a different application. However, considerable effort has been made to recognize and define innovative and emerging technology. The established criteria for successful clean-in-place is well defined, explained, and illustrated. A major goal of this book is to guide all readers to the development of the need for *clean-in-place* rather than *clean-in-part*.

Clean-in-place technology has been developing constantly for more than 50 years in dairy, beverage, and food applications for cleaning processes that produce fluid, semi-fluid, and dry granular products. Early development in these industries was generally guided by a small group of user personnel with intimate knowledge of the chemistry, bacteriology, and cleaning needs of the product, who were willing and able to seek outside support regarding the application of developing clean-in-place technology, sometimes referred to as "in-place cleaning" or "recirculation cleaning" in early literature. Much of the technical know-how was acquired during the early period of retrofitting clean-in-place to existing processes, primarily in dairy and brewing applications. A few well-qualified vendors of clean-in-place systems, sprays, and associated components served the needs of the entire market with "off-the-shelf" components and systems that were quite similar in the operating characteristics of flow and pressure. The regulatory agencies concerned with the application of this new technology were participants in the projects and members of the 3-A Sanitary Standards committee that quickly recognized the need for standards and practices to guide the application of the developing technology.

As the biopharmaceutical industry became clean-in-place users, the major application was to new projects. Most of the manufacturing companies lacked architects, engineers, and construction managers, and the overall design responsibility was necessarily delegated to large engineering companies. The selected firm often determined how clean-in-place would be applied, developing voluminous documentation describing its design, fabrication, and operation. Each new project was treated as a unique, different, and demanding application. During this same period, the industry recognized the need for "validatable cleaning" and clean-in-place was quickly recognized as a useful tool in meeting that need. The validation requirements for biopharmaceutical processes quickly spread to include all components of the clean-in-place system, and this has exacerbated the cost and complexity of clean-in-place in today's industry.

Clean-in-place technology is a powerful cleaning process when applied to a well-engineered "CIPable" process, and properly controlled and monitored to achieve the required combination of time, temperature, and concentration for the specific circuit and soil load encountered. The desired results are best assured by a combination of engineering design and end point control. A focus on program performance alone rather than the intricate details of how that program is made to occur has been demonstrated to be effective in dairy, food, and beverage applications for four decades. A change in the criteria for assuring validated cleaning—to apply what is necessary, rather than what is possible—could have great impact on the cost of the hardware and software required.

Dale A. Seiberling

Acknowledgments

During the final years of our life together, my first wife Jean often suggested that I give up the excitement of travel, new projects, and new problems and stay at home and write a book. She supported and encouraged my second love through fifty-four years of continuous activity in a variety of industries. Her ability and willingness to manage the family and home, combined with my use of personal aircraft for expedited travel, enabled my active participation in many phases of the development and commercialization of a technology that has fascinated me since I first wrote a term paper on the subject in 1949. I dedicate this book to Jean F. Seiberling. I also want to thank Informa Healthcare for the opportunity to share a lifetime of wonderfully satisfying experiences. My life's work has touched all peoples in the developed nations of the world in a manner they will never know, let alone comprehend. It is a privilege to help expand the successful application of clean-in-place to biopharmaceutical processes.

Contents

Contributors

Barry J. Andersen Seiberling Associates, Inc., Beloit, Wisconsin, U.S.A.

Jay C. Ankers LifeTek Solutions, Inc., Blue Bell, Pennsylvania, U.S.A.

Gerald J. Cerulli Integrated Project Services, Somerset, New Jersey, U.S.A.

Lyle W. Clem Electrol Specialties Company (ESC), South Beloit, Illinois, U.S.A.

Simon E. J. Forder JM Hyde Consulting, Inc., San Francisco, California, U.S.A.

John W. Franks Electrol Specialties Company (ESC), South Beloit, Illinois, U.S.A.

David M. Greene Paulus, Sokolowski & Sartor (PS&S), LLC, Warren, New Jersey, U.S.A.

Joachim Höller Boehringer Ingelheim Austria GmbH, Vienna, Austria

Albrecht Killinger Uhde GmbH, Biotechnology Division, Leipzig, Germany

Charles Lankford PharmaSys, Inc., Cary, North Carolina, U.S.A.

Samuel F. Lebowitz Electrol Specialties Company (ESC), South Beloit, Illinois, U.S.A.

James P. Norton Eli Lilly and Company, Indianapolis, Indiana, U.S.A.

Linda Rauch CH2M Hill, Boston, Massachusetts, U.S.A.

Johannes R. Roebers West Coast Engineering Biogen-idec, Inc., Oceanside, California, U.S.A.

Dietmar Rosner Ecolab GmbH & Co. OHG, Düesseldorf, Germany

Sally J. Rush Seiberling Associates, Inc., Beloit, Wisconsin, U.S.A.

Dale A. Seiberling Electrol Specialties Company (ESC), South Beloit, Illinois, U.S.A.

1 Introduction and Historical Development

Dale A. Seiberling
Electrol Specialties Company (ESC), South Beloit, Illinois, U.S.A.

INTRODUCTION

Many technical reference books begin with a review of the development of the subject material. Your editor, and the author of this chapter, has followed that practice in many prior publications. However, clean-in-place (CIP) did not start and was not initially developed in the pharmaceutical or biotechnology industries. It began more than 55 years ago, on the dairy farm, as the means of cleaning the newly introduced Pyrex® (Corning Incorporated, Corning, New York) glass milking system pipelines. During the subsequent 15 years, it was adopted by the dairy, beverage, brewery, winery, and food processing users. Only during the past two decades, has CIP been widely adopted by the biopharmaceutical industries.

My recent teaching experience has suggested that the technical personnel of the industries to which this book is directed are more interested in the current CIP technology than its historical development and application. Therefore, *clean-in-place for biopharma ceutical processes* will start with a question, "What is CIP?" in the biopharmaceutical industry at this time? A generic two-tank process and transfer line will be discussed in depth to provide the reader with an overview of the technology and some direction to those chapters in which the full detail can be found.

But first, it is necessary to consider the broad subject of "cleaning." Batch manufacturing processes require cleaning as one of many manufacturing steps. Whereas cleaning has long been recognized to be an important procedure, until recent years it was performed manually, typically by the least experienced and least trained employees under little, if any, supervision. The equipment and supplies provided to accomplish cleaning were also, in many cases, substandard. Cleaning in the biopharmaceutical process, however, is as important as the production of the active ingredients, the formulation and filling of those ingredients, and the sterilize-in-place (SIP) of the process equipment. Effective cleaning is the most important precedent to SIP as sanitization/sterilization requires contact between steam and microorganisms that will not achieve the desired time-temperature conditions if product residue insulates and/or protects the microorganisms. And it is well understood that chemical residue contaminants must be removed as well. It is also well understood that sterile filth in a pharmaceutical product is no more desirable than unsterile filth.

Cleaning can be accomplished by disassembly of the equipment followed by manually washing and rinsing, or transfer of the parts to a cleaned-out-of-place (COP) system. Or, it can be accomplished "in-place," by the CIP procedure. Tunner (1), when writing about validation of manual cleaning, states, "The most important difference between manual and automated cleaning can be summarized as the human factor, with the inherent variability of personal training and commitment to quality." He cites DeBlanc and coworkers' (2) observation that "cleaning is not the last step of a batch manufacturing process and has no impact on the quality of the batch for which the cleaning is performed." Rather, it is the first step in the manufacture of the next product batch and can greatly impact the safety and efficacy. If the process owner, validation department, and, above all, the cleaning operators view cleaning as the first step of manufacturing, the importance of a properly designed, validated, and followed cleaning procedure is obvious. By comparison, Tunner recognizes that "The effectiveness of an automated cleaning process depends primarily on the equipment and program design, which is confirmed by proper validation and maintained by an effective maintenance program. Operator training, while important, is secondary to the process and equipment design."

In today's manufacturing processes, Roebers (3) suggests that there is a mandate to apply automated CIP, noting that "If you cannot clean your process equipment and piping in a robust, reproducible way, don't even think about making a biopharmaceutical product!"

WHAT IS CIP?

The acronym CIP refers to a complex technology that embraces: (*i*) processing equipment designed and fabricated to be CIPable, i.e., capable of being cleaned in place, (*ii*) permanently installed spray devices to the maximum extent possible, (*iii*) CIP supply (CIPS)/return piping (CIPR), (*iv*) a CIP skid with chemical feed equipment, and (*v*) a control system to run the CIP skid and deliver the cleaning solutions in the correct sequences at the required composition, temperature, pressure, and flow rate. The process control system must operate process equipment pumps, valves, and other appurtenances to ensure the efficacy of the CIP process. When applied to biopharmaceutical processes the general application is to clean the process equipment and piping, generally in sequence, following each period of use.

The train may be comprised of individual tanks or equipment items, or groups of different sized tanks of similar function in series, the common element of the CIP process being the transfer line from the first tank to the next. SIP logically follows CIP, and the design of a CIPable process fulfills the major needs for SIP also, via the addition of clean steam (CS) and condensate drain connections.

Components of a Generic CIPable Process

Figure 1 illustrates a typical two-tank train supported by the requisite CIP equipment and piping. The train could extend to include multiple systems to the maximum extent possible in the applicable CIP circuits. The train may be individual vessels in sequence, per this example, or several in parallel in sequence with others in parallel via multiport transfer panels (TPs). The major components are identified by number in Figure 1 and include the following.

Vessels

Tanks T1 and T2, (1) each fitted with permanently installed fixed spray devices, a vent filter, manway, agitator, and (not shown) a rupture disk or equivalent overpressure protection device for the American Society of Mechanical Engineers (ASME) code vessel. The vessel may include heat transfer surface in one or more zones, and is commonly insulated for process and sterilize-in-place (SIP) reasons.

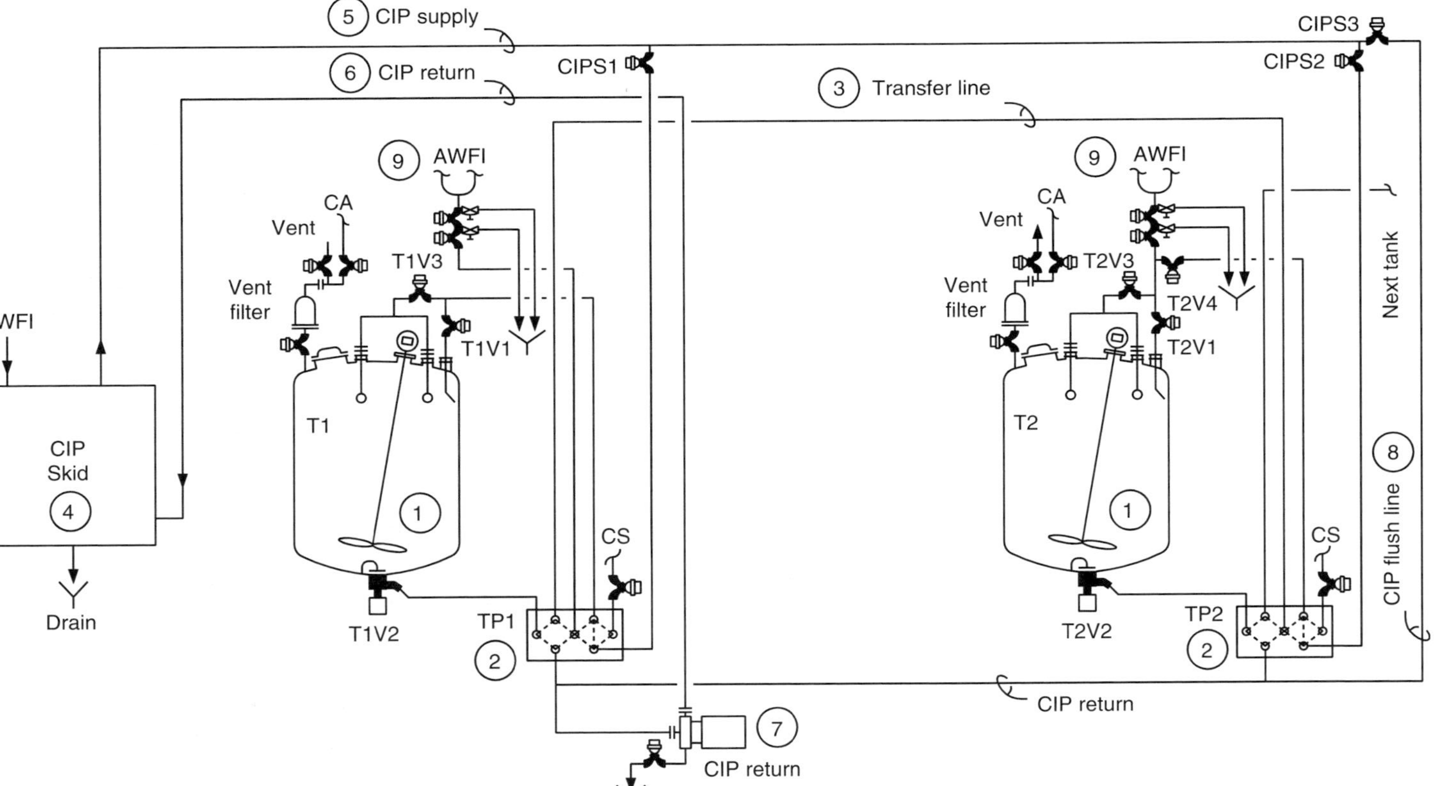

FIGURE 1 This generic two-tank process train illustrates a CIPable and SIPable design, with major components identified by number for reference in the narrative.

The vessel capacity may range from 30 L to 16 m^3 or more. A majority of CIP-cleaned processes are automatically controlled and the vessel and its dedicated piping will include air-operated valves to fill (T1V1 and T2V1), empty (T1V2 and T2V2), supply flush, wash and rinse solutions to the circuit (T1V3 and T2V3), and to subcircuits such as dip tubes and sample connections. Valves for the vent filter, sampling, etc., may be manually operated or automated, this decision being made on the basis of process automation requirements. An outlet valve of adequate size to accommodate CIP flush, wash and rinse solution flow rates is mandatory and a flat plate vortex breaker above the outlet opening is highly recommended to minimize air entrainment and prevent cavitation of any required return pumps. Additional criteria for CIPable design will follow later in this chapter.

Figure 2 is a photo of typical tank top piping as shown schematically in Figure 1 including a vent filter with a manual valve and a rupture disk. The two air-operated valves upper left control flow from a common line originating at a TP to either the inlet, or the spray, or both (for CIP) and are equivalent to T1V1 and T1V3 on T1 in the figure. Figure 3 shows the spray balls to either side behind the manway rim, and a magnetically driven agitator and permanently installed flat plate vortex breaker on the bottom. The selection and application of spray devices is described in chapter 9.

U-Bend Transfer Panels

Multi-port TPs (TP1 and TP2) shown as (3) are generally used to organize installation and support of product and CIPS/CIPR piping, and facilitate connections for required production, CIP and SIP operations. The design and application of TPs is described in depth in chapter 12. TPs require manual operations to

FIGURE 2 This view of a media prep vessel illustrates the high manway collar and also the rupture disc, vent filter, and level indicator nozzles that create clean-in-place spray coverage problems.

FIGURE 3 A view through the open manway shows the two sprayballs near the top, and the permanently installed vortex breaker and magnetically driven agitator on the bottom.

configure the equipment for process, CIP or SIP functions, and may be replaced or supplemented with additional air-operated valves for a higher degree of automation in large production processes. The TPs in Figure 1 would be considered to be "low-level" panels, as they are located so that the vessels drain to port 4 (Fig. 4). This figure illustrates the port arrangement for this "generic" panel, and the associated isometric view illustrates how a "standard" U-Bend would fit between

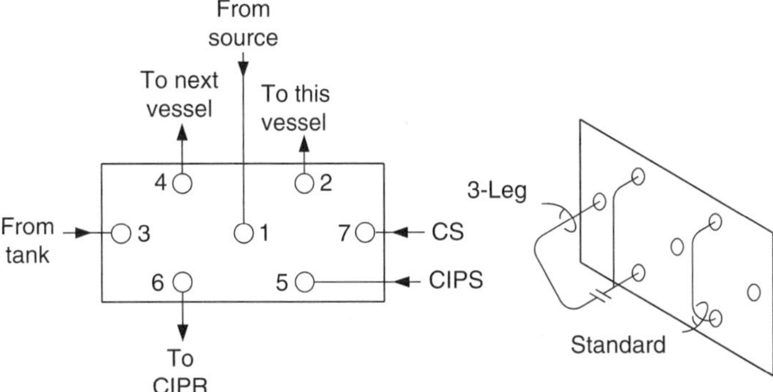

FIGURE 4 This detail of the low-level transfer panel shown on schematics in this chapter identifies the ports by number. Note the 3-Leg U-Bend. All U-Bends are of the same length, to permit full interchangeability. *Abbreviations*: CIPR, clean-in-place return; CIPS, clean-in-place supply.

any two ports. For purposes of this discussion, the 3-Leg U-Bend shown at the left side of the isometric view would connect any three ports, which is sometimes necessary for greater flexibility. The position of both types of U-Bends could be monitored by proximity sensors. Later references will be made to this figure for explanation of the use of the various ports.

TPs may be installed, supported, and accessed in many different ways. Alternatives include "through-the-wall" mounting as shown in Figure 5. The vessels in "gray space" behind the wall are accessed only from the clean room side, in this instance a high-level TP above the vessel head for the vent and product filters, and a low-level TP only partially visible behind the large filter housing. The design and application of TPs will be further discussed in chapter 13.

Transfer Line

Most early CIP systems in dairy, food, and beverage processes cleaned vessels and piping (lines) in separate circuits, as the vessels were often filled with product when the lines were cleaned on a daily or more frequent basis and the line circuits were generally large and complex, often including several arrays of air-operated valves. However, in many biopharmaceutical processes, the vessel and the associated transfer line (3) to the next vessel are available for cleaning at the same time following each sequential batch operation. Substantial savings in water, time, and chemicals can be made by cleaning a vessel in combination with the transfer line to the next process vessel. Since this line may be smaller than required for CIP flow, to reduce holdup volume, and because there is often no adequate CIP motivation in the process path beyond the vessel outlet, a preferred method is to

FIGURE 5 This view of a 1982 vintage sterile process shows U-Bend TPs installed through the wall between the clean room and the gray space in which the vessels and piping are located. Note use of TPs to mount filter housings. *Abbreviation*: TP, transfer panel.

clean the transfer line, in reverse flow, by sequencing flush, wash, and rinse solution from the spray CIP source to the transfer line on a repetitive, but intermittent, basis, i.e., 15 to 30 seconds of each minute. As the CIPS valve opens to the transfer line, the flow to the sprays is diminished, but continues at a reduced pressure and flow. The back pressure created by the sprays creates the driving force to cause flow through the transfer line and the supply flow will divide to create equal pressure drop through the two parallel paths for the periods of parallel flow. The flow rate and hence cleaning velocity in the transfer line can be easily estimated or calculated and/or measured. Though flow through this path is not continuous chemical cleaning action, continues during periods of no flow.

Automated CIP Skid

To ensure validatable CIP a fully automated CIP skid (4) is essential, including also the requisite chemical feed equipment, program control equipment, and instrumentation. Figure 1 provides a simple block diagram of one of several applicable CIP system configurations described in chapter 6, and supported by controls and instrumentation and chemical dosing equipment as described in chapters 7 and 8.

CIP Supply Piping

The CIP skid shown is interfaced to the two vessels via installation of permanently installed CIPS piping (5) to the TPs. CIPS piping is most commonly installed above the process vessels and TPs, and is commonly sloped to drain away from the spray.

CIPR Piping

Low-level return piping (6), preferably 15 to 18 in. (0.5 m) below the outlet of the largest vessel to improve return pump net positive suction head (NPSH) conditions with minimum vessel puddle will be installed from the TPs to a CIPR pump, or directly to the CIP skid if located below the process equipment.

CIPR Pump

Return pumps (7) may be used when the return flow must be elevated to high-level piping runs back to the CIP skid. Gravity return may also be used, and other methods of return flow motivation, will also be discussed in detail in chapter 10. Return pumps are generally fitted with casing drain valves and become the low-point drain for much of the circuit.

CIPR Flush Valve and Bypass Line

To eliminate dead legs in the CIPS and CIPR piping headers a small diameter CIPR flush line (8) is shown installed between the most distant ends of the CIPS and CIPR headers. This line also facilitates operation of the CIP Skid for test and maintenance purposes. Automatic control of valve CIPS3 enables brief repetitive pulses of flush, wash, and rinse solutions to partially bypass the spray devices and flush the full CIPS/R piping system. Deadlegs, their impact on CIP, and methods of elimination by piping design, will be further defined in chapters 10 and 13. This line and valve may also be used to bypass part of the CIPS/R flow required for proper line velocity when small outlet valves exist by intent, or more commonly, improper design and require vessels to be spray washed at a flow rate below that required for desired CIPS/R velocity.

AWFI (or Any Water) Supply

If the process requires the addition of product quality water to the process vessel, for product makeup, or perhaps through the sprays for rinsing following completion of the transfer, an ambient water for injection (AWFI) supply loop (9) may be connected to the vessel spray supply manifold. Two-ported diaphragm valves applied as shown in Figure 1 will isolate the AWFI loop from CIPS, and provide an AWFI sample port. The AWFI connection for initial batch makeup may alternatively be through the transfer panel TP1 to the first vessel (commonly used for mixing), and via a valve to the spray supply manifold on subsequent vessels in the train.

Operation of the Generic Two-Tank Process Train

Figure 6 will serve as the reference for description of the typical process required in the operation involving two tanks in a train connected by a transfer line.

Mixing and Transfer

As an example of mixing and transfer operations, consider T1 to be the initial mixing tank. TP1 would have a U-Bend between ports 1 and 2 for supply of AWFI from the facility loop via the isolation and sample valves shown. A U-Bend will also be required between ports 3 and 4 for the subsequent transfer and TP2 would requires a U-Bend between ports 1 and 2 for transfer to T2. AWFI would be supplied to fill the tank via path (F) under final control of valve T1V1, the quantity being controlled by a meter, load cells, or a probe (none shown). The vent filter vent valve would be open for the fill. The product ingredients (P) would then be added via the manway or by alternative means, and following agitation, transfer to T2 would be accomplished by first closing the vent filter vent valve and then applying pressure to T1 using compressed air (CA) (or other gas) through the vent filter. Valves T1V2

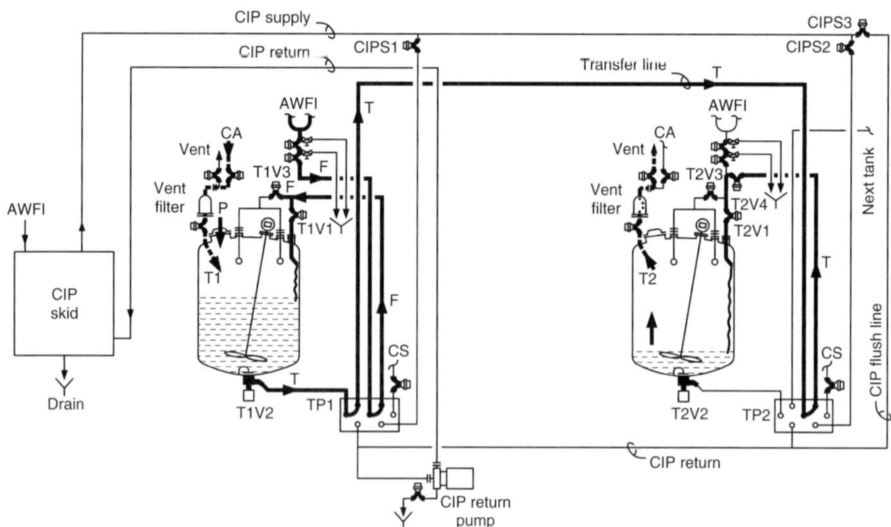

FIGURE 6 This schematic has been heavy-lined to define the movement of gases and the product during filling with AWFI, the addition of product (P) through the manway, and a top pressure motivated transfer from T1 to T2 (T). *Abbreviation*: AWFI, ambient water for injection.

and T2 V1 would be opened, and also the vent filter vent valve, to allow flow to commence from T1 to T2 in accordance with transfer path (T) arrow heads.

Product Rinse Forward

With the inlet line and sprays still connected to the AWFI supply, the loop valves and T1V3 would be opened briefly to flush the vessel via the spray device, thus recovering product from the vessel surfaces. If this step is necessary, the initial water volume will be reduced to allow for the dilution by this rinse volume.

CIP Cleaning of the Generic Two-Tank Process Train

Tank CIP

Following the transfer and AWFI rinse of T1 and the line, the TP1 U-Bend on port 2 would be relocated to the CIPS port 5 and the outlet line U-Bend on port 3 would be repositioned to port 6. The CIP system would then deliver flush, wash, and rinse solutions at the required conditions of time, temperature, and concentration, through the sprays, at a flow rate equivalent to 2.5 to 3.0 gpm/ft (30–35 lpm/m) of vessel circumference or 5 ft/sec (1.5 m/sec) in the CIPS/R piping, whichever is greatest. CIPS1 would direct solutions to TP1 and T1V3 would be open to the sprays for the full program, whereas T1V1 would be "pulsed" open for perhaps two to three seconds of each minute to clean the fill connection. The CIPR flush valve CIPS3 would also be "pulsed" a few seconds each minute to flush the CIPS header downstream of CIPS1 and the CIPR header upstream of TP1, thus assuring no "deadlegs" in the circuit. The flush, wash, and rinse solutions would follow the flow path designated by the (T) arrow heads in Figure 7 for the full duration of the CIP

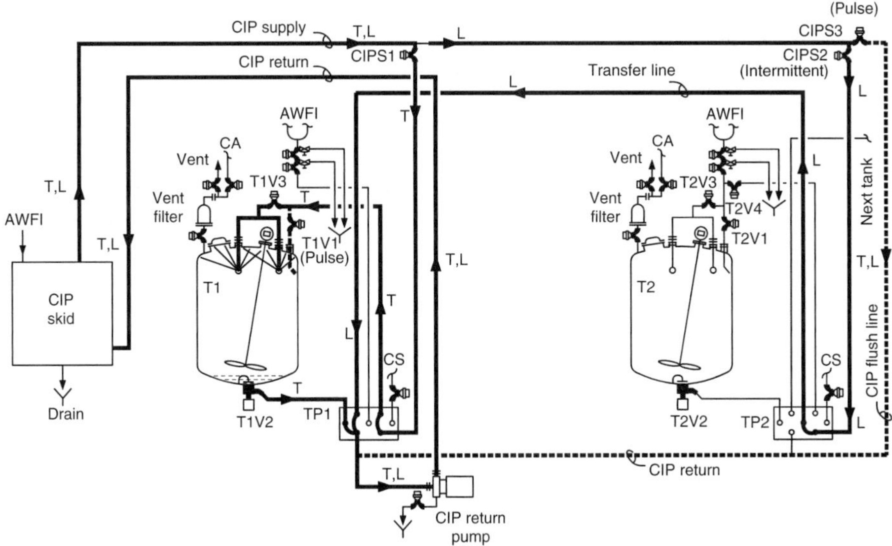

FIGURE 7 This schematic has been heavy lined to define the *tank* circuit flow path by arrow heads (T) and the *line* circuit flow path by arrow heads (L). The CIPR flush valve CIPS3 would be "pulsed" during both programs to clean the deadlegs of the CIPS/R headers. *Abbreviations*: AWFI, ambient water for injection; CA, compressed air; CIPR, CIP return; CIPS, CIP supply.

program, and through the "pulsed" paths by the arrow heads for brief intermittent periods. An air blow of the CIPS line will generally follow each program phase, after which, time will be provided for the return pump to evacuate the vessel. On completion of the program, low-point drain valves will open to assure drainage of the entire circuit.

Transfer Line CIP

Following completion of the tank CIP program, the TP2 U-Bend on port 1 would be relocated to the CIPS port 5 and the TP1 U-Bend on port 6 would be repositioned to port 4. The CIP system would deliver flush, wash, and rinse solutions at the required conditions of time, temperature, and concentration, through the piping path defined by arrow heads (L) in Figure 7, at a flow rate equivalent to or greater than 5 ft/sec through the largest diameter tubing in the circuit. CIPS2 would direct solutions to TP2 and the CIPR flush valve CIPS3 would also be "pulsed" a few seconds each minute to flush the CIPR header upstream of TP1, thus assuring no "deadlegs" in the circuit. The flush, wash, and rinse solutions would follow the flow path designated by the circuit arrow heads (L). When cleaning circuits with no vessel that accumulates a "holdup" volume, there is no need for an air blow of the CIPS line between each program phase, but the CIPS and CIPR lines will generally be evacuated by an air blow at the end of the program. On completion of the program, the low-point drain valves will open to assure drainage of the entire circuit.

Typical Cleaning Programs and Water Requirements

The effectiveness of mechanical/chemical cleaning is related to a number of variables including time, temperature, concentration, and physical action. Physical action is dependent on proper design and engineering, i.e., the selection and application of the correct sprays, supply and return pumps, and the sizing of CIPS/R and product piping to achieve the required flow velocity. There is no single "best way" to handle any particular cleaning program as the first objective must be to "do what is necessary to get the equipment clean" after which further adjustments giving consideration to limitations of temperature, time, or cleaning chemical cost may be completed. Dependency on exact or specific numbers (as part of the recommendation) is of no value *if* the equipment is still dirty upon completion of a cleaning cycle.

Four decades of experience have demonstrated that fat-, protein-, and carbohydrate-based soils encountered in most dairy, food, pharmaceutical, and biotechnology processes can be removed by one or a combination of several of the following treatments.

Prewash Rinse

Following completion of the batch operation and transfer, subsequent rinse forward, and reconfiguration of the process piping for CIP, any available cold water may be used to flush the remaining product from the equipment and piping surfaces. CIP skids are generally supplied with two different qualities or type of water [i.e., wash and rinse, soft and purified, reverse osmosis (RO) water, deionized water, AWFI, or hot water for injection (HWFI)]. The lowest cost water, or perhaps the lowest temperature water will be most often used for the prerinse, alkaline solution wash, and postrinse.

Alkaline Solution Wash

In its simplest form, this chemical solution may be nothing more than a mild solution of sodium hydroxide or potassium hydroxide, or a blended product which combines the base ingredient with other chemicals to enhance performance. Chemical concentrations may vary from as low as 800 to 1200 ppm of alkali for lightly soiled equipment to a maximum of 1% to 2% for heavily soiled equipment.

Cleaning temperatures are normally in the range of 70°F to 140°F (57–71°C), and exposure time (recirculation time at temperature) may vary from as little as 5 minutes to as much as 20 minutes, or more. Heavily soiled heat exchange equipment may require concentrations of 1.5% to 2% alkali and times of 45 to 60 minutes for effective results under all operating conditions. Lightly soiled equipment such as buffer prep tanks, or vessels used for highly soluble drugs may respond to rinsing alone.

Postwash Rinse

Following the solution wash, a minimum of softened water will be used to flush the soil and alkaline solution from the equipment surfaces. Neither sodium hydroxide nor potassium hydroxide is free rinsing, however, and whereas the simple check of conductivity at the discharge to drain may suggest that the alkaline material has been removed early in the rinse, sampling of the equipment surface may reveal a considerable alkali residual.

Acidified Rinse

The postwash rinse may be followed with a recirculated solution of soft or pure water lightly acidified with food grade phosphoric acid (or equivalent mild acid) to produce a pH of 5.5 to 6.0 (just slightly on the acid side of neutral). This solution, recirculated at the water supply temperature (no additional heating), will *neutralize* all traces of alkali residual on the equipment surfaces. Some users of CIP prefer to apply this step as an acid wash, at higher than suggested concentrations, and with the addition of heat for an increased period of time. Detailed discussion of chemical cleaning materials and programs is provided in chapters 4 and 5.

Post-Acid Rinse

The acidified rinse (or wash) will be followed with a rinse as described for the post-alkali wash.

Final Rinse

Pharmaceutical and biotechnology processes require removal of all traces of cleaning solutions from the equipment surface. This may be accomplished in one or two steps of rinsing straight through to drain with either purified water alone or soft water followed by purified water, to achieve the desired rinse test.

Typical Water Requirements

The volume of water required to prerinse a *piping circuit* is normally found to be 1.5 to 2 times the volume contained in that piping. To establish an understanding of water needs for CIP, consider the fact that 100 ft (30 m) of 1 in. (25 mm) diameter tubing will contain approximately 7 gal (26.5 L) and 100 ft (30 m) of 2 in. (50 mm) diameter tubing will contain approximately 14 gal (53 L). If the volume of the complete process piping circuit is, for example, 100 gal (380 L), then the total water requirement will be approximately 200 gal (760 L) for a pre-rinse plus 100 gal

(380 L) for the solution wash, plus 200 gal (760 L) for the post-rinse, plus 100 gal (380 L) for an acidified rinse, plus 200 gal (760 L) (perhaps more) for the final pure water rinse.

The total water requirement for a *spray cleaning program for a vessel* is related to (*i*) the spray delivery rate, (*ii*) the volume of the CIPS/CIPR, and (*iii*) the volume of the recirculation tank, plus (*iv*) the holdup volume required in the vessel being cleaned to achieve reliable recirculation. Prerinse and post-rinse times of 40 to 60 seconds are generally adequate. A final pure water rinse once through to the drain may require delivery at the spray design rate for two to three minutes, or more. A conservative estimate for the total water requirement for cleaning a tank at 80 gpm (300 lpm) in a system containing 200 ft (61 m) of 2 in. (50 mm) supply/return piping would include 80 gal (300 L) for the prerinse, 40 gal (150 L) for the solution wash, 80 gal (300 L) for the post-rinse, 40 gal (150 L) for the acidified rinse, and 240 to 360 gal (900–1350 L) for the final pure water rinse, for a single tank CIP unit which operates with no solution in the recirculation unit tank. Alternative multi-tank recirculating units may add an additional 100 to 150 gal (380–570 L) total in the solution tank and in the vessel being cleaned to achieve stable recirculation with pumped return. Deadlegs in the CIPS/R piping and excessive holdup in the vessel will dramatically increase the rinse volume required to achieve set-point resistivity.

Tank and Line CIP in Combination

Whereas vessels and lines are traditionally cleaned in separate circuits because of the different CIPS flow and pressure requirements, the careful consideration of the above numbers will reveal that the major amount of water used for the total program is to fill the CIPS/R piping to and from the circuit. For a CIPS/R length of 100 ft of 2 in. tubing, this will be 28 gal minimum. A very large vessel fitted with a vortex breaker may need only 3 to 5 gal more for reliable recirculation. A 50 ft transfer line of 1.5 in. diameter will contain only 3.5 gal of solution. The addition of the transfer line to the vessel circuit will increase the volume contained in the circuit by only 10% to 15% and washing the vessel and line in combination will effectively reduce total water for both circuits by 40% to 45%, as compared to washing them individually. There is generally no scheduling problem to wash the transfer line with the vessel as both are soiled and available for CIP at the same time following the transfer. The two individual tank and line circuits shown in Figure 6 can be combined by installing the 3-Leg U-Bend illustrated in Figure 4 on TP1, thus combining CIPR flow from the vessel (port 3) and the transfer line (port 4) to the CIPR pump (port 6). Tank T1 would be sprayed and the inlet line "pulsed" as described above for tank CIP, and CIPS2 would open for perhaps 20 to 30 seconds of each minute, allowing flush, wash and rinse solutions to pass through the transfer line in a reverse direction, with motivation being the back pressure caused by the sprays, commonly operated at 25 psi.

Addition of a Process Component to a Transfer Line

Seldom will a simple length of unobstructed tubing be installed between the two tanks of a train. The transfer path may include a pump, filter, heat exchanger, another process component, or a combination of several such devices.

Figure 8 illustrates a relatively simple product filter through which the transfer could be accomplished via head pressure as previously described. Following completion of the transfer, the filter housing would be opened to

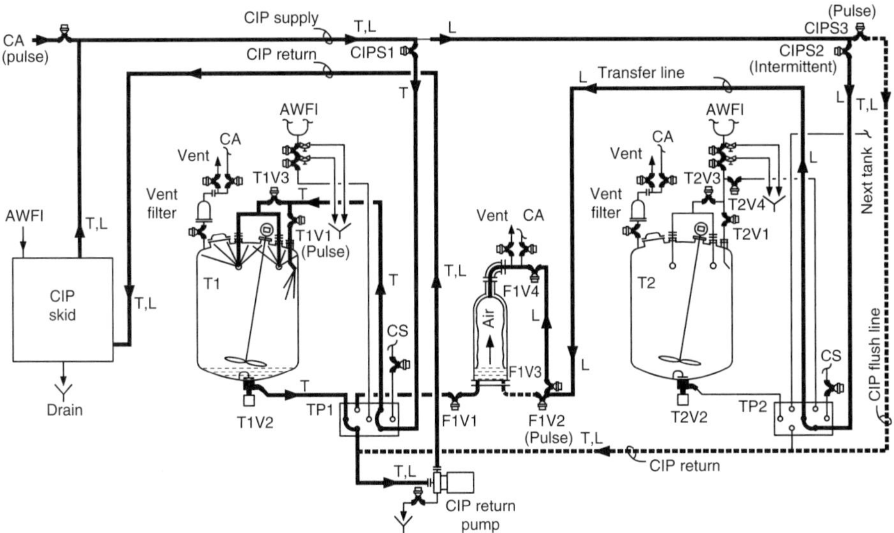

FIGURE 8 This version of the schematic includes a product filter housing in the transfer line. CIP of T1, transfer line and filter housing is described in the narrative.

remove the cartridge after which the housing would be replaced after manually cleaning the base joint and gasket area. The TP U-Bends would be installed as described for the more simple circuit shown in Figure 7. Filter valves F1V1 and F1V4 would be open full time for CIP. Tank T1 would be sprayed and the fill line "pulsed" as described above for tank CIP. However, when CIPS2 would open for perhaps 20 to 30 seconds of each minute, allowing flush, wash and rinse solutions to pass through the transfer line and filter housing in a reverse direction, F1V2 would be pulsed for two to three of the 20 to 30 seconds, allowing solution passage through the filter inlet line and valve. Then, when F1V2 closes, the reverse flow will spray the filter housing via the permanently installed ball or disc-type distributor (see chap. 9), the flow being caused by the back pressure created by the T1 sprays. The filter housing may be evacuated by the CIPR pump only, or may be assisted by the introduction of a controlled flow of CA to the CIPS line at the CIP Skid. The intentional injection of air to the CIPS flow will require a return system and CIP Skid that can handle retained air, and these will be discussed in chapter 6.

SIP of the Transfer Line with Filter Housing and Destination Vessel

Figure 9 illustrates one method of steaming the assembled components of the combination vessel and line CIP circuit (Fig. 8) for sterilizing purposes. Following CIP, the filter cartridges would be installed in the product filter and vent filter and the appropriate U-Bends would be installed on T1TP1 and T2TP2, with steam traps as shown. CS supplied to T2 via use of the 3-Leg U-Bend would then be cycled in sequence through the transfer line and process filter with the vent open to remove air, after which the vent valve would be closed and steam admitted at the top and bottom would continue to the trap on T1TP1. Then, with the T2 vent valve open, T2V2 would be opened to admit steam to the bottom of T2, driving air out the top.

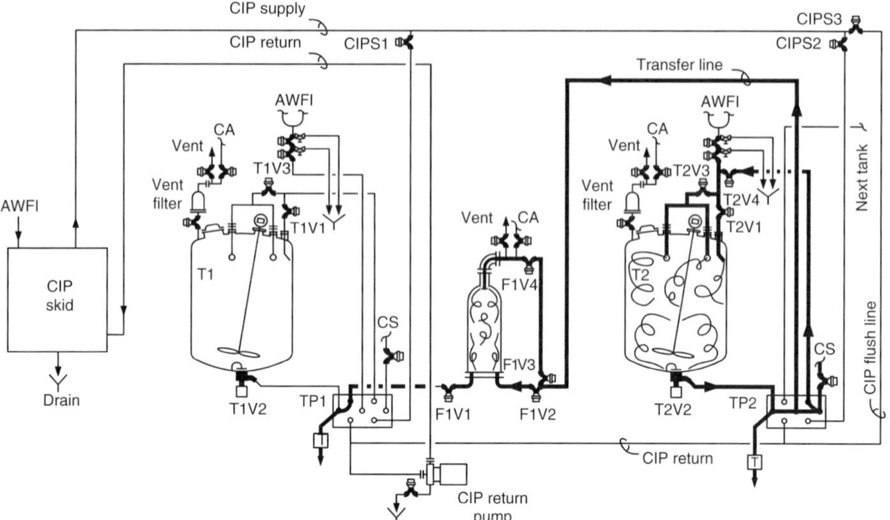

FIGURE 9 This version of the schematic illustrates the flow path for SIP sanitizing of T2 with the filter housing and transfer line prior to the T1–T2 transfer. The design would also permit hot water sanitizing. A full sterile process would require several added valves to maintain sterility during removal of the 3-Leg U-Bend after sterilize-in-place.

When steam appears at the vent valve, it would be closed and steam would be admitted to all of the remaining tank top piping to bring the entire system to temperatures and pressures required.

THAT IS CIP!

The selection, arrangement, and installation of the above-described generic CIP-cleaned process may vary in many ways, for example:

- Vessels may vary in capacity from 30 L to 16 m^3, or larger.
- Vessels may be fixed or portable, or a combination of both.
- Vessels may be simple in design and function (e.g., media prep or buffer prep), or complex (e.g., a bioreactor with multiple legs to be cleaned with the vessel).
- CIPS/R and product piping may be fixed, or fixed in combination with flexible hose final connections to portable equipment.
- Individual TPs may be replaced with larger top and bottom located multi-tank TPs.
- For the production facility, all TPs may be replaced with valve arrays of diaphragm, rising stem mixproof, or diaphragm mixproof type, to enhance control and minimize labor requirements for conversion from production to CIP and back.
- Transfer lines may be smaller (generally) than CIPS/R piping, and may include other equipment in the transfer path. Some components which permit flow in only one direction, i.e., a diaphragm pump, may require a bypass for CIP in reverse flow.

CIP and SIP Again Defined

As described above and practiced today, the CIP process is essentially chemical in nature, and generally requires recirculation to minimize water and chemical costs. Flushing, washing, rinsing, and (optional) sanitizing solutions are brought into immediate contact with all product contact surfaces under controlled conditions of time, temperature, and concentration, and continuously replenished. Vessels and filter housings are sprayed and piping is pressure washed. Steam-in-place (SIP) is the next logical step following CIP. The objective is to reduce the microbiological content in the equipment. Depending on the process requirements, SIP means to "sanitize" or "sterilize" the equipment. Both CIP and SIP can be applied to fixed or portable process vessels and holding vessels and process piping systems consisting of pumps, interconnecting piping, and valves. In addition to the basic tanks and piping, CIP circuits may include filter housings, membrane filters, homogenizers, centrifugal machines, heat exchangers, evaporators, dryers, congealing towers, screw and belt conveyors, process ductwork, and a variety of packaging machines.

The successful application of CIP requires that the technology be understood and accepted by all disciplines involved in design, fabrication, installation, commissioning, and validation of a project. DeLucia (4) suggested that "cleaning is (too often) an afterthought in the design of pharmaceutical facilities." Process development groups, design firms, and equipment vendors focus on their own process expertise and everyone assumes that "cleanability belongs to the CIP system and the validation department." The above-described equipment can be cleaned thoroughly and efficiently only if the cleaning requirements are integrated into the complete design process.

3-A Practice Revision

As this is being written, the 3-A Standards Committee is in the process of developing new definitions to be included in the next revision of the 3-A Practice.

B1.1 *CIP cleaning*. The removal of soil from product contact surfaces in their process position by circulating, spraying, or flowing chemical detergent solutions and water rinses onto and over the surfaces to be cleaned. Components of the equipment which were not designed to be cleaned in place are removed from the equipment to be manually cleaned. (CIP was previously referred to as mechanical cleaning.)

B1.2 *Manual (COP) cleaning*. Removal of soil when the equipment is partially or totally disassembled. Soil removal is effected with chemical solutions and water rinses with the assistance of one or a combination of brushes, nonmetallic scouring pads and scrapers, and high- or low-pressure hoses, with cleaning aids manipulated by hand, or wash tank(s) which may be fitted with recirculating pump(s).

The design principles on which the CIP and SIP procedures for the generic two-tank train have been described above constitute a "template" that has been successfully applied in many of this nation's newest biopharmaceutical facilities during the past 15 years. The projects have varied in size from as few as five fixed vessels, to as many as 65 vessels ranging in capacity from 30 L to 15,000 L (15 m^3). The process facilities have included pilot plant R&D and clinical trials production, blood fractionation, respiratory care products, the production of multiple biological products in a single facility, active pharmaceutical ingredient (API) processes, and even liquid creams and ointments. The physical arrangements have varied from single floor to five floors with CIP in a basement, maximizing the use of gravity CIPR. The degree of process automation has varied but all CIP functions have been fully automated, and the CIP programs successfully validated. There is virtue in considering the avoidance of all line CIP circuits by designing to clean every vessel in combination with its downstream transfer line and SIPing every vessel with its upstream transfer line, to achieve uniformity of design and operation, minimize the circuits required, and reduce water, time, and chemicals required for cleaning the total process.

CIP Cleanable Equipment and Process Design Criteria

Process equipment and piping that have been designed to be totally disassembled for manual cleaning, as used in the pharmaceutical industry until the recent past, is not suitable for application of automated CIP cleaning. The general design criteria for processes that handle fluid or semi-fluid products that must be maintained in a very clean or sterile condition include:

1. All equipment that will be contacted by cleaning solutions must be made of stainless steel, glass-lined, or equally corrosion resistant construction, and CIP-cleanable materials, sealed and closed with elastomers which are validated for the intended application.
2. The equipment must be designed to confine the solutions used for flushing, washing, and rinsing.
3. The entire process, consisting of the equipment and interconnecting piping, must be drainable.
4. A minimum radius of 1 in. (25.4 mm) is desirable at all corners, whether vertical or horizontal.
5. Mechanical seals must be used for agitators.
6. Projectile-type thermometer sensors are acceptable for use with filled tube or resistance temperature detector (RTD)-based temperature indicating and recording systems. Thermocouple(s) or RTD(s) installed so as to sense only the temperature of the tank surface provide an even more satisfactory installation from the standpoint of cleanability.
7. Automatic orbital welded joints are the most suitable for all permanent connections in transfer systems constructed of stainless steel.
8. Clamp-type joints of CIP design are acceptable for semi-permanent connections. An acceptable CIP design infers (*i*) a joint and gasket assembly which will maintain the alignment of the interconnecting fittings, (*ii*) a gasket positioned so as to maintain a flush interior surface, and (*iii*) assurance of pressure on each side of the gasket at the interior surface to avoid product build-up in crevices that might exist in joints which are otherwise "water-tight."
9. Dead ends and branches are prohibited, and all mandatory branches or tees should be located in a horizontal position and limited to a L/D (length/diameter) ratio of 2, or the branches shall be cleaned-through during CIP. Vertical dead ends are undesirable in fluid processes because entrapped air prevents cleaning solution from reaching the upper portion of the fitting.
10. All parts of the piping or ductwork should be continually sloped at 1/16 in. (5 mm/m) to 1/8 in. (10 mm/m) per feet to drain points.

11. The support system provided for the piping and ductwork should be of rigid construction to maintain pitch and alignment under all operating and cleaning conditions.
12. The process and interconnecting piping design should provide for inclusion of the maximum amount of the system in the CIP circuit(s). It is better to install one or two small jumpers than to remove and manually clean five or six short lengths of piping.
13. Mechanical/chemical cleaning is much more rigorous and is subject to better control than manual cleaning.

The Goals of Automated CIP

The goals of an automated CIP system include:

1. Elimination of human error and assurance of uniformity and reproducibility of cleaning, rinsing, and sanitizing not possible with manual procedures.
2. Prevention of accidental product contamination through operator error (by system design).
3. Improvement of the safety of production and cleaning personnel.
4. Improvement of productivity by reducing the production operation down time for cleaning. CIP-cleaned equipment generally requires less maintenance, thus also reducing maintenance downtime.
5. The automated CIP procedure enables modern computer-based technology to be applied to document the performance of the cleaning process when compared with the requirement.
6. The properly engineered CIP-cleaned process will generally contribute to substantial reduction of product losses.

The most effective and repeatable CIP operations are achieved by a high level of automation. Therefore, the highly automated process is generally more easily designed as a CIP-cleanable process when compared with processes which utilize manually operated pumps and valves, or considerable manually assembled product piping.

To achieve the most effective results, it is necessary to design the process and the CIP components and circuits simultaneously, giving equal consideration to the process requirement and the method of cleaning the process. CIP is seldom efficient as an after-thought.

Historical Development and Overview of CIP Technology

In-place cleaning was first applied in the dairy industry in the late 1940s, both at the farm level (pipeline milking systems) and in the early 1950s, in processing facilities. An early publication about CIP in the pharmaceutical industry by Grimes described "An Automated System for Cleaning Tanks and Parts Used in the Processing of Pharmaceuticals" (5).

The biopharmaceutical industry interest in CIP was confirmed in the late 1980s and early 1990s when the major professional societies began to develop educational programs. The first ISPE CIP seminar was presented in Chicago, Illinois in 1986. In 1990, the ASME bioprocess equipment design course included a two-hour session on CIP. This program developed rapidly and was soon an annual three-day and then a four-day CIP course. The Parenteral Drug Association (PDA) presented its first CIP seminar in St. Louis, Missouri, in 1992, and this program continued once or twice a year until the first aseptic process course was conducted in 1999.

The literature began to address CIP in the pharmaceutical environment as Seiberling (6) described a 1978 vintage large-scale parenteral solutions system that was fully designed and engineered for CIP but cleaned by rinsing with distilled water (rinse-in-place) and sterilized by steaming (SIP). In 1987, a book written for the pharmaceutical industry (7) reported a summary of the development of CIP in the dairy, brewery, wine, and food processing industries prior to 1976 as background leading to the application of this technology by pharmaceutical users. The 1987 book provided more detail about the large IV Solution process of 1978 vintage designed to CIP standards, but only rinsed and SIP. Attention was also given to a sterile albumin process and a blood fractionation process, both of early 1980s vintage, installed to 3-A (dairy) standards and cleaned with full CIP programs. These projects involved the first efforts to apply an integrated approach to piping design for the pharmaceutical process. A comparison of dairy and pharmaceutical piping processes and the unique problems involved with installation of CIP/SP systems in clean rooms were also discussed in this book.

Adams and Agarwal (8) contributed concepts in "CIP System Design and Installation" based on current experiences in 1990, and added information about integrated piping design was further described in "Alternatives to Conventional Process/CIP Design—for Improved Cleanability" by Seiberling (9) in 1992. A comprehensive review of current technology was offered by Seiberling and Ratz (10) in the chapter "Engineering Considerations for CIP/SIP," in *Sterile Pharmaceutical Products—Process Engineering Applications* edited by Avis in 1995, and this was perhaps the first major effort to use pharmaceutical and biotech design examples and installation photographs. This book gave attention to the large U-Bend TPs and mixproof valves than being placed into current projects, and also addressed the dry drug segment of the industry. In 1996, Stewart and Seiberling (11) reported a project which applied CIP very successfully to an agricultural herbicide process and provided all of the challenges of the current API processes. This project applied fully automated CIP via a CIP Skid to clean a solvent-based process with alkali and acid in essentially the conventional manner.

Engineering design approaches and validation were jointly described by Seiberling and Hyde (12) in (1997) in an article titled "Pharmaceutical Process Design Criteria for Validatable CIP Cleaning," an early effort to address the design of a validatable CIPable process in an exclusive publication by the Institute of Cleaning Validation Technology. Marks (13) further explored and explained "An Integrated Approach to CIP/SIP Design for Bioprocess Equipment" in 1999, and in 2002 Cerulli and Franks (14) compared the cleaning regimen currently employed in much of the API segment of the industry to the alternative CIP procedures now being applied, but generally under manual control. Greene (15), writing about "Practical CIP System Design" in 2003, has addressed in a scholarly manner the subjects of flow rate and pressure, and the kinetics of CIP. He observed that even with 15 years of industry experience, "Proper implementation of CIP appears to be a mixture of art and science," a valid observation from the editor/author's viewpoint also. Forder and Hyde (16) elaborate on everything taught in the early pages of this chapter under process system design for CIP, noting that "For the most effective use of CIP, circuits must be designed into the facility from the beginning" and not developed as an afterthought. This article mentions the use of potable water in some newer biopharmaceutical facilities for all phases of the CIP program except the final rinse and this cost reduction approach is noteworthy.

Substantial experience has shown that the effective application of CIP procedures requires some combination of the components described previously under section What is CIP? Seldom are two processes similar, and during the decades of the 1980s and 1990s and into 2005 many different types of pharmaceutical and biotech processes have been designed to be CIPable and successfully validated. Though applied at first to primarily processes which handled liquid products, the technology is equally applicable (via different procedures and components) to dry drug processes (chap. 14). And whereas it was first applied to the final products which required the highest degree of process cleanliness and/or sterility, the technology is now being used in the API segment of the industry (chap. 15). The individual or project design team member becoming involved with CIP for the first time can derive much benefit from a careful review of the most recent articles listed above, and in the *References*, specifically references 12 through 16 as these articles have been developed and contributed during the more mature stage of the application of CIP technology to the biopharmaceutical industry.

REFERENCES

1. Tunner RJ. Manual cleaning procedure design and validation. Cleaning Validation. Royal Palm Beach, Florida: Institute of Validation Technology, 1997:21.
2. LeBlanc DA, Danforth DD, Smith JM. Cleaning technologies for pharmaceutical manufacturing. Pharm Technol 1993; 17(10):118–24.
3. Roebers JR. Project planning for the CIPable pharmaceutical or biotechnology facility. In: CIP for Bioprocessing Systems Proceedings. San Francisco, CA: SBP Institute, May 23–27, 2005.
4. DeLucia DE. Cleaning and cleaning validation: the biotechnology perspective. Personal Correspondence, 1994.
5. Grimes TL, Fonner DE, Griffin JC, Pauli WA, Schadewald FH. An automated system for cleaning tanks and parts used in the processing of pharmaceuticals. Bull Parenter Drug Assoc 1977; 31(4):179–86.
6. Seiberling DA. Clean-in-place and sterilize-in-place applications in the parenteral solutions process. Pharm Eng 1986; 6(6):30–5.
7. Seiberling DA. Clean-in-place/sterilize-in-place (CIP/SIP). In: Olson WP, Groves MJ, eds. Aseptic Pharmaceutical Manufacturing. 1st ed. Prairie View, IL: Interpharm Press Inc., 1987:247–314.
8. Adams DG, Agaarwal D. CIP system design and installation. Pharm Eng 1992; 10(6):9–15.
9. Seiberling DA. Alternatives to conventional process/CIP design—for improved cleanability. Pharm Eng 1992; 12(2):16–26.
10. Seiberling DA, Ratz AJ. Engineering considerations for CIP/SIP. In: Avis KE, ed. Sterile Pharmaceutical Products—Process Engineering Applications. 1st ed. Buffalo Grove, IL: Interpharm Press Inc., 1995:135–219.
11. Stewart JC, Seiberling DA. Clean-in-place—the secret's out. Chem Eng 1996; 103(1):72–9.
12. Seiberling DA, Hyde JM. Pharmaceutical process design criteria for validatable CIP cleaning. Cleaning Validation. Royal Palm Beach, FL: Institute of Validation Technology, 1997.
13. Marks DM. An integrated approach to CIP/SIP design for bioprocess equipment. Pharm Eng 1999; 19(2):34–40.
14. Cerulli GJ, Franks JW. Making the case for clean-in-place. Chem Eng 2002; 109(2):78–82.
15. Greene D. Practical CIP system design. Pharm Eng 2003; 23(2):120–30.
16. Forder S, Hyde JM. Increasing plant efficiency through CIP. Biopharm Int 2005; 18(2):28–37.

2 Project Planning for the CIPable Pharmaceutical or Biopharmaceutical Facility

Johannes R. Roebers
West Coast Engineering Biogen-idec, Inc., Oceanside, California, U.S.A.

Dale A. Seiberling
Electrol Specialties Company (ESC), South Beloit, Illinois, U.S.A.

INTRODUCTION

The focus of this chapter is not to cover the vast subject of project planning, but rather the many considerations that must be given to planning a project that will include clean-in-place (CIP) as the major method of cleaning the process equipment. Successful integrating of CIP into process and facility design, and hence a "CIPable" facility, includes far more than adding CIP skids to the equipment and installing spray devices in process vessels. For CIP to be successful, it must be considered and integrated into all project phases of process and facility design.

Why Do CIP?

The current production of many active pharmaceutical ingredients and most protein-based biopharmaceutical drug products is essentially accomplished in liquid handling processes that require the cleaning of complex and costly equipment and interconnecting piping. The process steps involving solids or liquid/solid separation steps also involve complex equipment, which is not easily disassembled for manual cleaning, but susceptible, through appropriate redesign, to CIP on production scale. Manual cleaning of equipment is also considered inefficient and difficult to validate. In addition, regulatory agencies have increasingly scrutinized manual, nonautomated cleaning of process equipment. Hence, the need for robust, validated, and efficient cleaning drives the need for CIP.

If you cannot clean the process equipment and piping in a robust, validated manner, do not even think about making a pharmaceutical or biological product!

Why Focus on CIP During the Initial Part of the Life Cycle of a Facility Project?

The process and required-process engineering is typically well defined and gets a lot of attention from the very beginning of a process facility project. Essential process scale-up experiments will be planned and executed by process development departments. The critical process parameters from the process development work will be incorporated in the knowledge base developed for the project process design. Often material first produced in small-scale equipment is used in clinical trials and therefore the process needs definition. The scale-up to the large-scale production facility must be well understood and should be well documented. Regulatory agencies expect a well-documented technical transfer process as outlined, for example, by ISPE (1). Finally, there are many "process experts" available to provide the required assistance in completing all phases of the process development and scale-up activity. Many other aspects of facility and facility design must comply with the many applicable codes and standards [Canadian Standards Association (CSA), Factory mutual (FM), electrical, and mechanical] and therefore receive considerable attention.

However, CIP engineering and design in many instances receives little or no attention during scale-up and process development. Typically, no experiments are done for CIP-related design challenges. In addition, CIP is not often even considered in conceptual facility and process engineering. Then, as the project planning moves forward, especially for smaller scale production, clean-out-of-place (COP) is often considered good enough. COP procedures, however, fail to match the efficacy and reliability of CIP, and manual COP has been under increased scrutiny of regulatory agencies. The application of full automation to the COP process can easily assure compliance with the time, temperature, and concentration requirements for effective CIP, but the COP procedure seldom assures the effective application of uniform physical energy to all equipment surfaces. And, COP is applicable only to those smaller components that can be moved from the point of use to the cleaning equipment area, a labor intensive operation. However, the removal, handling, and reinstalling operations contribute to physical damage of equipment and equipment surfaces.

In other instances, CIP is an "after thought" to those lacking a full understanding of the intricacy of validated CIP cleaning. CIP may be attempted by use of portable equipment that can be "rolled in," or accomplished with a couple of portable tanks, a pump, and perhaps a few flexible hoses. The writer's personal experience suggests that even the common understanding of CIP may vary, based on the knowledge and experience of those involved. Some may consider the CIP design requirement as the addition of the components, specifically the spray devices for vessels, the CIP skids, CIP supply and return (S/R) piping, and CIP return pumps superimposed on an otherwise traditional smaller scale process design.

The Need for Integrated CIPable Design

The successful large-scale project requires the design of a CIPable process; i.e., a process which gives *equal* consideration to the process requirement and the means of cleaning all product contact surfaces via validatable CIP following each period of use. The definition will preferably be expanded to describe an *integrated CIPable process design*, one which gives equal attention to the process design as to the CIP design with the ultimate goal of achieving process excellence with efficient, robust, and validated CIP operation with minimal additions of valves, pumps, and piping.

Potential Causes for Failure to Achieve Best Possible Design for CIP

The Project Management Concern

The available literature on the subject of large-scale project facility design suggests that those involved in the supervision of this activity give little, if any, consideration to CIP as a significant, identifiable part of the overall task. A review of 28 articles published in *Pharmaceutical Engineering* (2) between 1994 and 2006 revealed many articles that fully identify the need for:

- Effective project management and personal skills, critical thinking, and leadership

- Organized approaches to commissioning and qualification
- Design/build as a project delivery method
- Process simulation as a method of confirming the design will achieve objectives
- The use of a technical program document to save time and meet budget and end result objectives

Odum (1999) provided "A Unique Look at Some Lessons Learned" (3) in a highly detailed review of a significant project, but mentioned only that "Process information regarding package equipment such as cleaning cycles or utility service requirements must be communicated to the equipment group..."

The CIP Equipment Designer/Supplier Concern

The review of a smaller number of more general articles about CIP in the same pharmaceutical engineering archive provided information about CIP system design, biopharmaceutical equipment design for CIP/sterilize-in-place (SIP), transfer panel design, and CIP cycle development. Each writer is focused on specific features rather than on the integration of the required components and technology in the overall process.

There is little recognition that successful validated CIP is less affected by the selection of the type of the spray device, CIP skid, or chemical cleaning agent, than it is by the *design of a cleanable process*. A report produced for the Joint Service P2 Technical Library (4), totally unrelated to pharmaceutical or biopharmaceutical applications, properly stated "CIP is more of a design method than a cleaning process. The CIP method works 'automatically' by eliminating the places where residue can accumulate."

A Need for a CIP Expert

A probable reason for the above findings is that there are few truly knowledgeable "CIP Experts" available; specifically individuals who understand the production process and operations, the fluid flow and hydraulics' issues of a complex CIP circuit, the mechanical attributes of the pumps and valves in the process, and the chemical and biological issues involved in the cleaning and sterilization or sanitizing processes, in sufficient depth to contribute fully to the basis of the design of the modern computer-based control system used for both production and CIP processes. Failure to fully recognize the CIP needs up front can lead to costly start-up delays. After the fact remedies may include the need to modify waste systems, increased water or chemical usage, and expensive computer software modifications.

The remainder of this chapter will define the attention that must be given to CIP, in all phases of the project, for the life cycle of the project. Various aspects of CIP design criteria will also be emphasized.

LARGE-SCALE FACILITY PROJECT LIFE CYCLE

Figure 1 illustrates the life cycle of a typical biopharmaceutical/pharmaceutical facility. The project duration before the facility becomes operational may vary from three to four years or more. This chapter will discuss only those activities following the feasibility study and site selection activities. The availability of water, in adequate quantity and quality, and the disposal of waste, however, are important considerations for site selection.

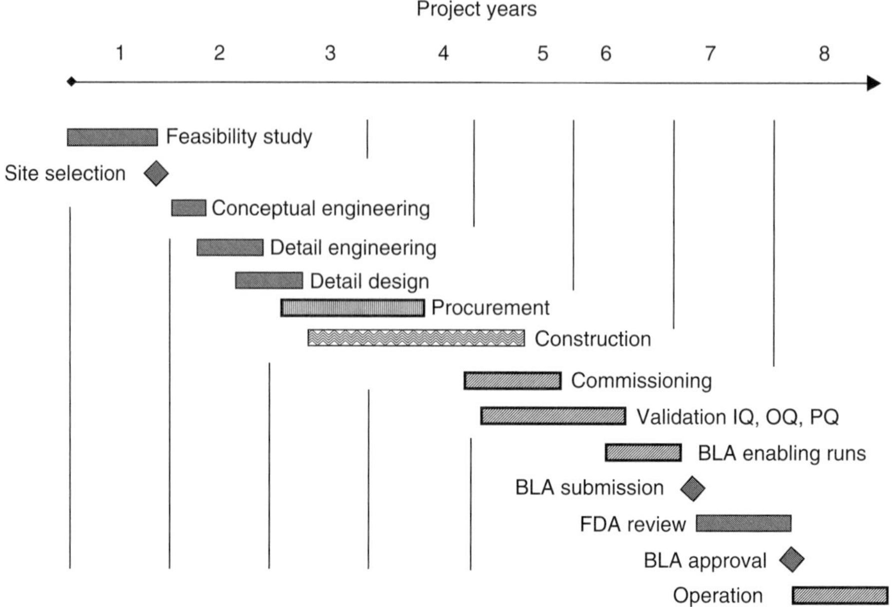

FIGURE 1 A project timetable showing the sequence and duration of major describable activity from the beginning of the feasibility study until start of production operations. *Source*: Courtesy of Genentech, Inc.

CIP CONSIDERATIONS IN THE CONCEPTUAL DESIGN PHASE

The major conceptual requirements of the CIP system should be second only to completion of the major equipment list. This approach will permit the establishment of locations, with respect to process loads, and the water-for-injection (WFI) or other high quality water supply. The number of CIP systems required may be determined by considering segregation requirements, scheduling of production and cleaning operations, and redundancy needs.

Segregation

A large-scale facility will comprise many functional areas of varying types of construction and finish ranging from warehouse space to clean room production areas. Separation may be governed by steps in the production process, and further refined by the functionality of the process zones, for example, media prep, cell culture, and purification in the biopharmaceutical process. Biological processes may also give consideration to separation of virus safety zones, i.e., pre- and post-viral inactivation.

Theoretically, a single CIP system and its associated CIP S/R piping, could clean an entire biopharmaceutical process, and some product development facilities have been validated on this basis. However, many considerations will favor complete separation of CIP systems by functional area. There is additional justification in doing this based on the size and geography of the process, related to the location of the process vessels with respect to the CIP skids. A typical CIP cycle consists of a prerinse, an alkaline wash, a post-rinse, and acid wash (or rinse), another post-rinse, and a final rinse to minimum required resistivity of the water going to waste at the end of the circuit. The circuit will normally be evacuated and refilled at least four times, following the alkaline wash, the acid wash or rinse, the last post-rinse, and on completion of the final rinse. The water required to "fill" the CIP S/R piping, and create a minimum puddle in the vessel being cleaned becomes greater as the distance between the CIP skid and the vessel increases. The first cost for a second CIP skid may be recovered quickly by reduced water cost for daily operation when the S/R piping contains significantly more water than the piping actually being cleaned.

Scheduling

Scheduling criteria must address not only the number of circuits, but also the frequency of cleaning. Consider, for example,

- How many lots are made on a daily basis?
- How many batch operations and transfers will be made daily?
- How many product changeovers occur in a given period of operation?
- How available are cleaning utilities, especially water?

A major consideration involves the philosophy of CIP to be applied. Alternatives include:

- Maximum flexibility, and the ability to clean any vessel, or any transfer line or other equipment item, individually, following each period of use
- Recognition that each vessel is used for a batch process associated with a subsequent transfer operation. Creative engineering design can make it possible to clean the vessel and all associated piping and downstream equipment involved in the transfer in a single circuit, thus reducing the number of circuits by a factor of as much as three, compared to the more conservative, flexible approach.

The writer recommends consideration of computer simulation of the chosen method as the means of reaching a final decision. Computer simulation can provide design data on which to base the decisions about the number of CIP systems required, the total time each will operate during a 24-hour day, the amount of water required, and waste to be generated, and the estimated chemical utilization. Consider also that CIP skids and associated CIP S/R piping used at a reasonable level of activity are less subject to developing contamination during long periods of nonuse. It is also not advisable to design for a CIP system usage of above 70% to 80% on average (Fig. 2).

Redundancy

A state-of-the-art CIP skid is a complex assembly of equally complex components which will occasionally fail, and require maintenance and the subsequent testing and documentation. Downtime considerations are essential, especially if multiple process areas are dependent on a single CIP skid. Redundancy can be achieved by two CIP skids, by routing CIP S/R lines through mix-proof valve arrays or transfer panels.

Careful consideration of segregation, scheduling, and redundancy needs, during the conceptual design phase, will enable the design team to define the system needs. Recognition of the value of the CIP system as a "start-up tool" is also

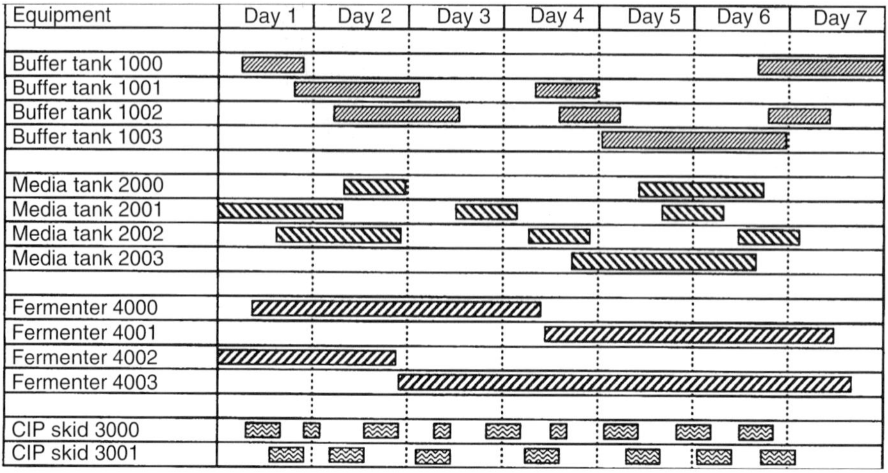

FIGURE 2 Bar graph showing computer simulation of the use of two CIP skids during a one-week period to clean buffer prep, media prep, and fermentation vessels. *Source*: Courtesy of Genentech, Inc.

essential. The CIP skids, and the operating software, are long-term delivery items, and also are the *first* major equipment items to be made functional at the beginning of the commissioning of all process systems.

Develop CIP Schematics

A major focus of many design teams is the development of individual equipment item P&IDs as the initial conceptual design effort. Experience in other industries suggests that the better approach is to develop one or more larger scale CIP schematics, drawings which focus on everything necessary to define the CIP circuits, flow rate, line size for process transfers, etc. Examples would include complete media prep or buffer prep functions, or a complete purification train; i.e., all equipment that might be cleaned from a single CIP skid. By assembling generic P&ID diagrams in small detail, in the relationship of the equipment on the conceptual layout drawings, the design team can quickly understand the significance of distance, air purge, drainage direction, etc., all very important to the fine tuning of the CIPable process design.

Layout of Process Facility for CIP

CIP Skid Distance to Process Circuits

The major reason for giving significant consideration to CIP early in the conceptual design process is to enable development of a facility layout favorable to minimum CIP S/R piping runs (Fig. 3). Water is the mostly costly ingredient in the CIP process, followed by time during which process equipment is unavailable for production, when being cleaned. Consider the following:

1. A large vessel, perhaps 15,000 L in capacity, can be spray cleaned at 80 gpm (300 Lpm). The 2″ CIP S/R piping, may be 100 ft runs each. The transfer line to the next vessel may total only 50 ft of 1.5″ line, another 50 ft of 1.5″ line may be an allowance for other vessel piping, dip tubes, etc.

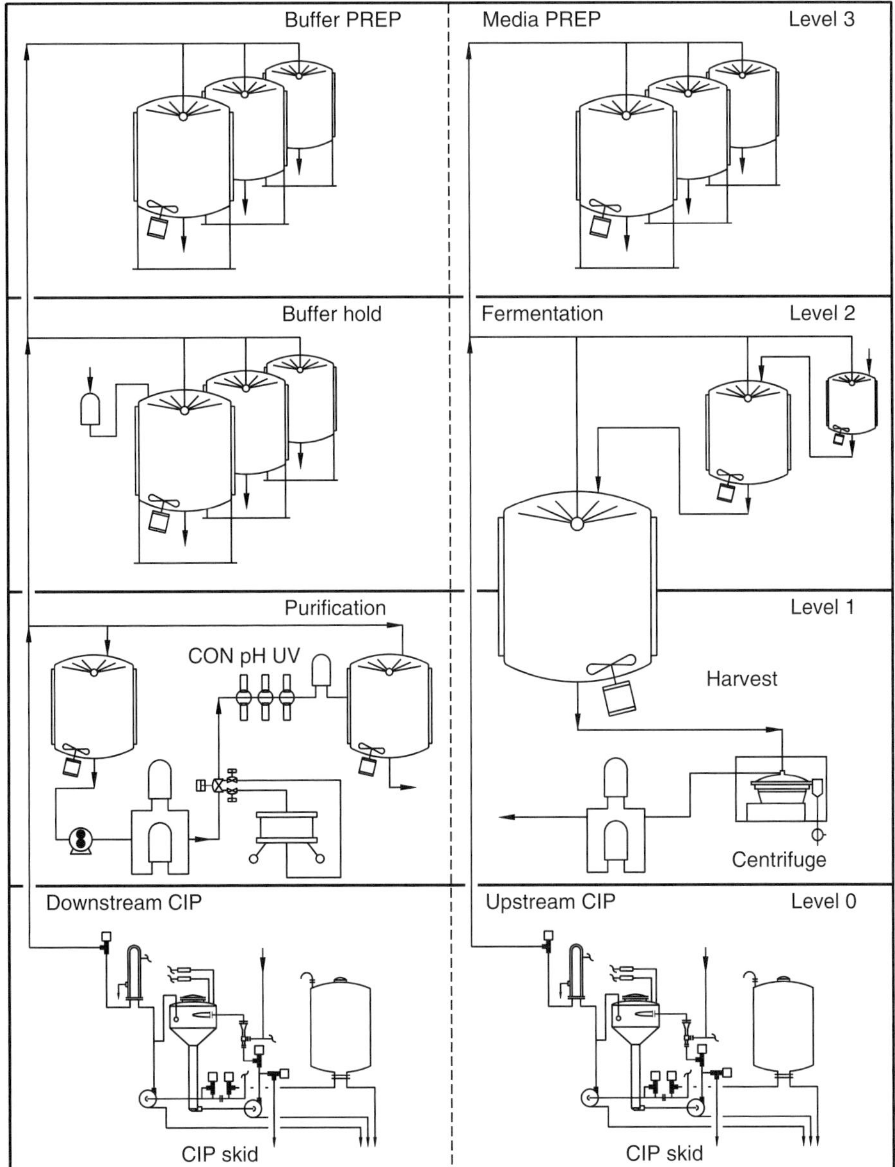

FIGURE 3 Simplified design schematic showing all equipment cleaned by each of two CIP skids. The final schematic would define individual circuits. *Source*: Courtesy of Genentech, Inc.

2. The water requirement for CIP would be 28 gal (106 L) to fill 200′ of CIP S/R piping, plus a 3-gal (10 L) puddle in a properly designed vessel with a flat plate vortex breaker, and approximately 7 gal (27 L) to fill 100 ft of 1.5″ tube.
3. If the CIP skid requires only 20 gal to fill skid piping, heat exchanger, and solution tank, then the total volume required for each recycle phase of the

program is 48 gal (180 L), of which only 10 gal (38 L) is in contact with the product contact surface at any given time.
4. The water required for the rinse phases will normally equal at least 1.5 times the line volume for pre- and post-rinses, and up to 5 times the line volume for a final rinse, perhaps 2400 to 2700 L for the total program.
5. If the vessel and the CIP skid are separated by only an additional 100 ft, the total volume will be increased to 3300 to 3500 L or more.

Whereas the above example suggested a 15,000-L vessel as the load, the total water requirement would be similar if the vessel was a small portable tank, or only the 50 ft of 1.5″ transfer line. This analysis demonstrates the need for maintaining the CIP skids in close proximity to the process equipment to be cleaned as compared to developing a centralized CIP skid area adjacent to the water supply. On multistory facilities, one should also consider the increase supply pressure that is required to transport the CIP supply from lower levels to equipment on higher levels.

CIP Return Flow Motivation

Lateral distance is only one factor to be considered. Gravity can be used to great advantage, to assist or create return flow motivation, and to facilitate rapid and complete draining of circuits following CIP operations. A basement location is desirable. A ground floor location is more common and requires CIP return pumps only for those vessels at the same level. A "top-down" process design may also use gravity to a great advantage in processing and facilitate vessel and transfer line drainage.

Accessibility

The layout of the facility should provide adequate accessibility and space around the CIP skid(s), valve arrays, and transfer panels. Around CIP skids great attention should be given to the maintainability of valves, pumps, and instrumentation. In addition, ports for sampling of CIP return solutions and sampling location for the water supply are essential in validation and operation.

The CIP S/R distribution scheme must give consideration to maintenance of required valve arrays, gasket change out, or perhaps transfer panels to establish routine connections or temporary connections to permit one system to be used as the replacement for another. The routing and installation of the CIP S/R piping should favor minimum footage and number of elbows, for capital cost and operating cost (volume to fill) reasons. The author strongly recommends establishing a design "hierarchy" in which CIP S/R piping and process piping is given preference over piping of support system such as steam, compressed air, water, and gases.

Whenever possible, CIP skids should be located in "gray space" immediately adjacent to (preferably under) the center of the CIP circuit load. CIP skids are not desirable in "clean/controlled spaces" for many reasons. Return pumps and valve arrays used for CIP S/R flow control should also be located in "gray space." Transfer panels may be placed in the "clean/controlled spaces" to avoid airborne contamination when making U-bend changes when such changes must occur frequently in operation. Campaign transfer panels may be located in either "clean/controlled" or "gray" space.

Cleaning Cycles or Programs

A typical CIP cycle is a sequential series of groups of events that can be controlled to achieve different cleaning or process objectives. The range of different cycles may include *rinse* only, a *full CIP* program only, a full CIP program plus steam sanitization, or a full CIP program plus steam sterilization. If the process equipment remains unused for longer than the validated clean hold time, a *rinse* followed by sanitization or sterilization or repeat of the full CIP cycle may precede its use. A typical cleaning cycle will start with prerinses of city water, distalled water (DI), reverse osmosis (RO) water, ambient water for injection (AWFI), or hot water-for-injection (HWFI) to drain.

The chemical wash, generally an alkaline solution, is preferably recirculated to minimize requirements for water, heat, and chemicals. Single pass chemical washes may be employed for systems that are used to process a variety of different products, to avoid the possibility of cross-contamination through development of residues of a specific soil throughout the system. Once-through cleaning is generally limited to circuits that can be cleaned at less than 50 gpm (190 Lpm) and chemical wash times of not more than five minutes.

A post-wash rinse with water may then precede either an acid wash or rinse. Evaluation of the final rinse discharge either by instrumentation or sampling and laboratory analysis will be applied to confirm removal of all traces of product and chemical agents.

Some process systems, i.e., water soluble buffer prep vessels, may require only a water rinse, followed perhaps by sanitization or sterilization.

The key in conceptual CIP engineering of cleaning cycles is to make good assumptions for time, water, and chemical needs for CIP cycles. These assumptions should be conservative enough to allow for potential extension of cleaning cycles as well as increased water and chemical use.

Cleaning Program Variables

The efficacy of CIP is well recognized to depend upon the control of time, temperature, and concentration. Prior experience may provide the best guide to establishment of a starting point. Cleaning trials and tests should be conducted if CIP technology is to be applied to a process used for a new, unfamiliar, product or new and challenging to clean equipment. Other chapters of this book fully address issues such as cycle development, the impact of hot rinses on protein residues, and the occasional need to cool a vessel following a hot rinse before starting the next batch.

Automation

Perhaps one of the most important and difficult decisions to be made during conceptual design is the degree of automation to be applied to both the process and the CIP system. High-volume production processes benefit from full process automation. Product development and clinical manufacturing facilities may be more flexible if the process is more dependent on human interaction; i.e., moving U-bends on a transfer panel to change the process configuration, or establish production transfer paths and CIP circuits. Fully automated CIP may be applied to either concept.

Transfer Panels vs. Mix Proof Valve Groups

Transfer panels have been used in pharmaceutical and biopharmaceutical processes since the very first applications of CIPable design to these industries in 1976. Seiberling (5–8) reported the use of both small-scale and large-scale panels using U-bends with proximity sensors in various early projects between 1977 and 1995. Louie and Williams (9) published an in-depth article in 2000 that further explored this technology. Transfer panels provide great flexibility at reasonable cost. However, all transfer panels require human activity at various steps in the production/CIP sequence, requiring labor, time and sometimes communication and documentation/verification of changes if proximity sensors are not included. The projects of the past decade, especially for high-volume production facilities, have favored the greater use of automatically controlled valves configured in valve arrays as illustrated in Figure 4 for both reliability and reduction of potential operator error considerations.

If the process is dedicated to one or a few products, or requires consideration of being fully contained, air-operated valves will be the choice. Rising stem mix-proof valves can be used nearly universally for CIP S/R flow control, and as isolation valves in non-sterile processes. Processes that must be sterilized require consideration of diaphragm-type valve assemblies as described by Chrzanowski, Crissman, and Odum (10). Other chapters in this book further describe these components and their use in the design process.

Whereas valves may have a greater first cost, factors including the reduction of labor, elimination of connection errors, increased operator safety, and less line hold-up volumes are considered by many to provide the best choice for the life cycle

FIGURE 4 A rising stem mix proof valve array used to control clean-in-place (CIP) supply distribution from two CIP skids to eight circuits. A similar valve array in the left background controls return flow. *Source*: Courtesy of Genentech, Inc.

of a facility. Air-operated valves may require additional maintenance but provide greater operator and product safety.

DCS vs. PLC or Combination

Though many highly automated large-scale biopharmaceutical facilities have been placed into production during the past decade, manually controlled processes still prevail. The highly automated process requires distributed control system (DCS) application, and when this elegant control is applied to the process, it is generally also extended to the CIP process. The main reasons for integrating CIP control into DCS are the elimination of control system interfaces, and the elimination of two operating, monitoring, and troubleshooting systems in later production operation.

Experience has demonstrated, especially in other industries, that a programmable logic controller (PLC) can control a CIP skid, and the sequencing of the necessary CIP S/R distribution valves in a highly adequate manner. And, there are CIP vendors that are knowledgeable about all aspects of the CIP process and are capable of developing and debugging the software and supporting commissioning and validation. An effective compromise may place all field (process-related) devices under control of the DCS, store all CIP recipes in the DCS, and download them to the PLC at the beginning of each program. Only minimal handshakes are then required for the PLC to control the CIP skid, and the DCS to control all process pumps, valves, and agitators. This book includes a chapter on CIP control that addresses this subject in more detail.

Special Cleaning Regimes

A typical highly automated CIP skid may be used to clean nearly any combination of biopharmaceutical vessels and interconnecting piping, and much of the related process equipment. Ultrafiltration systems require special control of flow and pressure of their membranes in operation and in cleaning. In addition, special nonalkaline solutions may be required for membrane cleaning. Chromatography columns must be bypassed for CIP and cleaned and regenerated with special solutions. Centrifuges can be subjected to the same time, temperature, and concentration criteria as the associated process piping and tanks, but often require dedicated control to handle the "shoot" sequence and often operate at substantially lower flow rates than other CIP operations. Filter housings greater than 4″ in diameter can be spray cleaned in combination with the transfer line in a typical CIP circuit. Large plate and frame filters can benefit from CIP of the large surfaces and small internal passages via secondary high-volume recirculation in the transfer line from a vessel. Powder mixers, i.e., Tri-Blenders™ (Alfa-Laval AB, Lund, Sweden) can be incorporated as both process pumps and CIP return pumps and effectively cleaned with minimal manual preparation. Portable tanks and intermediate bulk containers can be cleaned at portable tank wash stations connected to the CIP S/R lines from a fixed skid.

All of the above CIP issues need to be addressed during the conceptual design phase of the project and the selected scheme should be a part of the CIP schematic design package or document, subsequently used for development of the P&ID drawings for specification and purchase of each item of equipment. The project design team should investigate the opportunities to conduct cleaning tests and trials with vendors of both the specialized equipment and a knowledgeable CIP consultant.

Other Conceptual Design Issues

Other subjects of concern in the early stages of a large project, and the subject of complete chapters in this book, include the code and safety issues with hazardous cleaning chemicals and cleaning validation requirements and concepts.

Cleaning chemicals may be purchased and stored for use in several manners (Fig. 5). Following the determination of the quantities required (see sections entitled "Scheduling" and "CIP Simulation"), early decisions must be made about the method of delivery, sampling and testing, storage, containment, and also cleaning chemical and product disposal regulations in the jurisdiction of the facility location.

A cleaning validation plan that clearly outlines the validation requirements and approaches, including pass/fail criteria, should be well underway near the end of the conceptual design phase. Early commissioning of the CIP skids and distribution piping is beneficial to completion of start-up and passivation of the piping system. Due to the longevity and changes associated with large-scale manufacturing facilities, future expansion considerations need to be addressed at this time. The conceptual layout should provide the space for additional CIP skids, valve arrays, and transfer panels. Utility headers and should be sized to support such expansion.

FIGURE 5 Externally mounted bulk storage tanks for caustic and phosphoric acid cleaning chemicals handled by bulk delivery and storage. Distribution to the individual CIP skids was by chemical headers, valves, and magnetic flow meters. *Source*: Courtesy of Genentech, Inc.

CIP Conceptual Design Support

One of the most difficult decisions involves the method of obtaining CIP design services. Any project requires the support of an architectural and engineering firm(s) to handle the site planning, conceptual design, detail engineering and design. This requires the effective integration of many engineering disciplines. The process definition, however, must be provided by the owner. The CIP design may be handled by the A&E or by another consultant more highly experienced in CIP. However, it is essential to

- Develop basic CIP schematics
- Summarize the conceptual design decisions discussed above in a CIP basis of design
- Obtain quality assurance/quality control (QA/QC) as well as user group acceptance and approval for the defined CIP concept.

CIP CONSIDERATIONS IN DETAIL ENGINEERING

Develop CIP Schematics

This activity was identified as an early step in the conceptual design, before the development of P&IDs, to provide the means, in combination with the conceptual layout drawings, of defining all CIP circuits. The spray flow rate requirement for the *largest vessel* or *largest line* cleaned by each CIP skid will control the diameter of CIP S/R piping. The flow rate for process transfers will establish process line sizes. Flow distances can be obtained from the layout drawing $\pm 10\%$ and preliminary consideration of piping routing can enable a forecast of elbows in both CIP and process piping runs. This information, properly tabulated for all circuits, will permit calculation of line volumes, friction loss (for CIP pump sizing) and when combined with CIP cycle data, total water usage requirements for CIP. If it has not been done before beginning the detail design it is imperative to complete this step now, using the preliminary P&IDs as the basis for CIP schematic development for each CIP skid and associated process development.

Complete CIP-Related P&IDs, Drawings, and Specifications

CIP Functional Requirement

Following successful integration of all CIP Schematics into the individual process equipment P&IDs, the CIP skid P&IDs, and specifications should be completed to enable purchase of the these long lead time items. In most cases, a CIP vendor will provide the detail design of the CIP skids, based on detail specifications by the owner or A&E. The detail engineering can then continue by completion of drawings and specifications for transfer panels and CIP S/R valve arrays, all of which are needed for reference to prepare the basic CIP Functional Requirement.

CIP Utility Requirements

On completion of the CIP functional requirement, and full circuit and cycle information, a final review of utility requirements may include city water, RO water, WFI, clean air, steam, and estimated discharge to waste.

Preliminary CIP Equipment and Piping Layout

The CIP skid drawings supplied by the selected vendor are required to add CIP skids to the detail layout drawings, and this permits design of the utility connections.

Return pumps, valve arrays, and transfer panels, based on vendor drawings, must be located. The integration with other design disciplines (process, mechanical, electrical, HVAC, control system) can now move forward and preliminary pipe routing can be followed by a review of line sizing and pressure drop calculations to confirm or revise the estimates developed in the conceptual design phase. On completion of the preliminary pipe routing drawings, a careful review for potential deadlegs and low points should be made followed by appropriate remedial action.

CIP Automation
As the detail engineering nears completion, the CIP functional requirement should receive a final review and approval. This document will be the primary interface of the software integration vendor to the process design.

CIP CONSIDERATIONS IN DETAIL DESIGN

A valuable aid in guiding the CIP aspect of the project is the preparation of a *CIP design guideline* for reference by all project participants from this point forward. This can be done jointly by a CIP Consultant and the A&E or by the A&E. It should address all aspects of the CIP design in detail, including cleaning and Bioburden control methodologies, CIP skid, system and circuit configurations, sterilization or sanitization requirements (if any), and design criteria for valves, transfer panels, major processing equipment, spray devices, piping, and instrumentation.

The above document should be used for a final review and update of the CIP schematics, to ensure agreement with CIP skid and process equipment P&IDs. This work can be followed by the addition of valve numbers to the schematics. Extractions of the large schematic can be prepared to define each circuit, to show the primary flow paths for the flush, wash, and rinse solution, all subcircuits, and all air-purge paths. Line sizes and flow rates should be added and the schematics should be updated to reflect coordination with other engineering disciplines. These individual circuit definition and drawings should be supported by valve sequence matrix charts and submitted with the CIP Functional Requirement to the selected software integrator.

The CIP equipment layout, including the CIP skids, return pumps, valve arrays, and transfer panels should be fully defined at this point of the detail design. Equipment drains, CIP skid drains, and low point drains may require a final check. Also containment of CIP skids and valve groups should be considered. At this point the design drawings should fully integrate process P&IDs and CIP schematics and finalize process and CIP pipe routing. Full agreement and continued updating of both the P&IDs and the CIP schematics is recommended until completion of validation and in operation. Orthographic drawings as shown in Figure 6 should be developed by the piping designers or the piping contractor before production of piping isometrics to allow for review of all piping runs for proper sloping, low points, and shortest route.

The CIP automation detail design specifications should now be completed and must be brought into compliance with all CIP detail design drawings.

CIP CONSIDERATIONS IN PROCUREMENT

CIP equipment procurement may be on the basis of competitive bidding, or sole source negotiations. As with all specialty process equipment, there is great benefit

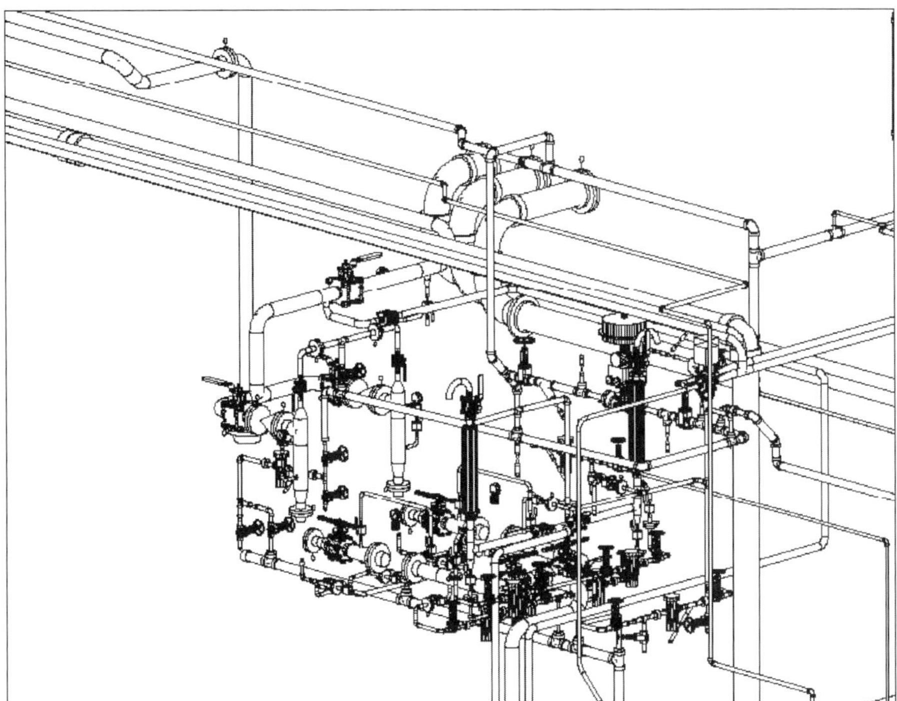

FIGURE 6 An orthographic drawing of area containing process and CIP system piping installed in gray space for ease of maintenance. *Source*: Courtesy of Genentech, Inc.

in purchasing such equipment from vendors that have the best experience, and perhaps a good previous track record. It is not advisable to make purchasing decisions solely on cost which is especially true with CIP equipment.

Spray devices for custom biopharmaceutical or pharmaceutical processes tanks require spray coverage verification which require custom design of every spray to ensure coverage of all tank nozzles. Procurement is best accomplished by providing the A&E guidance about tank top piping criteria, following the establishment of nozzle locations for vessel procurement, and then submitting the tank head drawings with a request for a design/fabricate bid from a selected spray ball vendor. The spray balls can then be sent to the vessel vendor for spray coverage testing in the tank shop. Neither spray device design nor spray flow rate decisions can be made by tank fabricators unfamiliar with the total project CIP concept. Vessels must be sprayed at a specified flow rate and pressure which is compatible with the CIP S/R piping design and ensures compliance with cleaning flow velocity criteria for each tank.

Standardization of pumps, valves, and instrumentation on CIP skids is beneficial from both procurement and long-term maintenance perspectives. This can be accomplished by including a preferred component and instrument list with requests for bids for skids. Consideration of lead time is important and it is desirable to prioritize deliveries of skids, valve arrays, and transfer panels. Early procurement of all CIP equipment will greatly enhance the ability to finish piping routing and design. Fabrication Inspections are of great value and may include pre-FAT Inspection, FAT planning and protocol development and FAT execution. The FAT procedure provides the opportunity for CIP design team members to become well acquainted with the design and intended operation of the CIP skids. While spray ball coverage has become a standard test at FAT, the author highly recommends repeating this test on site in commissioning or installation qualification (IQ).

CIP CONSIDERATIONS IN CONSTRUCTION

Assuming that the project has received the care and consideration during conceptual design, detail engineering, detail design, and procurement planning as suggested above, the construction process should create few, if any, problems. Remember Murphy's Law—"If anything can go wrong, it will!" Instruct contractors to bring potential problems to your attention as the owner as they are discovered. If the discrepancy is serious, it should be corrected before it is welded into place, even though this may require drawing and document revisions. It is strongly recommended that the design team reviews all isometrics before fabrication conducts a final check for additional low points. In addition, access and maintenance of skids, sample valves, valve arrays, transfer panels, and pumps should be diligently checked in construction.

The review of the piping isometrics should confirm that the final design complies with the previous assumptions used for circuit pressure drop calculations. The piping installation also needs to be reviewed for proper piping support especially if air-blows are used as part of CIP cycles. The start of construction is the last opportunity to review chemical solution containment at the CIP skid as shown in Figure 7 and the containment of CIP skid itself. A final hazard and operability (HAZOP) review of all CIP operations and chemical handling would be beneficial.

Insulation of CIP piping (as required for operator safety) should only commence after completion and verification of all piping installation work. Physical protection of CIP skids, valve arrays, and transfer panels during transportation, rigging and installation, proper protection in shipping and storage, planning of the installation path and protection following installation are all significant tasks to be planned by the procurement, rigging, and installation teams.

CIP CONSIDERATIONS IN COMMISSIONING

This section will be brief as another chapter of this book describes the subject in detail. The first step is to develop a CIP commissioning plan, followed by review and acceptance of CIP-related turn over packages. The review and general approval of a cleaning validation master plan may follow. Before beginning the commissioning process it is necessary to complete training of the CIP operator and maintenance personnel that will perform the required work. Obviously, nothing can be done until after the required CIP utilities have been commissioned and are operational. Then, the hard work follows, and will include, but is not limited to the following for a typical CIP circuit start-up:

- Piping walk down
- CIP skid operation check (level, temperature, flow, and conductivity control loops)

FIGURE 7 Eductor-assisted CIP skid with single water-for-injection tank located within a curbed area to confine minor leakage. Access for maintenance is excellent. *Source*: Courtesy of Genentech, Inc.

- Dry run valve check of CIP S/R distribution valves and all process system control valves that must be sequentially controlled for CIP
- A wet run and check of hydraulic balancing during valve sequencing will follow
- Preliminary chemical cleaning runs will then follow, with checks for the variables of time, temperature, and concentration, which determine the efficacy of the CIP program on each different soil residue
- Spray device CIP coverage test, generally with riboflavin, using the assigned CIP skid to provide the rinse water

Once CIP circuits are operational they may be used for convenient passivation of process and CIP piping by an owner team or outside consultant qualified to do this work.

Depending on owner company policies and guidelines, consider using most Commissioning work also for completion of IQ.

CIP CONSIDERATIONS IN VALIDATION

The cleaning validation function is covered in depth by another chapter in this book. Following is a discussion of the important elements of the task that must be addressed during facility planning.

The cleaning validation operation qualification must verify all parameters of each CIP cycle as documented in the detail design specification. verification of alarm and alert set points and messages is critical. Prior to cycle development, sampling considerations such as rinse tests, swab tests (microbiological, protein, and chemical residue), and visual examination criteria must be finalized.

Cleaning cycle development may be accomplished with simulated soil or expired or rejected product. However, it is critical that the simulated soils and the process conditions are as closely matched to the actual soils and actual process conditions as possible. The final setting of cleaning parameters (time, temperature, and concentration) must be based on a cleaning validation performance qualification with actual product soils. Regulatory and operational requirements make it necessary to establish robust *clean* and *dirty* hold times.

Large biopharmaceutical projects with hundreds of CIP cycles may generate thousands of CIP samples during cycle development and validation work. Therefore, the collection, identification, handling of the samples, analysis, recording of data, and communication of QC data must be carefully planned. It is not unusual to generate tens of thousands of cycle development and cleaning validation data points. Consider the need for CIP team training in sample handling and analysis activity, and involve your QC department as early as possible in the planning process. Attention should also be given to special analysis requirements.

CIP CONSIDERATIONS IN OPERATIONS

CIP provides many benefits to the user in addition to providing a means of achieving validated cleaning of all equipment, after each period of use. The advantage of full CIP automation is the capability to monitor operations on not only a daily but also hourly basis—program after program. The acquisition of key information about flow, pressure, temperature, conductivity, what is being cleaned, and who is operating the system, can all be tabulated and plotted against time. This in turn makes it possible to create *trend CIP circuit performance reports*, and monitor cleaning chemical usage and utility usage. Cleaning optimization is possible to reduce water, chemical, and time needs even though such work will require revalidation of each changed CIP cycle. CIP skids and circuits may also be used to repassivate process piping and equipment on a regular basis. Finally, another chapter in this book will suggest how CIP operational data can be used as the basis for troubleshooting and maintenance procedures.

CIP INTEGRATION AT THE GENENTECH OCEANSIDE PRODUCT OPERATIONS

From project conception until the completion of the project, the owner emphasized the objective of developing an appropriate CIPable facility layout and process. Consideration of a basement was eliminated during the early stages of the conceptual design for cost and code reasons. The three-floor manufacturing building shown in Figure 8 maintained "top-down" process flow and gravity CIP return capability with CIP skids on the ground floor, and standard centrifugal CIP return pumps on the ground floor for ground floor and some second floor vessels.

Most CIP equipment including skids and CIP S/R distribution valve arrays were located in "gray space" for maintenance accessibility. During the conceptual design phase of the project, the project team was able to reduce the amount of transfer panels from 80 to 3. This was accomplished by reducing the degree of desired process flexibility and by the use of valve arrays. A thorough evaluation of transfer panels versus valve arrays by the owner's engineering, safety, and operation teams universally favored mix-proof valve arrays as a means of providing CIP flow control and CIP skid redundancy.

FIGURE 8 Aerial view of large multiproduct biopharmaceutical manufacturing facility. The CIPable process is located in the high building (upper left). *Source*: Courtesy of Genentech, Inc.

The process piping design was integrated with the CIP piping design from the start of project. Vessels of all sizes were fitted with outlet valves and fixed plate vortex breakers to allow a spray CIP flow rate compatible with the size of the CIP S/R piping. Transfer lines downstream of the vessels were cleaned in reverse flow, under pressure, and sized for transfer flow rates, to minimize hold-up volume. All process and CIP operations were controlled by a DCS System.

An owner's CIP engineering and design team was established at the very beginning of the project. This group of dedicated personnel worked with an experienced CIP engineering and design service and CIP expert consultant intensively during the conceptual design, and throughout startup and commissioning. Other significant project management decisions included the following:

- The product/CIP solution segregation philosophy was established early by decisions to use rising stem mix-proof valve for CIP and diaphragm valve assemblies for the process, much of which could be SIP'd.
- QA/QC buy-in was obtained early in the conceptual design phase.
- A cleaning validation master plan was developed early.
- Cleaning chemicals were defined as caustic and phosphoric acid, handled by bulk delivery and storage.
- A dedicated CIP hot water-for-injection (HWFI) loop assured the availability of adequate water for five (5) eductor-assisted CIP skids.
- Some unique equipment items were evaluated by cleaning tests conducted by vendors with owner design team personnel, to assure proper process and CIP capability.
- Single-sourced CIP equipment including skids, valve arrays, and spray devices.
- CIP commissioning and validation was accelerated by teamwork of owner engineering, operations, and validation personnel and CIP with support of validation consultants.

CONCLUSION

- CIP and process engineering and design must be fully integrated into all phases of a facility project.

- The majority of CIP considerations are in conceptual engineering.
- The owner should not miss the opportunity for a truly CIPable facility design by deferring CIP decisions until later in the project, or leaving them to others!
- If you are not confident and competent in CIP engineering and design, get a "true CIP expert" to help.
- Learn from CIP experiences in other industries, companies, and projects.

REFERENCES

1. Good Practice Guide Technology Transfer. ISPE Tampa, Florida, March 2003.
2. Pharmaceutical Engineering. Article Archives and Back Issues 1994 through 2006. Tampa, FL: International Society for Pharmaceutical Engineering (Accessed July 26, 2007 at www.ispe.org).
3. Odom JN. Facility design and construction: a unique look at some lessons learned. Pharm Eng 1999; 19(1):8–14.
4. The Clean-In-Place (CIP) method to eliminate hazardous waste. Joint Service P2 Technical Library. www.p2library.
5. Seiberling DA. Clean-in-place and sterilize-in-place applications in the parenteral solutions process. Pharm Eng 1986; 6(6):30–5.
6. Seiberling DA. Clean-in-place/sterilize-in-place (CIP/SIP). In: Olson WP, Groves MJ, eds. Aseptic Pharmaceutical Manufacturing. 1st ed. Prairie View, IL: Interpharm Press Inc., 1987:247–314.
7. Seiberling DA. Alternatives to conventional process/CIP design for improved cleanability. Pharm Eng 1992; 12(2):16–26.
8. Seiberling DA, Ratz AJ. Engineering considerations for CIP/SIP. In: Avis KE, ed. Sterile Pharmaceutical Products—Process Engineering Applications. 1st ed. Buffalo Grove, IL: Interpharm Press Inc., 1995:135–219.
9. Louie E, Williams B. Transfer panel design: aseptic solution handling in biopharmaceutical facilities. Pharm Eng 2000; 20(4):48–55.
10. Chrzanowski GA, Crissman PD, Odum JN. Valve assembly use: a case study in maximizing operational flexibility, cost management, and schedule benefits. Pharm Eng 1998; 18(6):70–7.

3 Water for the CIP System

Jay C. Ankers
LifeTek Solutions, Inc., Blue Bell, Pennsylvania, U.S.A.

WATER—THE MAIN INGREDIENT OF CLEAN-IN-PLACE

Water, in one or more levels of purity, is the main ingredient of each phase of a clean-in-place (CIP) cycle. It is used by itself for the first rinse and intermediate rinses, mixed with a mild alkali or acid solution for the wash steps, or drawn from a compendial water loop for the final rinses. Water is used because of its solvent properties and is convenient to use because it is already available in each area of a facility that is used for processes that are CIP cleaned.

With the size and number of biotech and pharmaceutical facilities increasing in every region around the world, their individual demand for water has increased. A typical large-scale biotech facility will use between 250,000 and 2,000,000 gallons of water per day. Where does all the water go? A majority of this water is used in the cleaning of the process equipment and piping. One of the next large uses of the water is the blowdown or waste from the production of compendial waters.

Water is used in the biotech pharmaceutical facilities as follows:

- Process/CIP⇒*the majority is used here*
- Blowdown or waste from water purification⇒*30% to 40% of compendial water produced*
- Heating (steam—10% make-up—90% returned as condensate)
- Evaporative cooling (this can be a major user in warmer climates)
- Humidification of conditioned facility air
- Human consumption/use—drinking, cooking, and janitorial

CIP system design as it relates to water usage *directly* affects the first three in the list above.

CIP systems and cleaning circuits can be designed to use water efficiently and save a significant amount of water each day in a biotech or pharmaceutical facility. The type of water that the systems washes or rinses with and the way that the CIP skid stores water or calls for water as needed has a significant effect on the total facility water consumption. This chapter will divide the water types into either the pre-treated water grades that are potable or softened and the purified waters or *compendial* waters that are used in most pharmaceutical and biotech processes today. Compendial waters are defined by the American Society of Mechanical Engineers BioProcessing Equipment Standard (ASME BPE) as: *water purported to comply with United States Pharmacopeia (USP) and/or any other acknowledged body of work related to the quality, manufacture, or distribution of high purity water* (1).

The design of the water distribution systems that are connected to the CIP skid also have an effect on the function of the CIP system and the duration of the CIP cycle. The water distribution systems in a facility need to be carefully designed with respect to the demand created by the CIP systems. The CIP systems are the biggest user of the water but frequently the design of two systems (CIP and water treatment) is completely independent of each other. Often the water systems, loops, flow rates, etc., are finalized before the CIP loads are even determined, in which instance the CIP skids are then fitted with multiple large, costly, space consuming tanks to match supply to demand.

The choice of what types of water a designer will feed a CIP skid, depends on the type of water used in the process being cleaned. The USP simply states that for parenterals the minimum requirement is water-for-injection (WFI). For oral dosages (solid or liquid), USP purified is acceptable. For active pharmaceutical ingredients (APIs) and other upstream process applications, softened water or even potable water is used for cleaning and rinsing. With that decision being made by the end user, it is the designer's responsibility to ensure that when the CIP cycle is completed, the only water that is left behind after the final rinse meets quality of the water used in the process being cleaned.

The various grades of water found in pharmaceutical and biotech facilities are typically generated in a continuous water treatment system (Fig. 1). The starting point is usually *potable water* that is fed to the site from a municipal water supply or local water well. Before the water can be further purified in a facility, for use in a pharmaceutical or biotech process, it must first meet the quality attributes of the Environmental Protection Agency National Primary Drinking Water Regulations or comparable regulations such as those of the European Union or Japan (2). As the first step, the end user may filter the potable water to remove any particulates that are carried by the water into the facility. Potable water is not usually a good choice for CIP in processes other than for intermediate products or APIs because it still contains dissolved solids and can have small particulates and dissolved solids that could contaminate the bulk or finished product.

A softener bed (Fig. 2) is used to remove substances such as calcium and magnesium (cations) from the water to minimize scale deposits in the plant utility systems and more importantly, the water purification filters and distillation units. But, before softened water can be further purified, it is passed through an activated carbon filter to remove oxidizing substances (e.g., chlorine and its compounds) and

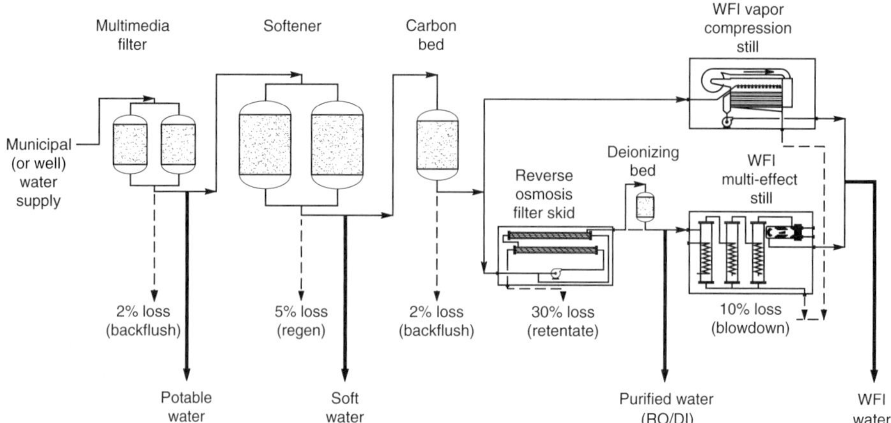

FIGURE 1 Typical water pretreatment and purification system. *Abbreviation*: WFI, water-for-injection.

FIGURE 2 Dual water softener system. *Source*: Courtesy of Lonza Biologics, Portsmouth, New Hampshire, U.S.A.

low molecular-weight organic materials before it is finally purified by reverse osmosis and/or distillation.

If USP purified, reverse osmosis/deionizing (RO/DI), or highly purified water is your compendial process water, a RO filter system with a DI filter bed is the most common way to meet the USP monographs for conductivity, pH, total organic carbon (TOC), and bioburden. These grades of water are generated and distributed around the facility at ambient temperature.

With WFI there are two distillation options that are dominant in the industry now: vapor compression still directly after the carbon bed (Fig. 1) or multieffect still fed by purified water. Both stills can produce water that meet the USP monographs for conductivity, pH, TOC, bioburden, and endotoxins. However, with the use of vapor compression technology, the retentate water that is lost during the filtration of the still's feedwater, is now conserved (Fig. 3). Some facilities are producing WFI or *highly purified water* by filtration alone in lieu of distillation. WFI is generated and distributed around the facility hot (65–80°C), chilled (20–25°C), or ambient temperature. Each one is acceptable for CIP but hot WFI is preferred because it does not require much heat to be added by the CIP skid.

For more information on compendial waters, when to use them, how to pipe them throughout the facility, and what validation and testing is required the designer can refer to the latest editions of the USP, ASME BPE, or the International Society for Pharmaceutical Engineering baseline guides. This chapter limits the discussion of these systems to their design as they relate to the CIP systems and their design for efficiency.

FIGURE 3 Vapor compression water-for-injection (WFI) still. *Source*: Courtesy of Lonza Biologics, Portsmouth, New Hampshire, U.S.A.

In the end, the grade of water the CIP system uses for the pre-rinsing and cleaning the equipment and piping in a given circuit is up to the designer. Softened water, USP purified, or USP WFI are the typical choices in the U.S. highly purified water or RO/DI water (that are not necessarily tested to the USP compendial monographs) may be the water of choice for cleaning equipment that is used for manufacturing APIs and oral dosage forms.

CHOOSE YOUR WATER WISELY

As water usage increases and the cost of water increases, the designer needs to carefully select the grade of water for both the wash and rinse phases of the CIP cycle. It is typical to choose the purest water available in the facility and justify its usage for all phases by saving money on capital costs such as piping and controls. But, at U.S. $1 to U.S. $2 per liter for WFI and other highly purified waters, you pay for the additional capital cost of connecting to another, lower cost water *in the first couple of cycles*. CIP systems also tend to "compete" with the process users for the compendial water due to the higher instantaneous flow rates required in CIP systems occurring at the same time a large process user is drawing from the loop. This can cause process-scheduling problems that can extend the batch times. Save the compendial waters for the final rinses.

For example, softened water is a good choice for the cleaning and intermediate rinses in most biotech facilities that produce parenterals. Potable water is very common as the primary cleaning solvent and rinse water in APIs and solid oral dosage process. The latest edition of the USP states, in the section "Water for Pharmaceutical Purposes" that "Drinking water may be used in the early stages of cleaning pharmaceutical manufacturing equipment and product-contact components" (2). In any case, a minimal amount of piping, filtering, and periodic sampling at the CIP skid will certainly save the end user money, water, and energy required to purify and validate the water to meet compendial monographs. Compendial water does not clean better than soft water when you are mixing it with alkali or acid and the additives that improve sheeting and contact time. Why not use softened water for all the pre-rinse, wash, and intermediate rinse steps? You won't save water in the cycle itself, but you will save the 15% to 30% blowdown that flows from the filter skid or distillation unit that generates the compendial water.

For processes where WFI or USP purified water is the primary ingredient/excipient, the final rinse cycle (as a minimum) should be supplied from the compendial system with proper point-of-use design and regular sampling. It is necessary to supply the compendial water to the CIP skid with the same control that a designer would supply a process skid. With the CIP system being used regularly and requiring a high volume of water in each cycle, the point-of-use piping gets regular use and flushing.

Hot WFI is a popular choice for all phases of a CIP cycle. It helps to heat the CIP skid and associated circuit piping so it does not require as much additional heat to get all the piping and equipment in a circuit up to temperature. Hot WFI also has the ability to sanitize the water tank on the CIP skid with the first couple of burst rinses of that tank.

GETTING THE WATER TO THE CIP SYSTEM

Softened water or potable water systems are usually piped in copper tubing, PVC tubing, or in some cases 304 stainless steel pipe. The piping is run in a branched system where the water only flows when a user of the water opens a valve. Both softened and potable water are distributed through a facility at ambient temperature. In colder climates or where the facility is fed from a well, the water temperature may be cold enough to exceed the heating capacity of a CIP skid if not planned for in the design.

Depending on the piping system used, it is advisable to filter the softened or potable water before it enters the tank to remove any particulate that may shed from the piping system or softener beds over time (Fig. 4A). In some cases softener resin beads or filter media is able to pass through the water pretreatment system and should be filtered before it is pumped through the process piping and equipment during the CIP cycle.

Another precaution when using softened water for pharmaceuticals and biotech processes: depending on the efficiency of the softener beds and the chemical composition of the water, softened water can have various levels of *stability*. A soft water that is excellent for feeding a stainless steel WFI still may have a low stability that can corrode copper tubing systems quickly. The designers of the water and CIP systems should be aware of the piping material requirements of soft water just as they would for high purity water.

Compendial waters are delivered from recirculating loops that are designed with features that eliminate branches with low flow and provide turbulent flow through the whole system to minimize bioburden. The loop circulates the compendial water continuously and is validated to maintain the water's temperature, pressure, and chemical quality. The constant velocity and in some cases, high temperature (65–85°C) help to maintain the water system with less biofilm for

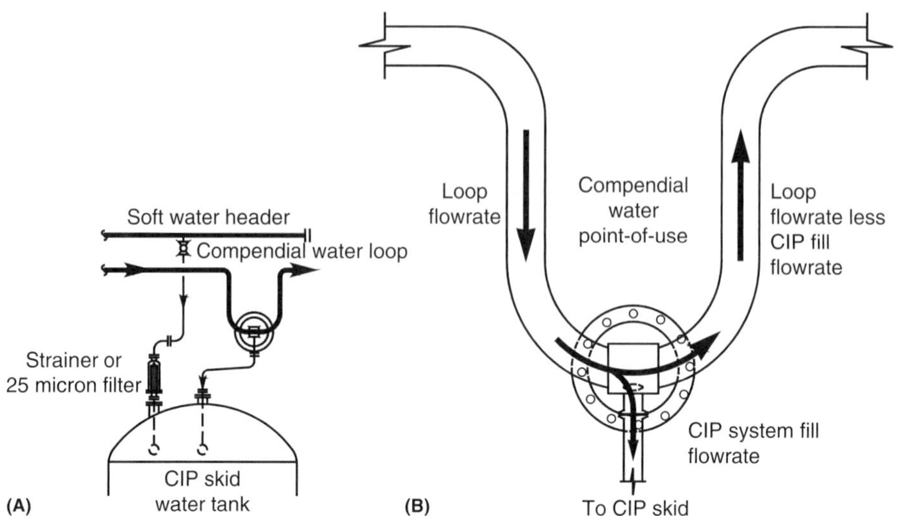

FIGURE 4 (**A**) Typical CIP skid water connections. (**B**) Compendial water point-of-use water flow paths.

longer periods between sanitization. It is important to note that the ability to sanitize a water system, including the point-of-use piping, by heat or chemical means is as important as the design for flow and pressure within the loop. The designer needs to be aware of the compendial water system's parameters and sanitization program because they may need to be incorporated into the point-of-use piping and water tank on the CIP skid.

Special point-of-use valve designs are used in loop systems to deliver the water from the loop without compromising its integrity and reducing downstream flow (Fig. 4B). The valves can be equipped with interlocks or flow restriction devices that prevent too many users from affecting the proper flows and pressures in the system. (Fig. 5) With flow and pressure in a looped system being inversely related (with a centrifugal pump), a large user like a CIP skid calling for water can reduce the system pressure or downstream flow below an acceptable level.

The ASME BPE standard provides two examples of point-of-use piping designs that are appropriate for use at a CIP skid (Fig. 6A,B) Figure 6A shows a simple direct-connect from the point-of-use valve to the top of the tank with a short run of sloped piping. This is the most common design for feeding hot WFI or ambient USP purified water to a CIP skid. Figure 6B shows a point-of-use piping design that allows for a longer run of piping that can be blown dry or steamed as needed. This design coupled with a steamable water tank on the CIP skid could complete a water tank design that is SIPable. This design would be appropriate for ambient WFI where additional microbial control might be needed for the point-of-use piping and CIP water tank. Both of these designs are drainable and intended to be left empty after filling a tank or flushing.

With a large user like CIP connected to a water system, the designer needs to consider the higher demand on all associated water systems for generation capacity as well as distribution capacity and pressure. Increasing the storage and distribution capacity of the system well beyond what is required for plant utilities and

FIGURE 5 Point-of-use valve assembly at CIP skid: Hot WFI loop, automated valve (interlocked), integral sample valve, flow restriction orifice. *Source*: Courtesy of Human Genome Sciences, Rockville, Maryland, U.S.A.

personnel use will need to be considered. Large storage tanks, pumps, and piping diameters should be considered for supplying the higher flow rates required by CIP systems (Fig. 7). If the other users of these water systems are considerably smaller, it may be advisable to use a dedicated branch or loop that feeds the CIP skids from a larger pump that can adjust to the large variations in flow demand.

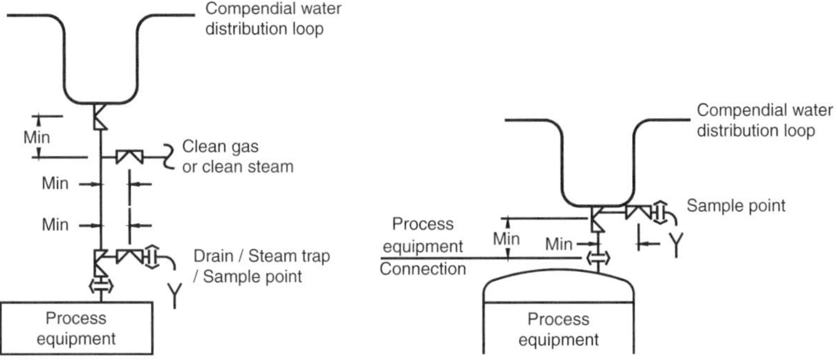

FIGURE 6 Point-of-use valve assembly. Courtesy of American Society of Mechanical Engineers Bioprocessing Equipment (ASME BPE)-2005.

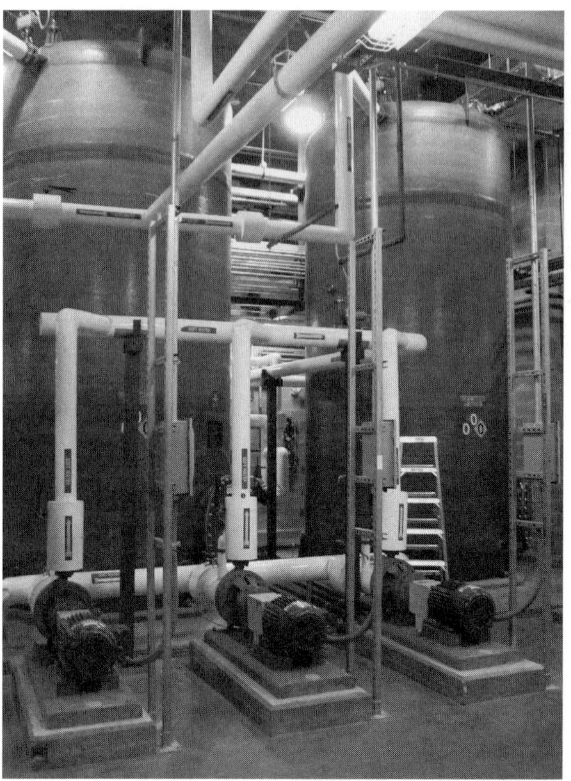

FIGURE 7 High-capacity soft-water storage and distribution: 304L stainless steel piping, fiberglass reinforced plastic (FRP) storage tanks, insulated for sweat control, and installed spare pump. *Source*: Courtesy of Lonza Biologics, Portsmouth, New Hampshire, U.S.A.

The designer of the water system can install several features into a compendial water loop that helps it adjust between the normal recirculated flow rate and the peak demand flow rates imposed by several CIP systems and process users drawing water from the loop at the same time.

First, select a pump with a flatter pump curve that can deliver a wider flow range with less pressure change. Within that model, size the pump to deliver the peak-circulated flow rate at 90% of the motor's maximum rpm. This well-chosen pump installed with a variable frequency drive can respond to the widest range of flow demand while operating at high efficiency. As with any pump selection, it is usually better to purchase the pump that is rated for your required flow and pressure, with the impeller that trimmed to approximately 85% to 90% of the largest diameter to allow for additional capacity with the purchase of a larger impeller later (spend the extra money and purchase the pump with the motor that will turn the largest impeller within the 1.0 service factor). These are general guidelines for pumps that run continuously (such as in a compendial water loop, Fig. 8) and are required to run reliably between planned maintenance. But, these pumpsizing guidelines also help the designer supply water over a greater flow rate range.

FIGURE 8 High-capacity, compendial water distribution pumps. *Source*: Courtesy of Lonza Biologics, Portsmouth, New Hampshire, U.S.A.

HOW WATER IS USED IN CIP SYSTEMS

CIP skids can be designed to fill the circuit and recirculation tank with water in two ways. One involves a second tank on the skid that is always kept full of water to draw from as needed. The other fills the circuit and recirculation tank directly with water before each phase of the wash. Each design has its pros & cons but both designs are used frequently.

The CIP skid with a tank favors a smaller water system that has process schedule or flow constraints. With a water tank mounted on the CIP skid, the cycle does not have to wait for water from the distribution system at each filling phase of the cycle. It merely feeds the CIP supply pump from the skid's dedicated water tank. The water tank levels the load on the water system over time and allows the water-use schedule to be de-coupled from the processing schedule when there is limited flow rate to work with. The greater the flow constraints the larger the tank needs to be. Obviously, the addition of any tank to the CIP skid adds to the size and cost of the skid, so it's not just as easy as putting a large tank on a CIP skid to solve all process schedule problems.

A dedicated water tank on the CIP skid may actually increase or decrease the amount of water used per cycle depending on how the cycle fills and flushes the tank. The cycle may call for the vessel to be flushed to drain excessively before each fill. The tank is usually flushed to drain before it is filled the first time each day but a designer may choose to flush each tank on the skid before each cycle thereby using significantly more water. On the other hand, the water tank may be used to save water if the leftover final rinse water from one cycle is utilized for the pre-rinse of the following cycle. The final rinse water is usually compendial water and in most cases would be hot. The designer should set a time limit for how long the water can be used after a cycle is completed. The time could be determined from the time it takes for the water to cool from 80°C to 65°C thereby remaining sanitized.

If a dedicated water tank is mounted on a CIP skid, it must be designed to maintain the quality of the water that it is stored for a validated length of time. Features like insulation and even a steam-heated jacket may be required for maintaining the temperature of the water for up to a day or two. If the tank is storing compendial water, it is necessary to put a vent filter with a 0.2 μm (absolute rated) filter element on the tank to vent the tank during filling, draining, or thermal changes. During validation and regular operations, sampling of the compendial water from the tank is typically required. Make the job easier and safer by purchasing the tank with a lower sidewall mounted sample valve. Finally, periodic steaming of the water tank is sometimes required, if there is a provision for porting clean steam to the top of the tank and installing a steam trap below the drain valve this operation can be completed quickly and safely when needed.

The other option for filling the CIP circuit with water is directly to the CIP recirculation tank or CIP supply pump depending on the skid design. This skid design is the smallest and lowest capital cost but puts higher instantaneous demand on the facility's water systems. If the designer chooses this method of filling the CIP circuit, the water system(s) need to be designed for the higher flow rates and pressures. The benefit of this design is shorter CIP cycle times because the system only fills with the volume of water required for one phase of the CIP cycle. With a direct fill CIP system it is a good time to consider a dedicated loop or branch on the water system to supply the CIP skid. See Figure 9 for an example of a dual-loop water system that has a loop dedicated to the CIP skids that runs independently from the process loop. With this arrangement, neither the CIP skid(s) nor the process users see any drop in pressure or flow when the other is drawing from their loop.

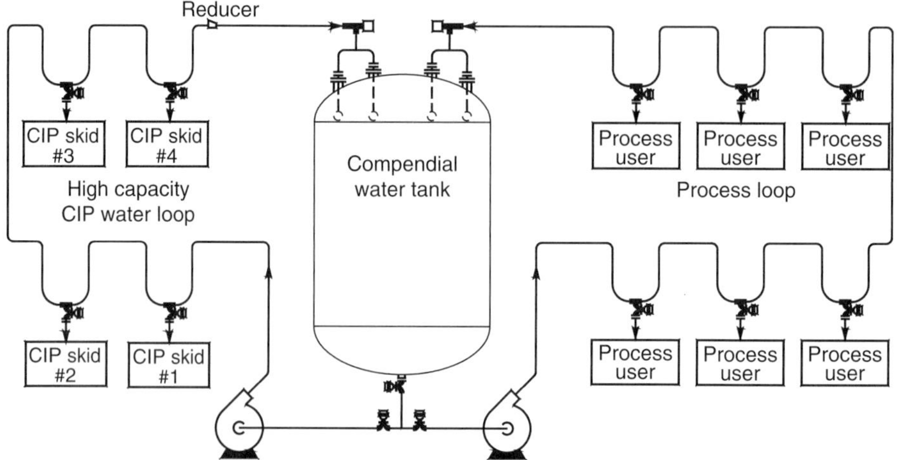

FIGURE 9 Dual compendial water loop with dedicated CIP water loop.

WATER USAGE IN FUTURE CIP SYSTEMS

As a designer responsible for a new CIP system, it is important to understand that *more* water in a CIP cycle does not clean *better*. The design concepts found through this book along with simple water conservation methods can actually clean better than just flushing a circuit with more water. A well-designed CIP *program* (code) uses water and energy efficiently and therefore has a big effect on minimizing the operating cost of a new facility. Also, designing piping and equipment for drainability and ensuring drainage takes place between CIP steps will minimize water usage during the rinse steps because there is less process fluid or chemical wash solution left behind.

The cost of water is increasing both in its potable state as delivered to the facility and also in its purified state after various levels of pre-treatment, filtration, and distillation. Increased municipal water costs (and wastewater treatment costs) are making water not as "low-cost" a cleaner as it once was. After figuring capital, operating, energy, and quality assurance costs, compendial waters can cost $1 to $2 per liter depending on the facility. Remember, it is not necessary or recommended to use the best quality (or compendial) water available in the facility for all steps in the CIP cycle. In fact, it is only necessary to complete the final rinse with the compendial water if that is the grade of water used in the associated process equipment and piping.

During start up and testing of a new system, the circuits should be optimized to minimize the amount of water used at the same time that they are validating the cleaning capability. Optimization of CIP circuits is rarely planned for in the projects schedule or is forfeited for the sake of making up time lost for other reasons. Optimization can be as simple as not keeping the water tank on the skid full through the cycle just to dump it to drain when the cycle is complete. A well-designed water tank on a CIP skid will maintain the water quality for 12 to 24 hours and deliver the necessary quality of water for the first rinse steps in the next CIP cycle. Other water optimization steps include choosing spray devices that clean by impingement force in lieu of flooding the soiled surface. Or, choosing valves that are easier to clean thereby reducing the number of rinse steps required to meet the final rinse water TOC and conductivity. A cleaner piping design will save water.

REFERENCES

1. The American Society of Mechanical Engineers, ASME BPE-2005, Bioprocessing Equipment Standard, Three Park Avenue, New York, NY 10016.
2. USP28, General Chapter <1231>, Water for Pharmaceutical Purposes. (See note.)

4 The Composition of Cleaning Agents for the Pharmaceutical Industries

Dietmar Rosner
Ecolab GmbH & Co. OHG, Düesseldorf, Germany

THE CLEANING PROCESS

Cleaning is a complex process based on both chemical and physical principles. The primary task of cleaning is the detachment of soil from a surface. Soil can be attached to surfaces by a combination of three physical effects as illustrated in Figure 1, including van der Waal's forces, electrostatic effects, and mechanical adhesion.

The sum of these effects can be termed as soil adhesion. In the cleaning process, soil adhesion has to be subdued by providing forces counteracting the adhesion. These forces can be reduced to the four basic parameters of any cleaning process.

1. The cleaning temperature
2. The cleaning time
3. The cleaning kinetics (mechanical effects)
4. The chemical activity of the used cleaner

The chemical activity depends primarily on the efficacy of a specific cleaner regarding a specific type of soil. It is a very complex phenomenon based on direct chemical reactions as well as the physical principles of desorption, dispersion, emulsification, and dissolution. The detergent concentration, as often stated, is only a factor of secondary importance. The higher the chemical activity, the lower the concentration required!

The four basic parameters of the cleaning process are all dependent on each other. Reducing or increasing any of these components will alter the balance of the other three; however, all four segments are required. If only one of them is omitted, then no cleaning occurs. Figure 2 illustrates the cleaning circle according to Sinner, which demonstrates these cleaning factors and their interaction.

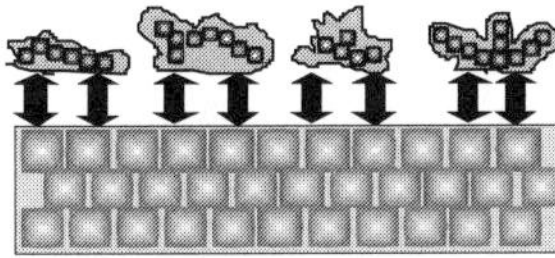

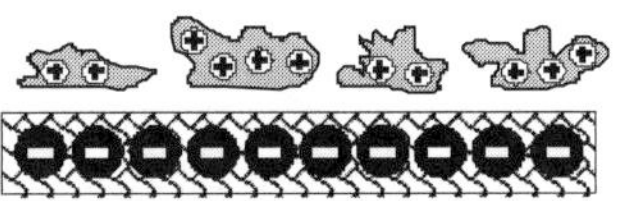

FIGURE 1 Soil adhesion on surfaces.

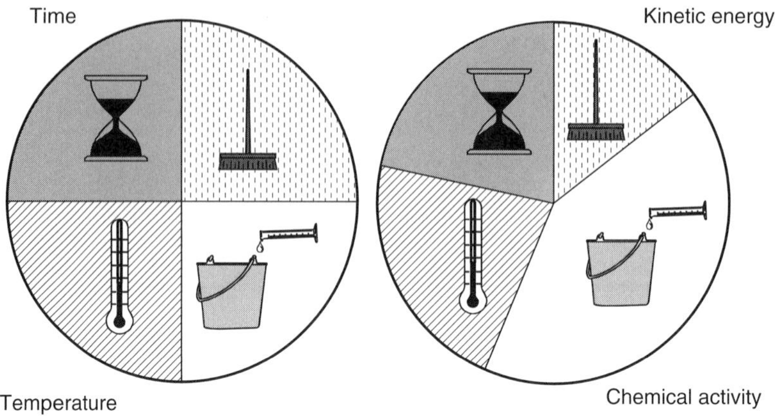

FIGURE 2 The cleaning circle.

The chemistry of cleaners with reference to the specific soil in pharmaceutical processes will now be addressed.

SOILS IN THE PHARMACEUTICAL INDUSTRIES AND THEIR INFLUENCE IN THE CHOICE OF CLEANING MATERIALS

As previously mentioned, the chemical activity of a cleaner has to be selected depending on the particular soil to be removed. However, the three other factors, time, temperature, and kinetics, will also influence the choice of the cleaner. If the cleaner is chosen correctly to match the demands of the soil, then any shortcomings in the other factors may be compensated by increasing the cleaner's concentration. In short, it is the actual soil, which is the main influence on the choice of the cleaning agent (and its composition).

The pharmaceutical industry faces a predominant problem with cleaning in the sheer number of raw materials used to manufacture a wide spectrum of different products. Many such raw materials are, for good reasons, kept secret. Table 1 is an attempt to place the product variants in a "cleaning hierarchy."

Formulating a universal detergent that will clean every plant is impossible because of the vast range of raw materials. Even using a wider range and

TABLE 1 The Influence of Soils on Choice of Cleaning Materials

Liquid forms	Semi-solid forms	Solid forms
Infusion solutions	Water/oil or oil/water emulsions	Tablet coatings
Syrups	Fats, oils, and waxes	Soluble in gastric acid
Tinctures	Polyacrylates	Permeable coatings
Carbohydrate solutions	Dyes	Powder preparations
Protein solutions		Granulates
Mineral solutions		Tablet bursting aids
		Gelatin
		Fillers
		Pigments

combination of different detergents, it takes a great deal of experience and numerous laboratory trials to find a specific solution to an individual cleaning problem. Care must be taken when processing similar products under different manufacturing conditions as this can alter the soiling and could call for a different cleaning procedure.

DEVELOPING A CUSTOMIZED CLEANING METHOD

For aforementioned reasons, the approach to developing a customized cleaning method for a pharmaceutical process is rather different and more empiric when compared with, e.g., food industries.

The development starts with a specimen obtained from the manufacturer. Depending on the information given, selective solubility test will be carried out under standardized conditions (temperature, volume, size of beakers, agitation, etc.) using alkaline, neutral, and acidic cleaning solutions. The testing needs permanent attention because the speed of dissolution has to be observed. At the end of the test, insoluble matter is retained by filtration and used for further solubility tests in a cleaning solution with opposite pH. Still insoluble matter is tried with using cleaning booster as oxidizers, surfactants, soil-dispersing agents, or others. This trial is aimed to provide basic information about how to tackle the soil.

Basing upon the information acquired in these tests, a second series of tests is started. The test material is applied to standardized stainless steel plates (steel type, surface roughness, and size). The plates with the applied soil are "aged" under controlled conditions and used in a simultaneous cleaning trial involving cleaning solutions that have proven most successful in the preliminary trial. This test aims to differentiate between cleaners of a similar type and to optimize temperature and concentration. Mechanical effects are kept strictly constant. At the end of the test, the amount of remaining soil on the test sheets is determined. These tests provide not only information about the quantity of soil removed at a time (cleaning velocity), but also about other effects important in a cleaning process as emulsification properties and prevention of soil redeposition.

The testing can result in a single-phase cleaning process (pre-rinse–cleaning solution–post-rinse) or a two-phase cleaning (pre-rinse–first cleaning solution–intermediate rinse–second cleaning solution–post-rinse).

The cleaning solutions may contain one, two, or in some cases even three formulated cleaners to achieve the best result.

The best results found in the laboratory trials are now transferred to practice. The procedure has to prove its performance in a practical trial to undergo finally a validation process.

The work and time required from the first testing to the validation process depends very much on information about the soil (product) available, on the existence of information about similar soil types (cleaning database), and the skills of the experimenter.

THE GENERAL COMPOSITION OF DETERGENTS

Cleaners may be classified by their physical appearance, liquid, solid or powder, or by their purpose of use. However, the most appropriate method to investigate into the composition of industrial cleaners is to start with a logical structure that divides cleaners into groups depending on their pH.

Classification of Cleaners by pH-Value or as Additives to Alkaline or Acidic Cleaners

The majority of industrial cleaners are either alkaline or acidic. Neutral products are mainly used for manual cleaning, sensitive surfaces, or as additives to alkaline or acidic cleaners (Table 2).

Acidic Cleaners

Inorganic soils are more soluble in acids than in alkalis. However, the choice of an acid base for a cleaner cannot be easily made like that of its alkali counterpart in alkaline products. The choice of acids may be restricted for a number of reasons, including the corrosion of equipment. Compatibility with other cleaning components and the effect on the environment should also be considered. Acidic cleaners may be divided into *acidic descalers* and *acidic cleaners*.

Acidic Descalers

These products serve the sole purpose to remove inorganic scale, as e.g., water scale. They consist mainly of one or a blend of two acids and a corrosion inhibitor, when intended to be used on acid-sensitive surfaces.

Acidic Cleaners

Since inorganic scale does often appear in combination with organic scale too, it would be advantageous to have cleaners that could deal with both types of soils. By combining surfactants with selected acids, it is possible to formulate acidic cleaners with a wider range of efficacy. The application of surfactants can cause foam problems, especially in clean-in-place (CIP) applications. This problem can easily be solved by adding foam controlling agents, so-called defoamers.

Neutral Cleaners

In most cases, the term neutral cleaner is used a more generously than the physical term "neutral" (pH 7!) permits. "Neutral cleaner" often stands for a mild cleaner in a pH range from about 6 (human skin) to about 10 (hand soap). The cleaning the efficacy of these cleaners is based on the detergency of surfactants supported by

TABLE 2 Classification of Cleaners by pH-Value

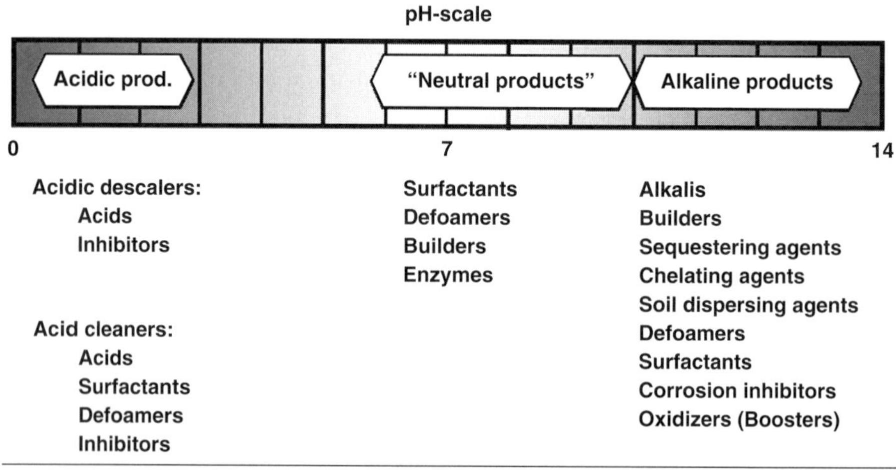

Acidic prod. (0)	"Neutral products" (7)	Alkaline products (14)
Acidic descalers:	Surfactants	Alkalis
Acids	Defoamers	Builders
Inhibitors	Builders	Sequestering agents
	Enzymes	Chelating agents
		Soil dispersing agents
Acid cleaners:		Defoamers
Acids		Surfactants
Surfactants		Corrosion inhibitors
Defoamers		Oxidizers (Boosters)
Inhibitors		

so-called builders. For applications that require foam control, defoamers can also be added. Such a product can be very efficient regarding hydrophobic soils and fat, but other organic matter like proteins might constitute a problem. To overcome this problem, there are neutral cleaners containing enzymes to deal with proteins and some other organic materials. Cleaners of this kind are used to clean pH-sensitive membranes used in cross-flow filtration processes.

Alkaline Cleaners

Strong alkalis, such as sodium hydroxide, can react with functional groups of organic matter and convert the organic soil into a water-soluble compound or change the chemical or physical nature of the soil so it can be more easily removed. In addition, some cleaning compounds show an increased efficacy in an alkaline environment. Sodium hydroxide, especially when dealing with organic residues, is the usual choice for a cleaner's base. In some cases, especially when formulation a strong alkaline products containing other cleaning compounds in a high concentration, it is advantageous to use more expensive potassium hydroxide. Potassium hydroxide also improves the freeze stability of alkaline cleaners.

However, sodium hydroxide (as well as potassium hydroxide) alone does not make a good cleaner, since it has the following drawbacks:

- Precipitates water hardness
- Very limited soil suspending ability
- Insufficient penetration into soil
- Reactions with organic soil, e.g., proteins, may form high-foaming substances
- No cleaning effect on mineral residues
- Removal of fatty substances not always satisfactory
- The cleaning effect of straight NaOH is very dependent on the type of soil and the NaOH concentration (Fig. 3).

For every type of soil there is an optimum concentration of NaOH for fastest soil removal. Too high a concentration, for example, actually has a negative effect on cleaning power as well as too low a concentration, because it can lead to unwanted reactions with the soil, such as coagulation, gelling, or hardening, making the soil more difficult to remove. Every type of soil has on an optimum concentration of NaOH for its

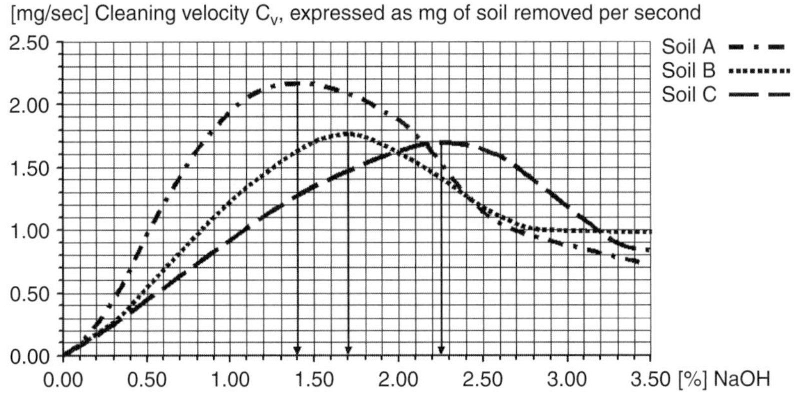

FIGURE 3 Cleaning velocity for NaOH.

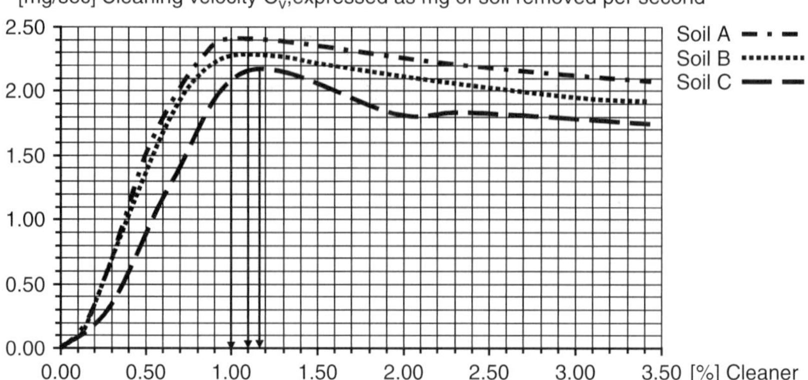

FIGURE 4 Cleaning velocity for a formulated cleaner.

fasted removal. Figure 3 shows the "cleaning velocity" C_V (mg/sec) for three different types of organic soil (A, B, and C) depending on the concentration of NaOH.

Figure 3 shows that soil A is removed best at 1.4% NaOH, soil B at 1.7%, and soil C at 2.25%. In case of a single type of soil, one could easily choose the concentration for an optimum cleaning. However, which concentration is best when a mixture of all three types of soil is present? Any choice would be a compromise prolonging the cleaning procedure for at least two types of soils.

By blending other detergent components into NaOH, it is possible to formulate a complete alkaline cleaner that does not display this characteristic and is effective against a wide range of soil types. The alkaline cleaner shown in Figure 4 has the same efficacy with all three types of soil and in contrast to sodium hydroxide alone, its cleaning effect is less dependent on the concentration.

Figure 4 also shows that with the formulated cleaner, a better result with all kinds of soil is obtained with even a lower concentration when compared to NaOH alone.

Now it becomes obvious that a good cleaner has to be formulated according to the requirements of the soil to be removed. A standard household detergent or even an industrial cleaner designed for dairy or food industries would not be the best choice for cleaning in the pharmaceutical industry. It is not only the soil, but also the pharmaceutical industries' requirements on a product's consistency, purity, etc., which demand an exclusively, dedicated product range for cleaning production equipment.

THE MAIN CONSTITUENTS OF CLEANERS AND THE PURPOSE IN THE CLEANING PROCESS

Builders

Builders are substances that enhance the cleaning effect and in particular the effect of surfactants. Builders also support the detachment of soil from surfaces. Builders belong to the group of chelating agents. The most important builder substances are polyphosphates, phosphonates, gluconates, citrates, nitrilotriacetate (NTA), and ethylenediamine tetraacetate (EDTA). Figure 5 demonstrates the effect of a builder.

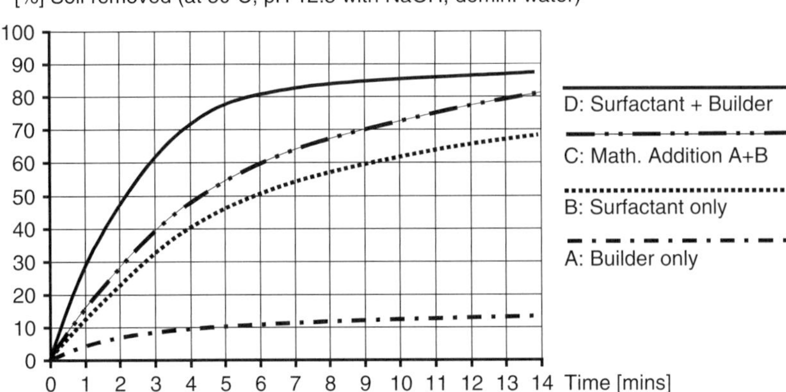

FIGURE 5 Cleaning velocity for NaOH plus surfactant and builder.

Under the given test conditions, the builder on its own could only remove about 12% of the total soil on a defined surface after 10 minutes of cleaning. Using the same conditions, a surfactant could remove 62%.

However, the combination of both surfactant and builder was able to remove 85%. This is even more as expected when adding both results obtained with the surfactant and the builder on their own (73% after 10 minutes). This clearly demonstrates that the components work in synergy to produce a superior and more efficient cleaning effect than when used individually.

Surfactants

Surfactants are substances consisting of molecules with a distinct physical polarity caused by the fact that their molecules contain both hydrophobic and hydrophilic sections. Taking a molecule of soap as an example, this can be clearly demonstrated. Soaps are well-known anionic surfactants produced by saponification of fats or vegetable oils with alkali (NaOH or KOH). During the saponification, the fat (or oil) is split into its base materials, glycerine, and fatty acid. The fatty acid immediately reacts with the alkali forming a "salt" due to a neutralizing reaction shown in Figure 6.

The soaps function as active cleaning substance is determined by its molecular structure as shown in Figure 7. The ionic bond contained within the molecule gives the soap certain water solubility thus the molecule section is hydrophilic. In contrast, the fatty acid tail is hydrophobic as the original fat. This polarity of the molecules causes them to react in a certain way when in aqueous solution since the hydrophobic

Oil		Glycerine	Soap
$H_2C\text{-}OOC\text{-}CH_2\text{-}(CH_2)_n\text{-}CH_2\text{-}CH_3$		$H_2C\text{-}OH$	$Na\ OOC\text{-}CH_2\text{-}(CH_2)_n\text{-}CH_2\text{-}CH_3$
$HC\text{-}OOC\text{-}CH_2\text{-}(CH_2)_n\text{-}CH_2\text{-}CH_3$	+ 3 Na OH	$HC\text{-}OH$	$Na\ OOC\text{-}CH_2\text{-}(CH_2)_n\text{-}CH_2\text{-}CH_3$
$H_2C\text{-}OOC\text{-}CH_2\text{-}(CH_2)_n\text{-}CH_2\text{-}CH_3$		$H_2C\text{-}OH$	$Na\ OOC\text{-}CH_2\text{-}(CH_2)_n\text{-}CH_2\text{-}CH_3$

FIGURE 6 Saponification of an oil.

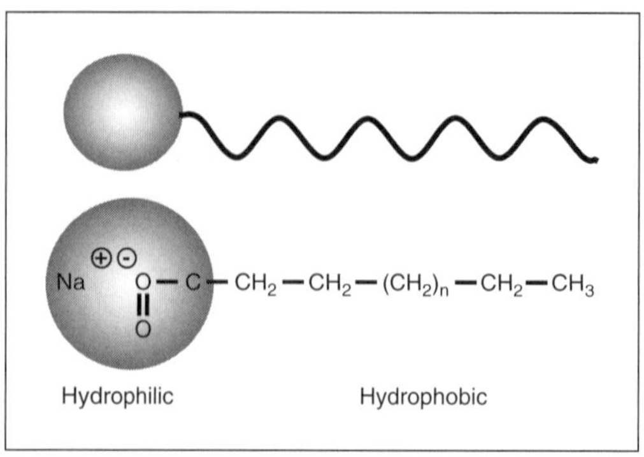

FIGURE 7 Molecular structure of a soap.

part is repelled by the water whereas the hydrophilic part is attracted. This means initially the surfactant molecules arrange themselves in the interface between the water and its surroundings, i.e., the air and the container walls. This is done in such a way that the hydrophobic section of the molecule is kept away from the water (Fig. 8).

With increasing surfactant concentration, all interfacial surfaces become occupied and the surfactant molecules begin to form globular bodies (micelles) with the hydrophobic sections of the molecules pointing inwards.

If the water contains hydrophobic particles such as oil droplets, these can be enclosed as the micelles are formed. This is due to the hydrophobic part of the molecules attaching themselves around the oil droplet (interface oil water). As the droplet becomes completely enclosed by the surfactant molecules (and its hydrophilic groups finally face the outside of the globules), the oil droplet can then be emulsified in the water and dispersed. Through this process, water is converted into a "solvent" capable of "dissolving" hydrophobic substances, such as fat and oil.

Types of Surfactants

There are different types of surfactants that are classified according to the ion characteristic of the cleaning-active group (Fig. 9).

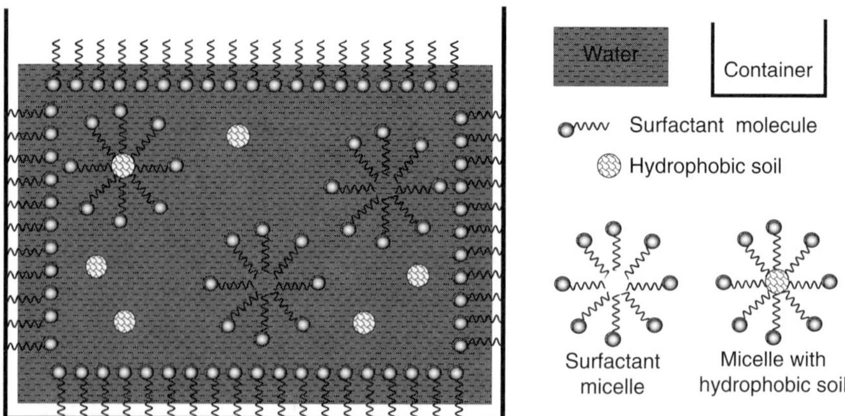

FIGURE 8 Behavior of surfactants in water.

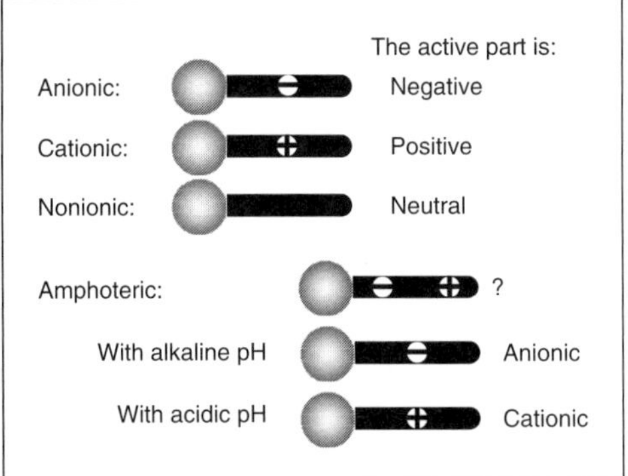

FIGURE 9 Classification of surfactants.

A distinction can be made between anionic (e.g., soaps), cationic (mainly quaternary ammonium compounds), and amphoteric surfactants. Its almost exclusively nonionic and anionic surfactants that are used as components for detergents, whereas cationic and amphoteric surfactants are used to formulate disinfectants due to their microbicidal effect.

The Chemical Structure of Nonionic Surfactants

Nonionic surfactants are produced by reactions of fatty alcohols or fatty amines (both made from natural oils) with ethylene oxide (EO) alone, EO plus propylene oxide (PO), or PO alone (Fig. 10).

Ethylene and PO are two very reactive gases. The ring-shaped connection of the oxygen with the two neighboring carbon atoms is extremely instable. When these substances are brought together with, for example alcohols under suitable conditions, the ring-shaped bonds are broken. The hydrogen atom of the alcoholic hydroxyl-group (–OH) binds to the oxygen of the ethylene or PO respectively.

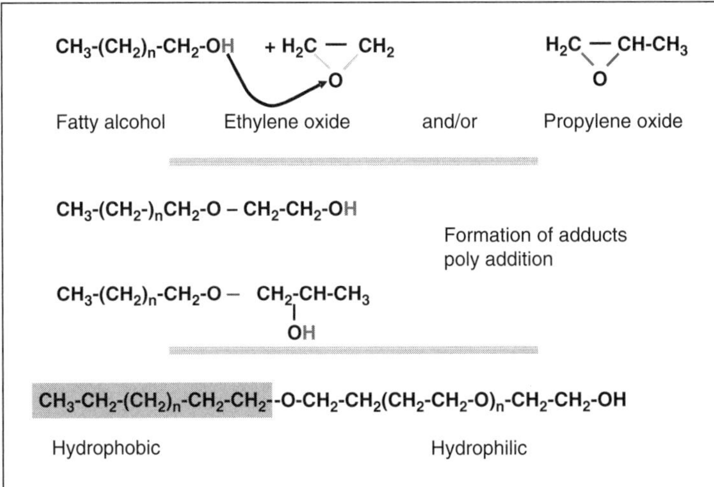

FIGURE 10 Nonionic surfactants.

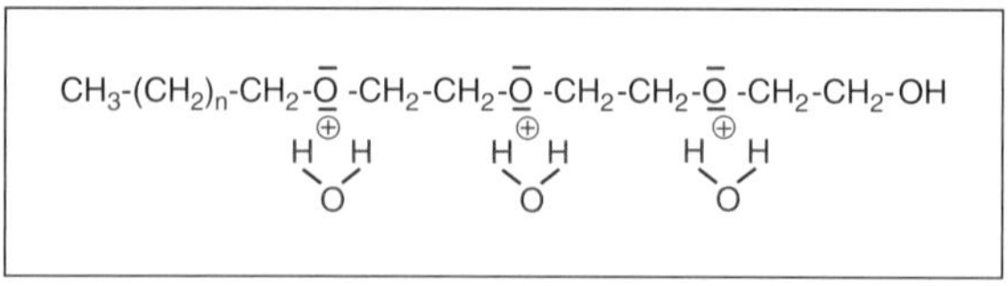

CH_3-$(CH_2)_n$-CH_2-$\bar{O}$-CH_2-CH_2-$\bar{O}$-CH_2-CH_2-$\bar{O}$-CH_2-CH_2-OH

FIGURE 11 Formation of hydrate shells on nonionic surfactants.

During this reaction, bonding electrons are released at both the EO (or PO) and the orphaned oxygen of the alcohol molecule. These form a new bond so that the EO (or PO) is added to the alcohol molecule. The newly formed molecule again possesses a terminal hydroxyl-group to which another EO molecule (or PO molecule) can be added. This reaction can be repeated time and time again or interrupted, when the desired end product is obtained. Such a reaction is called a "polyaddition."

The "ether-bridges" –CH_2–O–CH_2–, produced by this reaction can form hydrate shells in aqueous solutions, because the two free electron pairs of the oxygen can attract the positively charged side of water molecules. The molecule part with these ether bridges thus becomes hydrophilic (Fig. 11).

The water solubility and hydrophilic nature depend upon the number of ether bridges in comparison with the length and shape of the hydrophobic carbon chain of the fatty alcohol. The more ether bridges (added EO) the more hydrophilic and water-soluble is the final product. (When PO is used, however, ramified molecules are produced which are less hydrophilic).

The choice of the alcohol compound, the number of added ethylene or PO molecules (or a mixture of these two), allow very accurate control of the surfactants and their various properties (cleaning properties, solubility, foaming behavior, electrolyte tolerance, etc.). Since nonionic surfactants are easily tailor-made to fulfill a special function, they are the most commonly used surfactant type in CIP-suited formulations.

Effects of Surfactants in Cleaning Solutions

Surfactants offer more benefits in a cleaning process than just the emulsification of hydrophobic soil. They also reduce the surface tension of water. Surface tension results from the attraction of water molecules to other. In a liquid, a water molecule is surrounded by other water molecules, but the molecules at the surface have no neighboring molecules above (Fig. 12). This results in a stronger attraction towards each other and the molecules in the rest of the liquid, which results in the formation of globules. This phenomenon can be seen if water is poured onto hydrophobic surfaces (Fig. 13). Instead of leaving a continuous sheet of water (Fig. 14), several misshapen globular islands are formed.

During cleaning, the formation of droplets is an unwanted phenomenon since a cleaning solution should be evenly distributed over the entire surface. This enables a maximum contact area with the surface and the soil to be removed. Surfactants can counteract the surface tension of water due to their accumulation and orientation towards the interfaces (Fig. 15).

Figure 16 demonstrates the effect of an increasingly lower surface tension on the shape of water droplets on a hydrophobic surface. Alpha (α) is termed "angle of wetting" and is a measure to describe the wetting properties of a fluid on a surface. The decrease in surface tension actually relates very little to the surfactant's concentration, but is more dependent on the physical properties of the surfactant molecule.

FIGURE 12 Water molecules at the surface.

FIGURE 13 Droplets on surface.

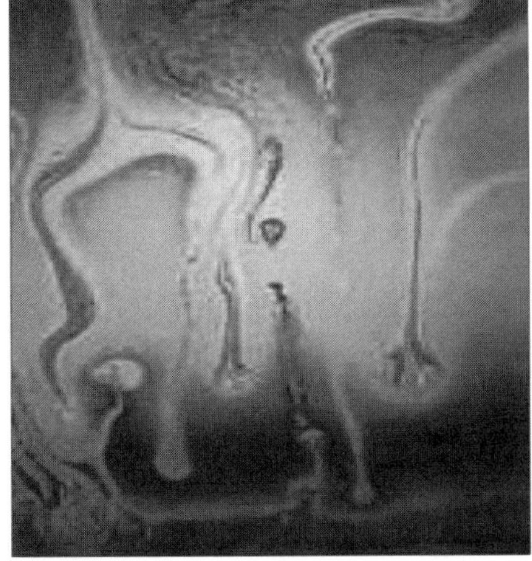

FIGURE 14 Continuous water film.

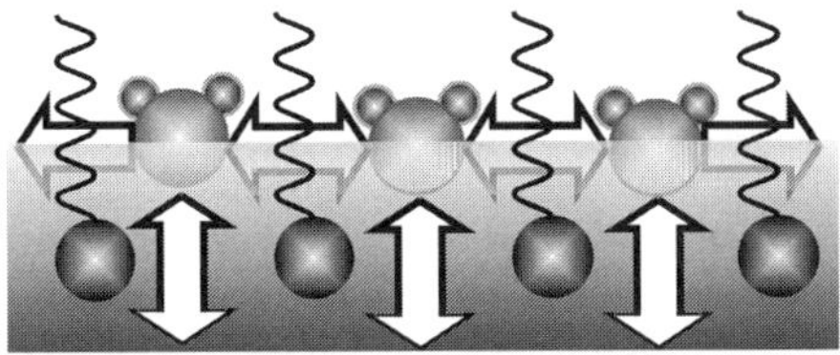

FIGURE 15 Surfactant molecules interfering with forces of attraction between water molecules at the surface.

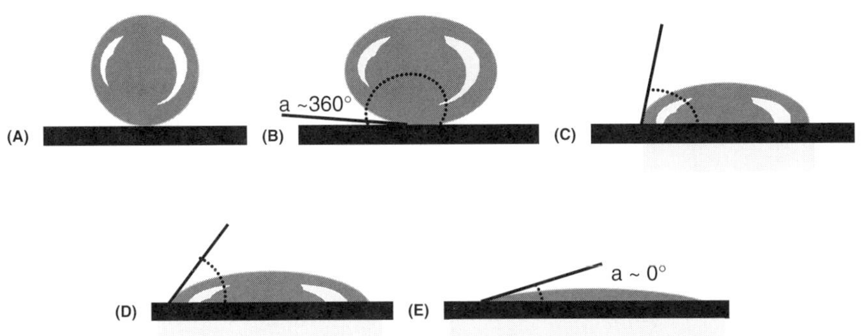

FIGURE 16 The effect of lowering the surface tension on a water droplet. (**A**) Without the influence of gravity, a water droplet would form a perfect sphere. (**B**) A water droplet under the influence of gravity. (**C**) Decreased surface tension by addition of a surfactant. (**D**) Further decrease by adding a more efficient surfactant. (**E**) Complete wetting using a highly efficient surfactant.

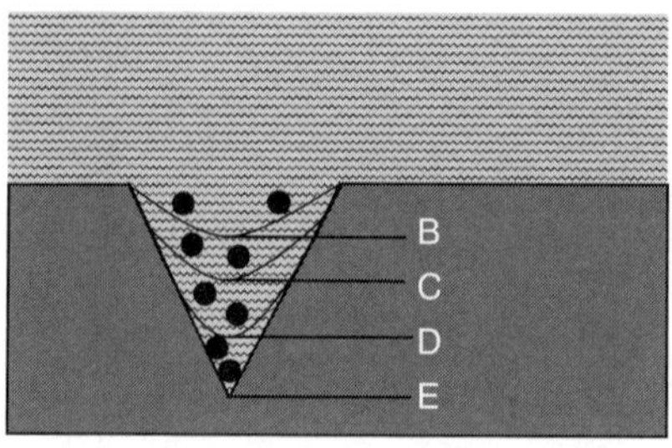

FIGURE 17 Capability of penetration.

The decrease of surface tension allows the cleaning solution to penetrate into crevices and small depressions within the surface to be cleaned. The lower the surface tension of the cleaning solution is, the better its penetration (Fig. 17).

However, there is no direct relationship between the ability to reduce surface tension and cleaning ability. There are examples which clearly demonstrate that, although some surfactants reduce the surface tension dramatically, they actually show very little cleaning efficacy. The cleaning properties of surfactants are largely determined by such factors as size and concentration of the micelles and the affinity to the soil in balance with the affinity to water.

Figure 18 demonstrates that there is no relationship between the surface tension and the cleaning power of a surfactant solution, even if a series of chemically very similar surfactants is chosen. Surface tension therefore cannot be used to measure cleaning power. Thus, the contribution of surfactants to the properties of a cleaning solution is influenced by both the decrease in surface tension (wetability) and how effectively the soil is emulsified and removed.

Chelating Agents

In alkaline cleaning solutions, water hardness has an adverse effect on the efficacy of the cleaning solution, especially on surfactants. In addition, alkalinity has very little effect on mineral deposits and inorganic components of soils found in manufacturing equipment. On the other hand, if an acidic cleaner suitable for inorganic soil removal is substituted then organic soil is not efficiently removed. By adding

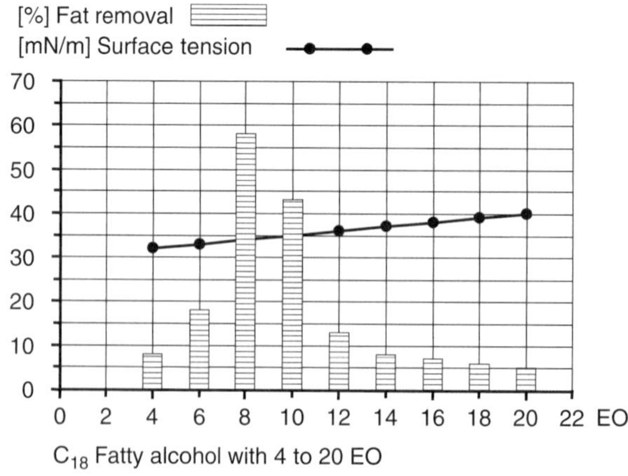

FIGURE 18 Fat removal and surface tension of surfactants at 25°C, 250 ppm concentration, and pH 7. *Abbreviations*: EO, ethylene oxide.

chelating agents to alkaline cleaners, cleaning effectiveness regarding inorganic soil and overall performance of the cleaner can be greatly improved.

In alkaline solutions, chelating agents can bind metal ions in strong complex bonding thus screening the ions from other influences. Typical chelating agents in industrial cleaners are EDTA, NTA, imido disuccinate, and gluconates (Fig. 19).

EDTA is the most powerful chelating agents regarding calcium and magnesium ions and forms very strong chelate bonds with many types of metal ions.

However, the same efficiency is not expected from alkaline detergents containing chelating agents as from acid cleaners. This is certainly true when dealing with large amounts of inorganic soil or thick layers of scale. Chelating agents simply will not work fast enough. In such cases, an additional acid cleaning step after the alkaline clean is recommended.

Chelating agents work in strictly stoichiometric rations (e.g., 1 mol of calcium requires 1 mol of EDTA for complete chelation). When using a product containing EDTA, it is essential to take into consideration the water hardness of the make up water preparing the cleaning solution. There must be sufficient EDTA to deal with the water hardness, there must be still a sufficient excess to deal with the mineral compounds of the soil during the cleaning process.

One problem that unfortunately cannot be solved by using chelating agents in cleaner is the formation of water scale during the rinse cycles after alkaline, especially hot alkaline cleaning.

During the rinse cycle, the remaining film of the cleaning solution on equipment surfaces is diluted thus reducing the concentration of all components contained. Since the rinse water is invariably hard (unless softened or demineralized water is used), the power of the chelating agents is fast and greatly reduced.

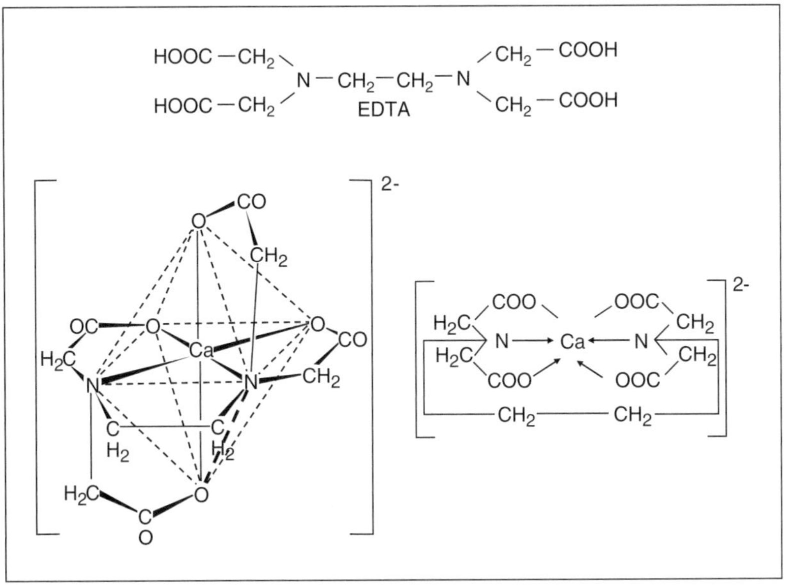

FIGURE 19 Structure formula of ethylenediamine tetraacetate (EDTA) and the EDTA-calcium complex (*left*, simplified representation). *Abbreviations*: EDTA, ethylenediamine tetraacetate.

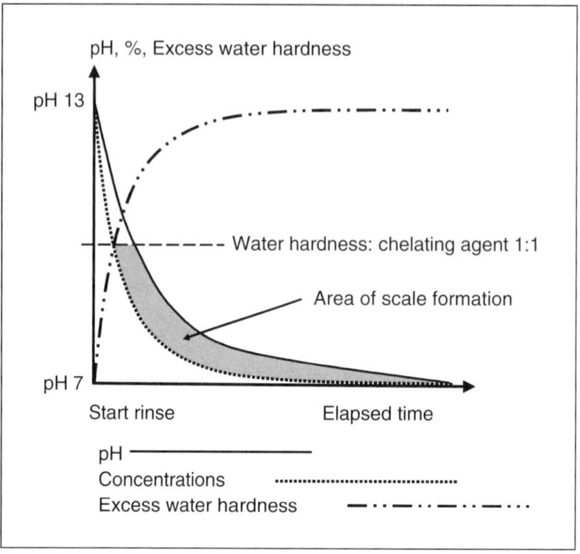

FIGURE 20 Reasons for scale formation during rinse.

Within seconds, the stoichiometric ratio between water hardness and remaining chelating agent is largely in favor of water hardness (Fig. 20).

On the other hand, the surface to be cleaned remains at the cleaning temperature and a residual alkalinity long enough to cause water hardness to precipitate. It has to be considered that the relationship between pH-value and alkali concentration is a logarithmic one, while the relationship between concentration of a chelating agent and water hardness is a simple linear pattern.

Sequestering Agents

Sequestering agents are chelating agents with one additional property. By disturbing the growth of crystals, e.g., calcite crystals that are responsible for the formation of water scale, the development of deposits is prevented even when used in sub-stoichiometric concentrations (Fig. 21).

Well-known sequestering agents are polyphosphates and phosphonates. Polyacrylates give similar results but in a slightly different way. They prevent the formation of water scale by dispersing the water hardness crystals (Fig. 22). The water hardness continues to precipitate during the rinse cycles, but no calcite crystals are formed (Figs. 23 and 25). The precipitate itself is a fine, powdery residue. This fine and light powder cannot leave any hard deposit is easily taken away by the flowing rinse water.

Sequestration at Work

Depending on the conditions during the cleaning process (such as water hardness, cleaning temperature, concentration of sequestering agents, etc.), the acid cleaning step may only be necessary every third, fifth, or even tenth cleaning operation (Figs. 24 and 26).

Defoamer

The generation of foam during automatic cleaning operations should be avoided for several reasons. Foam residues are extremely difficult to rinse off and this lends to

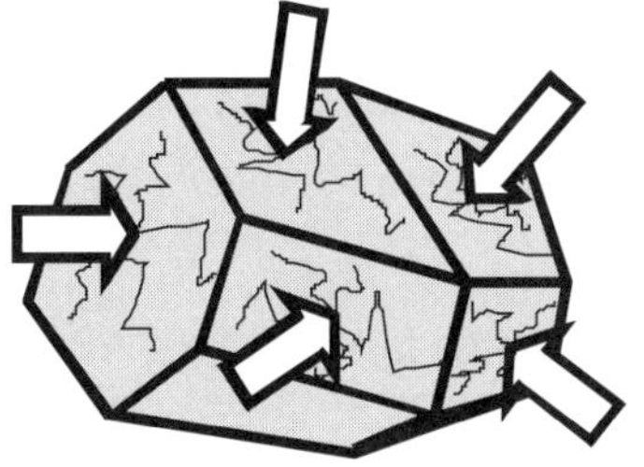

FIGURE 21 Sequestering agents disturb the surface of crystals during their growth phase.

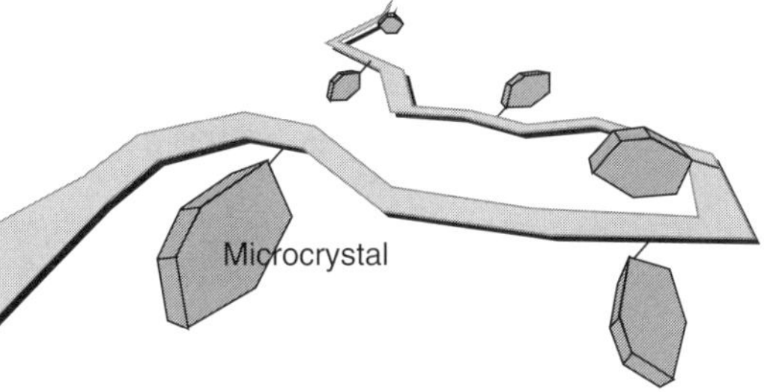

FIGURE 22 Polyacrylates bind microcrystals thus retarding the formation of scale.

Without sequestration

Hard, adherent crystals

FIGURE 23 Undisturbed calcite crystals.

With sequestration

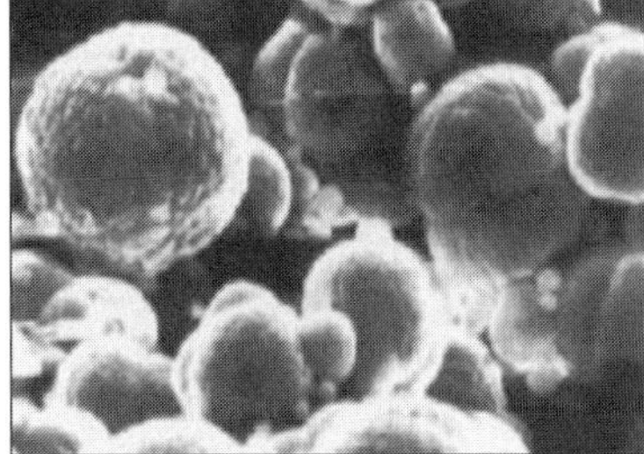

Soft, powdery "sludge"

FIGURE 24 Crystallization disturbed by presence of sequestering agent.

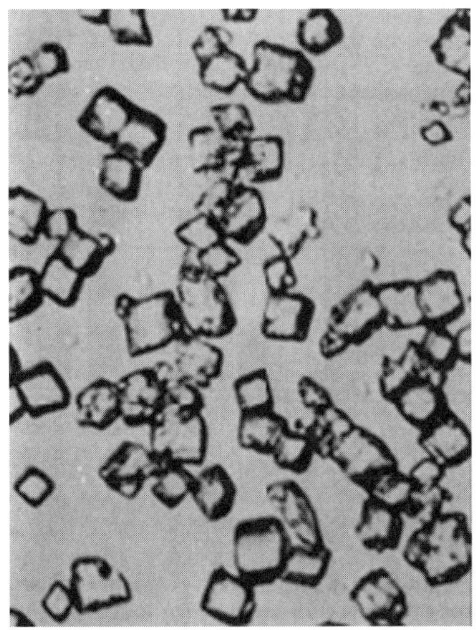

FIGURE 25 Water hardness in the form of calcite crystals on a surface.

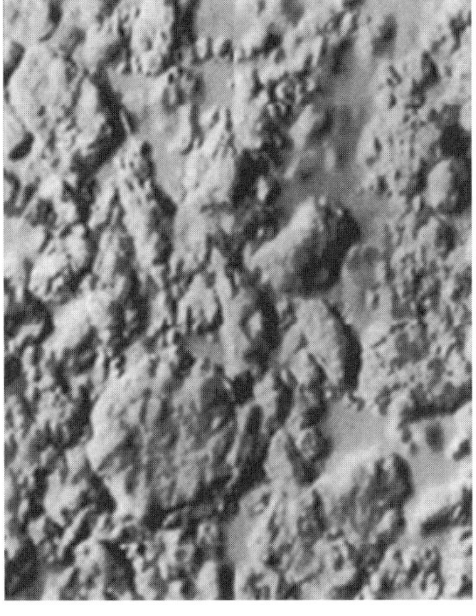

FIGURE 26 Amorphous water hardness sludge on a surface.

dramatically increased rinsing times and high-rinse water consumption. The formation of foam also can affect level control equipment and can cause severe damage by cavitation in centrifugal pumps.

Foam is created either by foaming surfactants, production residues, or from the reaction between cleaner and soil. Defoamers could be used to alleviate the problem. They are predominantly hydrophobic substances and work at the interfacial areas of the cleaning solution, preventing the formation of foam. However, ironically, their disadvantage is their hydrophobia for they are not water soluble, difficult to integrate into cleaner formulations, and a problem to rinse. Typical substances used as defoamers are paraffin oil, silicones, fatty acids, and fatty alcohols with very short EO or PO chains, etc.

However, the last substance class offers a solution to the problem. By combining longer hydrophilic molecules with relatively short EO-chains or by introducing PO into the EO-chain, it is possible to produce surfactants that reach their insolubility (so-called "cloud point" because the clear solution becomes cloudy at this temperature) at a temperature far below the intended cleaning temperature.

For example, a surfactant with a cloud point at 30°C will become increasingly hydrophobic at temperatures above 30°C. This effect produces foam-controlling properties and the surfactant does not foam anymore. Even more, it acts increasingly as a defoamer for all other substances that might be present and cause foaming. When finally rinsing with cold water (<30°C), the substance becomes increasingly more soluble thus facilitating its removal from surfaces (Fig. 27).

Oxidizing Agents

Oxidizing agents, such as peroxides and hypochlorites, assist in decomposing carbonized residues and insoluble high-molecular compounds. They also can destroy dyes or discoloring agents providing a bleaching effect. When considering the use of such chemicals as cleaning boosters added to a cleaning solution, it is essential to be aware of the risk of chemical attack.

Although hypochlorites are extremely efficient, they are also highly corrosive towards certain materials and can cause pitting corrosion on stainless steel as well as attacking such sealing materials as rubber and plastics. Oxidizing agents based on hydrogen peroxide show a good efficacy, but they are less likely to attack equipment.

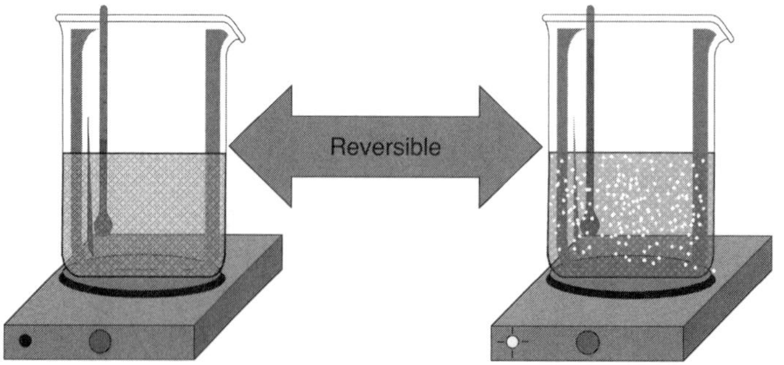

FIGURE 27 The cloud point of a nonionic surfactant.

Corrosion Inhibitors

Production equipment that consists of nonferrous metals, tin plated or galvanized steel, mild steel, or aluminum requires specific corrosion inhibitors. The choice of a corrosion inhibitor is dependent on many parameters such as the substrate to be protected, the nature of the detergent, and the pH-value of the solution. Silicates, modified carbohydrates, and phosphonates are normally used.

VALIDATING THE CLEANING PROCESS

Validation, in general, intends to establish documented evidence that a specific process will consistently meet its predetermined objectives. In the case of cleaning processes, the objective is that the next batch of product, processed in the cleaned equipment does not become contaminated from any chemical or microbiological source.

Providing documented evidence that the cleaning does indeed meet this objective includes selecting an efficient and effective sampling method, deciding on sampling frequency, finding appropriate assay methods, and setting acceptance criteria.

This article will not address the issues of properly documenting the cleaning validation; however, it will address how to achieve the objective of a cleaning process.

Avoiding contamination of a production batch focuses primarily on preventing carryover of product residues from a previous batch. This is especially important when more than one type of product is produced in the same process equipment. Acceptance criteria for carryover are based on considerations such as potency and concentration of the active component, toxicity, etc. How can the level of previous batch product residue be reduced to the acceptance limit by a cleaning process? As outlined in this article, the success of a cleaning process depends on four basic factors (choice of detergent, cleaning kinetics, solution temperature, and contact time); unfortunately, there is no "mathematical miracle formula" and in practice very much depends on previous experiences and test results. When setting up a cleaning procedure, there is usually an "overkill" cleaning program proposed to be on the safe side. If proven successful, individual parameters can be down-scaled to result in an optimal and "fine tuned" cleaning program that meets acceptance limits. Cleaning performance tests typically include swab tests and ringing water analyses. Residues left by the cleaning chemicals are yet another possible contamination source that should be considered. As with batch product residues, acceptance criteria for carryover of chemical residues must be determined. An essential element to consider is whether cleaning chemical components stick to equipment surfaces or are easily rinsed off. This depends on physicochemical properties of both the chemical components and the equipment surface (stainless steel and gasket material).

Because of the vast number of variables, it is difficult to give a general outline; however, an attempt will be made: to give some indications concerning the rinseability of cleaning agent components. Most ingredients of cleaning agents have an ionic character. Therefore, they are readily water soluble and can be easily removed from surfaces by a water rinse. However, depending on the affinity of a detergent to the surface material, it must be considered that a monomolecular layer of that detergent is attached to the surface by adsorption and cannot be removed by dissolving and dilution. In this case, the adsorption has to be counteracted by desorption. If the affinity is very high, desorption with water is not always possible. If after the cleaning process, this surface comes into contact with a product with higher desorption capacity, the residues are released from the surface and replaced by the new product.

Surveying adsorptive layers is extremely difficult because a cleaner is never a single substance, but a mixture of several components. These components have different affinities to the surface and compete in the formation of an adsorbed layer.

For example, when treating a surface with nitric acid, a nitrate layer is formed on a steel surface. If a mixture of phosphoric and nitric acid is applied, the phosphate layer is preferentially formed, since the phosphate ion has a higher affinity to stainless steel. These monomolecular layers arc not accumulative, which means they will not build up to form layers with increasing thickness (1).

Surfactants are a different category. These molecules always have hydrophilic and hydrophobic groups. The hydrophobic group has a higher affinity to a steel surface than to water. Therefore, surfactants are of major concern when examining residues of cleaners. A survey has shown that in case of surfactant residues of 2 mg/m^2 may occur on stainless steel (2).

For example, a small tank of 1-m diameter and 1-m height has a volume of about 785 L and a surface of about 4 m^2. This means about 8 mg of surfactant may remain on the surface. When filling the tank with a product that completely desorbs the residues, there will be 8 mg of surfactant in 785 L, which is equivalent to a concentration of 0.00255 ppm. When increasing the tank size, the ratio of volume: surface will increase, decreasing the potential contamination. The most critical residual substances are surfactants. Nonionic surfactants have a better solubility in cold than in hot water. When using cleaners containing such surfactants, the rinse water should always be cold. In addition, the practice of using demineralized water for rinsing is questionable. Alkaline residues are much easier to rinse with normal potable water containing some water hardness. In case of demineralized water, the rinse water consumption may be three times higher than with hard water.

The adherence of residues to the equipment surface is a physical process that cannot be avoided. Therefore, it is important to evaluate the impact of the residues on production. Cleaning agents for the pharmaceutical industry contain only substances with noncritical LD_{50} values. This fact and the extremely low contamination level permit classification of the potential residues as noncritical.

Besides the adhesion of cleaning agent components to the equipment surface, other factors can also play a role in the minimizing of residues. Those include the hygienic design of plant equipment, appropriate surface material finish, internal geometry, cleanability of equipment, etc.

One has to bear in mind that, whatever theoretical assumptions are made beforehand, extensive cleaning performance testing and documenting will always have to be carried out in practice, and successful cleaning processes rely heavily on previous experiences and chemical supplier expertise (3,4).

ENVIRONMENTAL EFFECTS OF CLEANING AGENT CONSTITUENTS

Several factors must be considered when developing cleaning agents. Not only are effectiveness and performance important, but also wider issues such as toxicity of the product and its environmental effects. Additionally, the chemical industry has voluntarily agreed to renounce the use of certain chemical products.

The optimization of the cleaning process by such means as process control, dosing and control technology, as well as process engineering can make an important contribution to environmental protection. Modern cleaning agents themselves are increasingly environmentally compatible so the use of these products rarely triggers specific problems during wastewater treatment.

Phosphates and nitrates are nutrients for such plants as algae and therefore contribute to the eutrophication of water. They can, however, through certain processes be largely removed in sewage treatment plants. In fact, the amount of these substances entering the watercourse through chemical products is minimal when compared to the discharge resulting from agricultural and communal discharges. In total, phosphates from chemical products represent a proportion of less than 6% and nitrates less than 1% of the total loading.

Like sodium chloride, sulfates contribute to the salt loading of the watercourse. The proportion of the total loading resulting from chemical products is nevertheless far lower than 1%. Under aerobic conditions, damage to concrete can be caused by corrosion due to the discharge of high levels of sulfate. Under anaerobic conditions, concrete corrosion can also be caused by hydrogen sulfide formation in the case of low sulfate concentrations.

Hypochlorites (with special emphasis on sodium hypochlorite), although highly effective as cleaning boosters, have increasingly attracted attention because of their formation of adsorbable, organic halogen compounds. The EU Detergent Legislation has set down a framework with minimum requirements for biodegradability of surfactants used in cleaning agents, which means they must be biodegradable in sewage plants and watercourses.

For anionic and nonionic surfactants, a primary degradation (i.e., loss of the surface-active characteristics) of at least 60% measured as methylene blue-active substance and/or bismuth-active substance according to the Organization for Economic Cooperation and Development (OECD) screening test is required. The biodegradability of surfactants and de-foaming agents represents an important influence in product development for the cleaning agent manufacturer. Today, for the most important surfactant groups, there are comprehensive tests available on their effect upon the environment.

SELECTION OF CLEANING AGENTS WITH RESPECT TO SOILING

The variety of raw materials, manufacturing processes, and manufacturing equipment used in the pharmaceutical industry means the selection of the most suitable cleaning agent can be extremely complicated. Further considerations to remember must also be the type and degree of soiling.

Exact specifications on cleaning can only be evaluated during practical trials, but before this is carried out, the basic conditions needed for successful cleaning can be emulated in laboratory trials using samples of the soil to be removed.

The following recommendations are only intended to be general guidelines as a basis for cleaning trials:

- Cleaning procedure in the case of water/oil emulsion: Alkali and surfactants
- Cleaning procedure in the case of acrylic polymers as residues: Alkali and chelating agents and oxidizers
- Cleaning procedure in the case of gastric juice soluble polymers: Acids and surfactants

- Cleaning procedure in the case of zinc oxide residues: Alkali and chelating agents and surfactants
- Cleaning procedure in the case of pigment-containing soil: Alkali and chelating agents and oxidizers

Details, like the specific type of cleaning material, the cleaning temperature, the concentrations, and the need for a secondary acidic clean have to be evaluated in practice.

REFERENCES

1. Puderbach H, Grosse-Boewing VV. Analyse von Adsorptionsschichten auf Edelstahlblechen. Fresenius Analytische Chemie 1984; 319:627–30.
2. Schluessler HJ. Rueckstaende von Reinigungs- und Desinfektionsmitteln bei der Getränkeherstellung. Chem Ing Technik 1980; 52(3):246–7.
3. Hambuechen T, Jezek H, Serve W, Thomas U. Reinigung und Desinfektion von Mischern und Homoginisatoren. Journal for Cosmetics, Aerosols, Chemical and Fat Products, No. 9/1080, SOFW, Augsburg, 1980.
4. Goerd T, Hambuechen T, Serve W, Thomas U. Massnahmen zur Verbesserung der Produktionsbedingungen bei der Herstellung von Salben, Gelen und Emulsionen aus hygienischer Sicht. Journal for Cosmetics, Aerosols, Chemical and Fat Products, No. 10/1980, SOFW, Augsburg, 1989.

5 Cleaning Cycle Sequences

Sally J. Rush
Seiberling Associates, Inc., Beloit, Wisconsin, U.S.A.

INTRODUCTION TO CLEANING PROGRAMS

Three commonly applied clean-in-place (CIP) programs include aqueous-based, solvent-based and solvent-assisted CIP procedures. The biotech industry primarily employs aqueous-based cleaning, using alkaline and acid solutions to remove the product soil, returning the equipment to its original use condition. Pharmaceutical chemical and bulk active pharmaceutical ingredient (API) facilities may employ solvent-based CIP cleaning, as these potent product soils are soluble in a process solvent, but may have minimal solubility in aqueous solutions. To counteract the solvent-based cleaning issues, such as explosive atmospheres and waste handling of potent compounds, "solvent-assisted" cleaning programs use a process solvent pre-rinse to remove the bulk of the soil load, followed by an aqueous-based CIP program. A process solvent post-cleaning flush may be used to remove residual water prior to reestablishing the equipment back to the process mode.

Cleaning Cycle Development

Howard and Wienceck properly note that "CIP cycle development (CD) is a systematic approach...that promotes rapid and successful execution of Cleaning Validation activities" (1). The author's company, based on the teaching of this book's editor, has always advocated that CIP CD (CIP program and circuit definition with preliminary recipe development) must start well in advance of programming, installation and commissioning. The foundation for effective cleaning programs start with the definition of cleaning criteria in the user requirements, continues with specification of cleanable process equipment and definition of the means to achieve the desired results in the functional requirements.

The functional design specifications must provide sufficient detail to ensure the planned cleaning recipes are configurable with equipment and recipe parameters to permit circuit optimization during commissioning.

The execution of chemical testing on the process soil and the definition of CIP circuits and cleaning cycles should be initiated during the earliest part of the conceptual design. This early effort to determine the CIP cycle establishes the foundation for an accurate assessment of processing schedule, utility and chemical requirements. The CIP impact on water-for-injection (WFI) requirements can impact the WFI still and loop capacity, thus the number of WFI stills required to generate adequate water for the peak loads. The conceptual design data and assumptions must be confirmed in detail design, firming up the foundation for the effort to program, commission and qualify the facility. The project's commissioning effort can then focus on the start-up and optimization of defined CIP cycles, with the effective cleaning chemistry already a known entity. The preplanning of CIP operations will ensure there are no surprises during qualification, with the effort being performed relatively "hands off" by recording observations, doing swabbing and rinse sampling, etc. to confirm the successful optimized cycle.

CIP PROGRAM PHASES

Cleaning programs are made-up of "phases," the building blocks used to create a CIP "batch," also referred to as a "master recipe." The foundation for phase determination is founded by the internationally recognized American National Standards Institute/Instrumentation Systems and Automation Society (ANSI/ISA) S88 (Parts 1–3) batch control standard. The S88 standard uses identical and/or similar operations that are repeated during a CIP batch to define a phase. For example, drain phases may be repeated multiple times during a CIP program.

There are a number of CIP phases or control module operations that are inherent in every CIP program, but are not described below. These operations include CIP circuit field device sequencing, water surge tank charging, CIP unit level, flow and temperature control operations. These CIP operations run in parallel with the phases noted below and are applied to nearly every CIP program, but are tailored to the CIP unit and CIP circuit design. CIP programs use the following basic phase building blocks to create cleaning programs:

CIP Program Initiation Phase

Establishes the CIP circuit boundary and acquires devices (pumps, valves, equipment) within the boundary, ensures adequate utilities and chemicals are available for the CIP program and checks permissives and interlocks.

Rinse Phase

Rinse water (or) solution is supplied from the CIP unit, and this phase is used to flush the CIP circuit of all free rinsing soil and chemical solutions to waste. The rinse phase may be repeated multiple times during a CIP program as a pre-flush prior to a chemical wash, or after a chemical wash as a post-wash rinse. Although the majority of rinse phases are performed with water, the rinse may also be performed with a base, acid, saline or solvent solution.

Gas Blow Phase

This phase uses compressed, filtered gas to clear the CIP supply and process piping to the process equipment being cleaned to enhance a clear transition between dissimilar phases, when necessary.

Intermediate Drain Phase

Used between CIP phases to actively transfer spent wash and rinse solutions from the circuit boundary to enhance a clear transition between dissimilar phases.

Chemical Wash Phase

This phase performs the chemical cleaning duty and may be either single pass through the circuit to waste or recirculated through the CIP circuit. The chemical wash may be an alkaline or acid solution in aqueous cleaning programs. Solvent-based cleaning programs may apply process solvents, such as alcohol, methanol or dimethyl sulfoxide (DMSO) to remove the process soil.

Final Rinse Phase

The final rinse is performed with the highest quality water available within the facility, meeting the quality of water used for process operations (2). A CIP return resistivity or conductivity sensor may be used to monitor the final rinse purity to confirm the removal of soil and chemical.

Final Drain Phase (Gravity)

This phase begins by opening all drain valves, with CIP return pumps and equipment gas blanketing "off," providing for gravity drainage to and from all CIP circuit low points.

Program Complete Phase

Establishes the circuit boundary as "clean" and returns all devices to their safe "stand-by" state, ready to be acquired for the subsequent process or steam-in-place operation.

CREATION OF PHASE-BASED SINGLE-PASS OR RECIRCULATED CIP PROGRAMS

To confirm the relative demands for water, chemical, heat and time during a detail design effort, it is necessary to sequentially link phases to create a CIP master recipe or CIP "program." The basic CIP programs are often classified as *single-pass* or *recirculated* CIP procedures. The single-pass CIP program makes up chemical wash solution at the CIP unit by batching or in-line mixing and supplies it to the circuit, with spent solution discharging to process waste. The recirculated CIP program uses single-pass rinse phases, but recirculates chemical wash solutions between the CIP unit and the circuit, permitting an extended chemical wash exposure, while minimizing water, heat and chemical cost (3) and the waste volumes discharged for treatment.

Why Single Pass?

Single-pass CIP systems require significantly more water and chemical, and therefore a greater capacity for waste handling. In spite of the utility burden, a single-pass CIP program may be desirable and an evaluation of the user requirements will aid in determination of the requirement. Multiproduct facilities may elect to use single-pass CIP to avoid extensive validation to prove no cross contamination of dissimilar product through the CIP system. Soils with "undesirable" insoluble particulates, for example broken glass vials in a freeze dryer, are single pass cleaned to avoid the distribution of particulates throughout the process. Single-pass cleaning may also be desirable when the CIP unit can only be located in a position elevated above the process to take best advantage of gravity flow.

Definition of a Generic Circuit

For purposes of comparing relative merits of single-pass versus recirculated CIP programs, a generic CIP circuit is defined below to permit a review of typical recipe parameters, CIP program durations and utility usage. Consider the following "typical" CIP circuit:

- A 1000 L mixing vessel, having a 4′ diameter with a 1-1/2″ fill line and 1″ reagent addition line. The tank discharge line is 2″ in diameter, and the outlet is equipped with a vortex breaker. Although the tank could be cleaned with spray devices operating at 115 to 135 lpm, the tank is equipped with spray devices specified at 165 lpm to accommodate for proper flow velocity in the 2″ discharge line.
- An evaluation of the process/CIP piping reveals 200′ of 2″ CIP piping, with 100′ of 1-1/2″, and 10′ of 1″ sanitary tubing. The following breakdown of piping lengths with line diameters noted: The CIP unit recirculation tank requires 60 L of water to charge prior to establishing circuit operation and recirculation. Based on this data, the total system fill volume is estimated roughly at 200 L.
- The CIP circuit device sequence involves the following sub-paths, which repeats every 60 seconds:
  Step 1. Spray tank—40 seconds
  Step 2. Spray tank and pressure wash reagent inlet—10 seconds
  Step 3. Spray tank and pressure wash fill line—10 seconds
  Return to Step 1

Recirculation CIP Programs

Table 1 describes a typical phase based on recirculating CIP program using cleaning chemicals "A" and "B." The table describes the phase, phase function and water usage to clean the example circuit. The time requirement does not include operator time to assemble the CIP Circuit.

Single-Pass CIP Programs

A comparable single-pass CIP program may begin with a chemical wash to provide a flush, which includes the benefits of chemical cleaning, while reducing waste generation by elimination of the normal prerinse. However, a water prerinse may be required if a potent soil must be flushed to an alternate drain destination prior to the chemical wash. Table 2 denotes the linked phase sequence for a typical single-pass CIP program utilizing cleaning chemicals "A" and "B," with phase function, water usage and duration noted.

Schedule and Utility Comparison—Single-Pass vs. Recirculated Cleaning

These comparable CIP programs applied to our typical circuit illustrate the difference between the single-pass and the recirculated CIP programs, enabling a comparison of program time, water usage, and waste generation.

- Single use, recirculated CIP cleaning—2860 L of water required and waste generated over a 64 minute program.
- Single-pass CIP program—4600 L of water required and waste generated over a 46 minute program.
- Single-pass CIP program with prerinse—4900 L of water required and waste generated over a 48 minute program.

It should be noted that gas blows between flow phases in both programs may be considered unnecessary as the CIP supply line will be flushed with the next solution. However, the gas blow and drain process are critical for circuits not readily drained by the CIP return system, as the process equipment puddle containing soil

TABLE 1 Typical Recirculating Chemical Wash Program

Phase(s)	Function	Water usage (L)	Time (min)
CIP program initiation	Confirms utilities, CIP boundary, permissives	NA	5
Rinse	Flush circuit of all free-rinsing soil	300	1.5
Intermediate drain	Drains return side, CIP supply side remains charged with water	NA	2
Chemical wash	Establish circuit recirculation, feed chemical A, confirm conductivity and temperature, wash for required duration	130	10
Gas blow and intermediate drain	Clears CIP supply of chemical A, drain circuit for effective minimum volume rinse	NA	2
Rinse	Flush circuit of spent chemical A	300	2
Intermediate drain	Drains return side, CIP supply side remains charged with water	NA	1
Chemical wash	Establish circuit recirculation, feed chemical B, confirm conductivity and wash temperature, wash for required duration	130	10
Gas blow and intermediate drain	Clears CIP supply of chemical B, drain circuit for effective minimum volume rinse	NA	2
Rinse	Flush circuit of spent chemical B	300	2
Intermediate drain	Drains return side, CIP supply side remains charged with water	NA	1
Final rinse with high quality water	Flush with high-quality water to defined end point, removing soil and chemical	700	5
Gas blow and intermediate drain	Clears CIP supply of water, drain rinse	NA	2
Final drain	Gravity drain of CIP boundary low points	NA	5
Program complete	Releases clean CIP boundary	Total volume 2860	Total time 64

and chemical will be rinsed by dilution, requiring significantly more water to accomplish the task.

DETERMINATION OF CIP PROGRAM RECIPE PARAMETERS

The critical factors that determine cleaning efficiency are the controllable variables of time of exposure to cleaning solutions, cleaning temperature, and chemical concentration. The physical action of flow rate, pressure and spray coverage, are controllable only by engineering design (4). In many cases, the required cleaning chemistry to remove the product soil is well known, although the definition of a CIP recipe to effectively apply the cleaning chemicals may be less easily determined. Generally, an increase in exposure time, cleaning temperature or chemical concentration will positively impact the cleaning efficiency...but there is no universal recipe for all cycles and soil types. An evaluation of the nature of the soil load will impact the chemical selection, cleaning temperatures and the required wash duration. The individual CIP circuit configuration will determine

TABLE 2 Typical Single-Pass Chemical Wash Programs

Phase	Function	Water (L)	Time (min)
CIP program initiation	Confirms utilities, boundary, permissives	NA	5
Chemical wash	Single-pass wash with cleaning chemical "A"	1650	10
Gas blow and intermediate drain	Clears CIP supply of chemical A, drains solution for effective minimum volume rinse	NA	2
Rinse	Flush circuit of spent chemical	300	2
Intermediate drain	Drains return side, CIP supply side remains charged with water	NA	1
Chemical wash	Single pass wash with cleaning chemical "B"	1650	10
Gas blow and intermediate drain	Clears CIP supply of chemical, drains solution for effective minimum volume rinse	NA	2
Rinse	Flush circuit of spent chemical B	300	2
Final rinse with high quality water	Flush circuit with high quality water to remove soil and chemical	700	5
Gas blow and intermediate drain	Clears CIP supply of water, drain rinse	NA	2
Final drain and	Gravity drain of all CIP boundary low points	NA	5
Program complete	Releases clean CIP boundary	Total volume 4600	Total time 46

the CIP flow rate set point, rinse volumes, gas blow pressure and duration, and drain times.

Laboratory Testing to Define Chemical Application and Exposure Time

"New" soils require special consideration and laboratory testing to determine the appropriate cleaning chemistry, and this work should be performed well in advance of the development of functional (software) design specifications. Chemical suppliers or the owner's laboratories can support this investigation by performing chemical testing on the processed soil to determine the proper application of time, temperature and concentration, and the sequence in which chemicals are applied. The test described in chapter 4 provides additional detail on this topic.

Impact of Cleaning Temperature

The removal of many soils are greatly enhanced by an increase in temperature, which can provide the following benefits:

- Decreases viscosity and increases turbulent action
- Improves soil solubility and increases chemical reaction rates
- Melts fats and decreases soil adhesion to the equipment surfaces

For any soil with a fat component, the minimum cleaning temperature should be several degrees higher than the fat's melting point.

CIP Circuit Evaluation to Determine Required Physical Action

The fourth critical cleaning parameter physical action, enhances soil removal through turbulence, dissolution and/or suspension, and is difficult to accurately simulate in a coupon laboratory test. The guidelines for physical action have been suggested in chapter 1, and will be further discussed in subsequent chapters. An evaluation of the CIP circuit configuration must be performed to determine the proper flow rate to achieve the desired physical action, i.e., spray coverage of a vessel or velocity in a line.

As noted in chapter 1, CIP circuits are typically classified as "tank," "line" or "combination" circuits. On tank circuits, a simple assessment of the spray device application will define the CIP supply flow rate recipe parameter. However, as illustrated with our typical circuit earlier, the CIP spray supply flow rate should meet or exceed the flow rate requirements for the largest line size within the CIP circuit boundary. Basic line circuits require a review of the maximum process/CIP piping diameter to determine the target flow rate set point. The determination of a combination circuit flow rate set point requires consideration of both the spray device flow specification and the maximum piping diameter. A combination circuit may present a challenge in hydraulically balancing CIP supply and return flow. These challenges can be addressed by designing circuits to provide intermittent flow for transfer lines in parallel with tank sprays, and when possible, return flow through the vessel, always at a fixed supply flow rate. The size of the vessel outlet valve will often be the greatest deterrent to achieving adequate flow rates.

DETERMINATION OF PHASE-BASED RECIPE PARAMETERS

Rinse Phase

Selection of Rinse Solution

The selection of the rinse solution is dependent on the solubility of the soil in aqueous solutions or solvents and the available water quality in the facility. The most common application of a rinse solution is high quality water, but not necessarily the highest quality water in the facility (4). If there are two water qualities available in the facility, the lesser quality water may be suitable for the rinse and chemical wash phases, with the process quality water being used in the final rinse phase, thus reducing production cost of high quality water in the facility.

Process solvents may be used in the rinse phase(s) for soils impervious to aqueous solubilization. However, the use of solvent rinses can complicate the CIP system installation and cycles, as the CIP boundary will likely require an inert atmosphere and oxygen monitoring may be required. The preparation of a solvent/water blend can alleviate the explosives concern, while providing the benefits of a solvent-based rinse.

Temperature Requirements

The recipe set point for the rinse temperature may require adjustment, heated or cooled, to tailor the phase to the cleaning needs. For example, a heated post-rinse is desirable to improve the solubility of alkaline solutions, reducing the rinse volume required to remove the spent solution. A cooled pre-rinse may be desirable when *(i)* flushing a protein soil to avoid denaturation, *(ii)* rinsing containment equipment to avoid thermally damaging viewing panels, and *(iii)* a cooled post-rinse may accelerate the turn around time between CIP and ambient or chilled process operations.

Rinse Volume (Duration)
The starting point for determining the rinse volume recipe set point is typically founded in the CIP circuits system fill volume; the volume required to fill the CIP supply, process and CIP return piping, and if required for CIP return pump priming, a small puddle (20–40 L) in the equipment being cleaned (2). The minimum prerinse volume is 1.5 to 2× the system fill volume, and the minimum volume for a postchemical wash rinse is 1× the system fill volume. However, if the CIP circuit involves a complex valve sequence of extended duration, the rinse may be extended to achieve at two to three passes through the device sequence to flush all CIP circuit subpaths (3).

Return Flow Check
A return flow check step is often included in the initial rinse phase, and is used to ensure that rinse solution being sent out to the circuit is returned from the equipment. The CIP return probe or flow switch is used to monitor return flow after a rinse volume equal to the system fill volume is supplied to the CIP circuit. At this point, CIP return flow should be established and the program will require an input from the return flow check device to indicate established return flow for a brief preset time. This step is a permissive for the phase to continue, and should fail the CIP program if an input is not received from the device. The failed test is indicative of a blocked return path, or may indicate that the rinse solution is being sent to an unintended destination.

Criteria for Rinse Phase Success
The success criteria are dependent on when the phase is applied within the CIP program. For example, the criteria for the success of an initial rinse is that the volume supplied flushes the CIP circuit of all free-rinsing soil, and the solution being discharged to waste runs "clear." The criteria for a rinse following a chemical wash is often misunderstood. These rinses are included to rinse out the majority of the spent chemical solution, but may not need to fully rinse all chemical out of the circuit (4). For example, in a CIP program that employs an alkaline wash followed by an acidified wash, the water rinse after the alkaline wash is not intended to totally remove the alkaline solution. The acid wash phase is included in the program to neutralize residual alkaline solution and remove mineral deposits, performing the dual duty to reduce water usage and minimize waste (4). A thorough post-acid rinse, which is much freer rinsing than the alkali, reduces the overall water usage. Typically, the post-chemical rinse phase volume is equivalent to one circuit volume, with two to three passes through the device sequence being accomplished.

Gas Blow

Phase Objective
The objective of the gas blow is to facilitate a clear transition between CIP program phases by clearing the CIP supply and process piping prior to supplying fresh solutions, thereby improving the cleaning effectiveness. The gas used is filtered process air or nitrogen for systems, which require an inert atmosphere. A gas blow diminish step is always included in the phase to allow the gas pressure to dissipate prior to continuing on with the CIP program. An intermediate drain phase always follows a gas blow phase to discharge the solution blown into a vessel for transfer to waste handling.

Requirement for Sustained Pressure and Volume

The gas blow phase effectiveness is dependent on the gas source's ability to supply a sustained pressure and volume for the step duration. The gas blow duration is influenced by supply pressure and volume, as well as piping routing and circuit volume. The gas blow pressure must be sufficient to overcome the friction loss of a full line at the blow down rate, the static head (if any) and any overlay or blanket pressure in the destination process equipment. The effective clearing of a transfer line of substantial length requires a gas supply equivalent to a flow-rate of 20 ft/sec in the line to be cleared. For example, to clear a 2″ transfer line, the gas volume must be equivalent to a liquid flow-rate of approximately 600 to 650 lpm at the beginning of the blow-down. The gas pressure must be sufficient to move the liquid through the line at that flow-rate. As the gas blow clears the line, the head loss due to friction rapidly dissipates and the gas in the transfer line expands very rapidly to the destination tank, further accelerating the movement of liquid through the piping.

Determination of Gas Blow Phase Duration

The recipe parameter for gas blow phase duration is the time sufficient to clear the CIP supply line to the CIP circuit high point, with the expanding gas pushing the solution in front of it to clear the line through the equipment inlet(s) and spray device(s). The optimal gas blow time is best determined at the CIP circuit supply high point with a stopwatch and a stethoscope on the piping to note the time when the air/water interfaces passes. An excessive gas blow duration, which blows mist through the sprays devices until clear unnecessarily extends the CIP program and risk over-pressuring downstream equipment.

Criteria for Success of Gas Blow Phase

The criteria for the success of gas blow phase are somewhat subjective. A successful gas blow will clear the CIP supply line into the equipment being cleaned, or on line circuits, push it directly into the CIP return path for solution removal by the CIP return system. Evacuation of the downstream portions of the circuit will occur via gravity and expansion of gas in the line. The CIP circuit will not be dry, and some solution may accumulate at piping low points after a period of time upon completion of the gas blow.

Factors impacting effectiveness of a gas blow phase include insufficient gas pressure and supply volume capacity, as well as piping configurations involving long vertical runs of piping. The volume of water idle in the piping low points after a blow will be increased on undersupplied or poorly designed systems. Gas blow duration is irrelevant if adequate volume and pressure cannot be maintained to push the gas/liquid interface in plug form. Extensive vertical piping runs can compromise the integrity of the water/gas interface, and may result in a core of gas tunneling through the water, reducing the gas blow effectiveness.

Gas blows designed for line clearing are not suitable for CIP circuit drying operations, and a separate drying cycle should be defined if equipment dryness is required. The drying cycle air should supplied at a lower pressure, and it is recommended that the CIP unit not be the source of the air, as the unit will be out of cleaning service for 30 to 60 minutes while the drying step is performed, impacting CIP system utilization. However, the portion of the CIP supply line for the circuit that is dedicated to the circuit may be used for supply of the gas blow.

Intermediate Drain

Objective

The intermediate drain objective is to remove previous rinse or wash solutions from the circuit prior to introducing the next solution. An intermediate drain is recommended after a rinse or chemical wash phase to ensure a clear separation between the previous solution and fresh wash or rinse solution to be supplied.

Determination of Drain Time Parameter

The determination of the drain time recipe parameter is a factor of CIP return flow capacity, and drain times may be different phase to phase within a CIP program. An early activity during CIP system commissioning is performing a test to evaluate the CIP circuit return flow capability. The drainage test charges a known volume of water into the vessel to be cleaned with CIP return valving closed. After the volume has been transferred, a drain phase is initiated and the accumulated water is sent to drain via CIP return, monitoring the equipment level via sight glass. For prompt drain operations and balanced circuit hydraulics, the CIP return flow capability should meet or exceed the CIP supply flow rate for that circuit. This information makes it possible to estimate intermediate drain times accurately, contributing to an efficient CIP commissioning effort and effective drain operations.

Criteria for Success of Intermediate Drain Phase

The criteria for the success of the intermediate drain are affected by the subsequent phase. If the next active CIP program phase is a single-pass rinse, then discharge of all solution from the process equipment outlet within the established drain time is adequate. However, if the next program phase is a recirculated wash phase, the previous solution should fully transfer back to the waste system within the timed preset; thereby avoiding potential contamination of the next wash phase with remnants of the previous rinse solution. The effectiveness of each individual drain in a CIP program should be confirmed during commissioning for CIP cycle optimization.

Chemical Wash Phase

Objective

The objective of the chemical wash phase is to expose all equipment surfaces within the cleaning boundary to the required physical action, time of exposure at cleaning temperature and chemical concentration to dissolve, suspend and remove the product soil.

Chemical Selection

The required cleaning chemistry is dependent on the nature of the soil to be removed. For example, if the soil is primarily a biological material made-up of carbohydrates, protein and fat, then an alkaline based cleaner is suitable for the primary chemical wash phase. However, if the soil load is primarily protein in nature, a hot acid wash may best serve the needs of the equipment to be cleaned. For additional detail on cleaning chemistries, please see chapter 4.

Conductivity and Temperature Monitoring

Conductivity monitoring of chemical wash solution ensures that the chemicals are added in the proper amount to achieve the required chemical concentration for the defined soil load. The conductivity instrument is preferably located on the CIP system supply side, and single-pass CIP units may use a static mixer to ensure the solution uniformity prior to measurement. Recirculated CIP systems may not immediately create a uniform solution and a chemical mix time recipe variable may be desired prior to confirming chemical concentration via conductivity. The chemical wash phase CIP return temperature is a critical parameter to be defined in the User Requirements, as the wash phase effectiveness is dependent on cleaning activity and is impacted by temperature variations.

Criteria for Success of Chemical Wash Phase

The criteria for success of the chemical wash phase, when applied for soil removal, is best determined at the conclusion of the rinse, blow and drain phases which follow the chemical wash phase. The success point is that the combined effects of the chemical wash supplied at the use of concentration and temperature, for the required duration and using the necessary physical action, results in the removal of soil. Typically, the chemical wash phase volume or time will be sufficient to ensure four to six passes through the device sequence (3).

Final Rinse Phase with High-Quality Water

Objective

The objective of the final rinse phase is to flush all equipment surfaces within the cleaning boundary free of all product soil and residual chemical wash solution. The final rinse water should be specified in the User Requirements as either: (*i*) the highest quality water available in the facility or (*ii*) the water quality equivalent to that specified for process operation. The termination point for the flush is typically related to CIP return conductivity or resistivity monitoring, or based on supply of a flush volume validated to achieve the required results. The most common approach is resistivity monitoring, and our phase description will cover this method.

Final Rinse Phase Recipe Parameters

The final rinse phase is usually comprised of two substeps, (*i*) a volumetrically controlled rinse followed by. (*ii*) rinse to resistivity set point near the incoming water supply. The initial rinse volume recipe set point should flush the CIP circuit with at least one CIP circuit system fill volume. After the initial volumetric rinse has completed, the final rinse recipe parameter set point is determined through CIP return line resistivity monitoring, confirming when the resistivity set point has been achieved for a recipe-specific timed preset. The resistivity should be maintained for a timed duration equivalent to at least one pass through the CIP circuit device sequence, ensuring all sub-paths of the CIP circuit are flushed. If the solubility of a chemical solution is dependent on temperature, a final rinse CIP return temperature requirement may be established and monitored.

Criteria for Success of Final Rinse Phase

The criteria for success of the final rinse phase involves confirming rinse monitoring meets the resistivity recipe set point for a circuit recipe specific duration. An alternative method involves establishing a final rinse volume, which flushes the circuit to meet an acceptable total organic carbon limit, indicative that the residual chemical and soil is below an acceptable limit. The samples are obtained at volumetrically recorded intervals, and a final rinse volume is established.

Final Drain Phase (Gravity)

Objective

The final drain removes, via gravity, the residual rinse at CIP circuit low point drains prior to concluding the CIP Program. As there are differences in the function, there are also device-positioning differences in the intermediate and final drain. The intermediate drain uses some form of motive force (pumping, gravity, eductor assist, or some combination thereof) to actively direct solution to waste. The final drain opens all low point drain valves and gravity drains off the minimal residual amount of final rinse water or solvent.

Final Drain Phase Recipe Parameters

The final drain phase is time based, and the determination of the drain duration recipe parameter is best determined in commissioning with field observations and a timer. The piping beyond the low point or casing drain valve is broken to permit timely determination of when all gravity-drained solutions have been removed from the CIP circuit. This recipe parameter set point is usually stated in minutes, rather than seconds as water that clings to the piping will slowly dribble to a low point for drainage.

Criteria for Success of Final Drain Phase

The confirmation of a successful final drain phase ensures that the drain duration recipe parameter is sufficient for residual water to gravity drain through CIP circuit low points. The CIP circuit will not be dry and droplets of water will adhere to the piping and equipment surfaces. However, no pooling or puddling should be noted in the equipment and piping.

TYPICAL CIP CLEANING MASTER RECIPES

Aqueous-Based Cleaning Program—Water Soluble Soils

Some soils are effectively removed by water alone, including alkaline and acid buffer solutions used on purification column or filtration operations. A full chemical cleaning program may not be justified on a lot-to-lot basis, and a single-pass water-rinse program could provide sufficient cleaning for campaign runs to minimize process turnaround time and utility use.

The program in Table 3 represents an effective single-pass water-rinse program for routine cleaning on easily solubilized soils with water.

The above-noted program is also suitable for a number of other routine facility operations and should be planned in facilities, with optional use for the following purposes:

- Spray performance (riboflavin) testing of spray devices.
- Single-pass rinse of an aborted chemical wash program due to equipment failure to remove chemical and facilitate safe circuit disassembly for maintenance purposes.
- Single-pass rinse of a failed chemical wash program, to facilitate running a complete chemical wash program without deviation.

The single-pass rinse program is not suitable for insoluble salts and does not provide sufficient cleaning for endotoxin or soil removal after completion of a maintenance procedure. As a result, the facility should also maintain the capability for a full chemical cleaning program on a regularly scheduled or post-maintenance

TABLE 3 Single Pass Water Rinse Programs

Phase	Typical parameters	Success criteria
CIP program initiation	None	Utilities and CIP boundary available, permissive confirmed
Rinse	375–500 L	Rinse is clear, 2–3 passes through sequence
Gas blow	30–60 sec	Gas–liquid interface at CIP supply high point
Intermediate drain	30–60 sec	Rinse drains from equipment into CIP return
Final rinse with high quality water	750–1000 L, megOhms	Resistivity set point maintained for period greater than 60 sec
Gas blow	30–60 sec	Gas–liquid interface at CIP supply high point
Intermediate drain	30–90 sec	Rinse discharge through CIP return to waste
Final (gravity) drain	5–10 min	All free-draining liquid is drained by gravity
Program complete	None	All utilities and CIP boundary are released

basis. The routine application of this program may also be supplemented with regularly scheduled passivation for aggressive removal of accumulated mineral salts and oxides.

Aqueous-Based Cleaning Program for Soils Susceptible to Alkaline Cleaners

Most soils related to biotech fermentation or cell culture and purification operations are effectively removed with an alkaline-based cleaning program. The program in Table 4 represents an effective chemical wash program for cleaning of soils susceptible to removal by alkaline solutions.

The primary cleaning duty is performed by the alkaline cleaner, although an acid wash phase is noted within the program as well. Of the two main alkaline cleaners, potassium hydroxide rinses more readily than sodium hydroxide, but neither is very free-rinsing, especially with cold or ambient water. To avoid generating a high volume of waste solution by fully rinsing the circuit, the acid helps to neutralize residual alkaline as well as remove mineral salts and oxides during every CIP cycle. The acid wash works to avoid the build-up that becomes a visual film, which is considered as "failed" cleaning. The acid solution is often prepared with phosphoric acid, and is in the pH range of 4.5 to 5.5 for neutralization, and lower pH ranges (pH range 2–3) may be used to provide for removal of tenacious mineral build up or rouge. The typical exposure time is a minimum three to five minutes exposure for all circuit sub-paths. Citric or acetic acid solutions have also been used to perform the same duty; however, not all acids have the same neutralizing capacity and the additional chemical feed volume must be recognized when planning the chemical feed scheme.

Aqueous-Based Cleaning Program for Soils Susceptible to Acid Cleaners

Many biotech purification and final sterile vial or cartridge fill operations are associated with protein therapies, which have unique cleaning concerns, and

TABLE 4 Recirculated Chemical Wash Program—Alkaline Cleaning

Phase	Typical parameters	Success criteria
CIP program initiation	None	Utilities available, all permissives confirmed, CIP boundary acquired
Rinse	375–500 L	Rinse is clear, 2–3 passes through sequence
Intermediate drain	30–60 sec	Rinse discharge through CIP return to waste
Chemical wash	Conductivity _to _mS temperature _C, ±_C wash volume _L	Soil dependent conductivity, temperature recipe requirements maintained for wash duration, 4–5 passes through device sequence
Rinse	250–375 L	1–2 passes through device sequence
Gas blow	30–60 sec	Gas–liquid interface at CIP supply high point
Intermediate drain	30–60 sec	Rinse discharge through CIP return to waste
Chemical wash	Conductivity_to _mS temperature _C, ±_C wash volume _L	Acid to neutralize residual alkaline or remove mineral deposits, temperature maintained for wash duration (3–5 min, and 2–3 passes through device sequence)
Gas blow	30–60 sec	Gas–liquid interface at CIP supply high point
Intermediate drain	30–60 sec	Rinse drains from equipment into CIP return
Rinse	250–375	Minimum 1 circuit volume, 1–2 passes through device sequence
Gas blow	30–60 sec	Gas–liquid interface at CIP supply high point
Intermediate drain	30–60 sec	Rinse drains from equipment into CIP return
Final rinse with high quality water	1st Rinse—1 circuit volume then, rinse to resistivity of _megOhms. Typical total rinse—3–4 circuit volumes	Resistivity set point maintained for period greater that 60 sec, resistivity set point facility and water quality specific. Time period is circuit recipe parameter
Gas blow	30–60 sec	Gas–liquid interface at CIP supply high point
Intermediate drain	30–60 sec	Rinse discharge through CIP return to waste
Final (gravity) drain	5–15 min	All free-draining liquid is drained by gravity
Program complete	None	All utilities released, CIP boundary "clean"

these soils may be effectively removed with an acid cleaning program. The program in Table 5 represents an effective chemical wash program for removal of soils susceptible to acid solutions, such as insoluble mineral salts and oxides.

Management of Protein Foams

Soils primarily comprised of protein, or a combined protein-carbohydrate components can create stable foam, which can interfere with the repeatable cleaning operations. Alkaline cleaning solutions act to reduce surface tension, which facilitates the removal of many soils, but enhances the formation of protein foams. This protein foam can result in the following cleaning complications, which can result in cleaning performance inconsistencies. The soil laden foam will rise and fall with the cleaning water levels in the equipment being cleaned, and will cling to the equipment sidewalls. Foam interferes with efficient CIP return pump operations resulting in unreliable drain phases. Foam and liquid levels are not readily measured in the CIP recirculation tank, and the foam may overflow the tank or block vent filters.

To counteract the impact of a protein foaming soil, the following measures may be considered:

- A cooled, single-pass acid pre-rinse will eliminate the complications of a denatured or foaming protein soil. The acid rinse should be discharged to a local process drain to avoid soil loading the CIP return piping, and volume flush all free-rinsing protein from the CIP circuit.
- Burst rinsing, rather than continuous rinse supply can be employed in combination with additional intermediate drains to assist in efficient foam removal with minimal waste generation. The CIP return system must be capable of efficiently returning an air-water mixture.
- Consider a CIP re-circulating tank with a tangential return design, which uses centrifugal forces to eliminate gas from the CIP return stream. The specified CIP recirculating tank level sensing instrument should not be affected by foam.

Again, the acid-based cleaning program is insufficient cleaning for soils with fat component, endotoxin or soil removal after completion of a maintenance procedure. The facility should maintain the capability for a full chemical cleaning program with alkaline cleaners, performed on a post-maintenance basis.

Cleaning Program for Denatured Protein Soils

The acid override cleaning program can be effectively applied on denatured protein and mineral laden soils, which may also include a fat component. Examples of this

TABLE 5 Recirculated Chemical Wash Program—Acid Cleaning

Phase	Typical parameters	Success criteria
CIP program initiation	None	Utilities available, CIP boundary acquired
Rinse	375–500 L	Rinse is clear, 2–3 passes through sequence
Intermediate drain	30–60 sec	Rinse discharge through CIP return to waste
Chemical wash	Soil dependent conductivity _to _mS temperature _C, ± _ C wash volume _L	Conductivity, temperature recipe requirements maintained for wash duration, wash duration to ensure 4–5 passes through CIP circuit device sequence. Wash durations range 10–40 min
Gas blow	30–60sec	Gas–liquid interface at CIP supply high point
Intermediate drain	30–60 sec	Rinse drains from equipment into CIP return
Rinse	250–375 L	1–2 passes through device sequence, minimum 1 circuit volume
Gas blow	30–60 sec	Gas–liquid interface at CIP supply high point
Intermediate drain	30–60 sec	Rinse drains from equipment into CIP return
Final rinse with high quality water	1st Rinse—1 circuit volume rinse to resistivity of _ meg. Ohms common rinse volume 3–4 circuit volumes	Resistivity set point maintained for period greater that 60 sec, resistivity set point facility and water quality specific. Time period is circuit recipe parameter
Gas blow	30–60 sec	Gas–liquid interface at CIP supply high point
Intermediate drain	30–60 sec	Rinse discharge through CIP return to waste
Final (gravity) drain	5–15 min	All free draining liquid is drained by gravity
Program complete	None	All utilities released, CIP boundary “clean”

Abbreviation: CIP, clean-in-place.

type of soil include fermenters or bioreactors that have been thermally inactivated prior to cleaning; especially challenging is the air–liquid interface where media and cell mass agglomerates and dries during the cell growth phase. Spray dryers and evaporators also present a cleaning challenge as the equipment surfaces are coated with a highly concentrated soil, often comprised primarily of denatured protein and mineral content.

Protein denaturation in these examples can result in a carbohydrate-protein-fat-mineral matrix, which can be very difficult to attack with an alkaline cleaner alone. The acid override program is a more extensive treatment than any of the previously described programs, and can be effective in removing the soils noted above. The program uses a strong, hot acid solution in recirculation for an extended wash period to attack the tenacious protein and mineral soils. The concentrated alkaline is introduced after the acid wash duration to "override" the acid solution and continue recirculation, without discharge of solution. The alkaline wash phase to attack the fat and carbohydrate components of the soil. This approach conserves water usage, and minimizes discharge of concentrated acids to waste.

The CIP program begins with a traditional rinse operation to send all free-rinsing soil to drain prior to establishing a recirculated chemical wash phase using a hot acid solution. The CIP circuit is next subjected to a hot acid wash to attack the protein component of the matrix, breaking down peptide bonds and minerals, exposing the fat and carbohydrates trapped within the denatured protein matrix. The acid solution may be recirculated for a period of 25 to 30 minutes prior to introducing concentrated alkaline cleaner to override the acid solution, the alkaline solution then working to remove the remaining exposed soils. The alkaline solution may be recirculated for a period 30 to 45 minutes prior to being flushed out with a subsequent rinse phase. The remainder of the program concludes in the same manner as the traditional alkaline-based wash program.

Solvent-Based Cleaning Programs

The application of a solvent-based cleaning program is primarily to bulk pharmaceutical chemical or API facilities, with organic and inorganic solvents employed for both process operations and cleaning the equipment. As noted in chapter 16, the process equipment and piping may not be hygienic in design or fully susceptible to CIP technology. The process cleaning is often performed in a contained manner to protect personnel from potent compounds and solvent exposure, as well as solvent release to atmosphere. The application of CIP cleaning can improve cleaning reliability while minimizing the operator exposure to potent compounds and reduce manual cleaning required to restore this equipment for process use.

These CIP operations typically occur between process campaigns, and cleaning frequency may vary as a campaign may be one week or one year in duration. However, between process lots, a process solvent flush may be used to prevent build up of process soil. The soils noted may not be susceptible to dissolution, suspension and removal by aqueous cleaning solutions. Organic process solvents rely primarily on solubility for the removal of the soil residue, and can perform their cleaning duty without introducing an external contaminant into the processing system (2). The solvent laden with residue is transferred to a waste handling system capable of handling the solvents and potent compounds. Some dedicated product facilities perform very infrequent cleaning, and may use process reactors equipped with spray devices and transfer piping to perform the duties of the CIP unit. When a CIP unit is used, the design will likely include solvent headers to supply the necessary cleaning solvents, for example acetone, methyl alcohol, DMSO, and dimethyl formamide.

Solvent-Based Cleaning Sequence Example

The primary phases of a solvent-based cleaning sequence, for cleaning between Drugs "A" and "B" campaigns are presented in sequential order in Table 6.

TABLE 6 Solvent Cleaning Sequence

Phase	Function	Success criteria
Process solvent A pre-flush	To process compound recovery and to reduce soil load prior to cleaning	All soiled process paths flushed clear
Process gas blow—nitrogen	Solvent recovery, line clear	Primary solvent and soil load pushed to solvent recovery
Process drain (complete process "A" operations)	Release process for cleaning	Residual solvent drained to solvent recovery
CIP program initiation	Acquires CIP boundary, initiates nitrogen purge and blanket	Boundary and utilities available, permissives met
Chemical wash	Expose all soiled surfaces to recirculated cleaning solvent	Chemical wash removes soil, confirmed later in program
Gas blow, intermediate drain	Remove soiled solvent from circuit	Primary solvent and soil load sent to solvent recovery
Rinse process solvent flush	Single pass flush of all soiled surfaces with cleaning solvent	Circuit flushed clear of spent solvent
Gas blow, intermediate drain	Remove flush solvent from circuit	Primary solvent and soil load sent to solvent recovery
Chemical wash	Recharge with solvent and recirculate for sampling	Sampling for previous product soil reveals residual below acceptance criteria (less than _ ppm)
Gas blow, intermediate drain	Solvent recovery, line clear	Sampling solvent to solvent recovery
Program complete	Releases CIP boundary, terminates nitrogen and blanket	Perform visual inspection or equipment breakdown for manual cleaning prior to release for processing
Process solvent B flush	To remove residual solvent A	All soiled process paths flushed clear
Process gas blow and process drain	Solvent recovery, line clearing	Primary solvent and soil load sent to solvent recovery
Ready for process "B" operations	Release for process B	

The pre-cleaning process flush with Solvent "A" is a key ingredient in the equipment cleanability. This post-process flush removes the primary soil load and will lessen the overall CIP cleaning load. Typically, the flush does not involve the CIP unit, but does use the equipment CIP spray devices and is considered a process, not a cleaning function. As such, there is no cleaning performance "acceptance criteria" assigned to the operation.

Combustion Concerns

Due to the combustion concerns related to the CIP system solvent utilization it is essential to consider the potential hazards and ensure the required safeguards are in place. Nonconductive solvents build up an electrical charge, and sprayed mist will be capable of containing sufficient total electrical charge to generate the spark energy needed to exceed the solvent's minimum ignition energy (MIE). If the spray device piping is grounded and bonded adequately, then these charges will go off to ground and cannot participate in the generation of a spark.

An ignition source with air present could initiate an explosion, and that ignition source could potentially be the sprayed organic mist. The CIP program must ensure the exclusion of a fuel source below the minimum oxygen concentration. A CIP circuit nitrogen purge to create an inert atmosphere may be required to avoid an explosive atmosphere. The CIP unit should include a filtered nitrogen supply and vent on to permit purging of the CIP circuit, and active nitrogen blanketing may be considered over the course of the CIP program. To determine the impact on the instrumentation selection and wiring installation, the solvent(s) selected and their MIE, as well as the expected use conditions must be evaluated to determine how best to respond and adapt.

Solvent-Assisted Cleaning Programs

Solvent-assisted cleaning programs are being applied with greater frequency in bulk pharmaceutical chemical or API facilities with both solvents and aqueous-based cleaners playing a role. Again, these CIP operations typically occur between process campaigns and may be infrequently scheduled.

The soils may be marginally vulnerable to aqueous cleaners, and a solvent process pre-flush is used to supplement the benefits of the alkaline and acid based aqueous cleaning programs. This approach reduces the solvent waste generated, minimizes the hazards associated with spraying solvents, and reduces the added cost for specifying and supplying a CIP unit capable of running supplying solvents under a nitrogen blanket.

PROCESS CONTROL TO ENSURE CLEANING PROGRAM SUCCESS

There are two (2) relevant timers associated with the control of a CIP cleaning boundary which aid in ensuring the integrity of clean processing operations, or the success of a validated CIP program under controlled soil conditions.

CIP Hold Dirty Expiration Timer

The CIP "hold dirty" timer may also be phrased with a more positive image as the "requires cleaning" timer. This timer is initiated at the conclusion of process operations, and ensures that the equipment is not held post-process for an excessive period of time where the soil load may alter state to a condition, which is more difficult to clean. For example, soil drying on equipment surfaces, concentrating the soil load, denaturing proteins, and resulting in greater surface adhesion. Typically the Requires Cleaning timer is of a time range of 1 to 24 hours, depending on the process soil, operating schedule, and the effort undertaken to validate an acceptable time period. If the timer is exceeded, a more aggressive cleaning regimen may be specified prior to renewing the equipment to a clean status.

Clean Hold Expiration Times

The "clean hold" time status is established at the successful conclusion of a CIP program and is used to provide an expiration time limit for cleanliness. This must be closely evaluated on process operations where post CIP microbial growth would impact integrity of a process operation or reduce the likelihood of success for a steaming operation. Again, time durations vary from 24 hours to several weeks as the process operation, operating schedule, and the effort undertaken to validate an acceptable time period factor into clean hold time definition.

SUMMARY

The described variations of CIP cleaning programs using aqueous-based, solvent-based and solvent-assisted CIP procedures have been proven effective on most soils found in the biotech and pharmaceutical facilities. The key to successful, repeatable CIP operations is an early effort to define CIP CD, start with the definition of cleaning criteria in the User Requirements, and with continued focus through all phases of the project to:

- Ensure specification and installation of cleanable process equipment and piping
- Perform timely laboratory testing on the soil to permit definition of cleaning chemistry and the basic CIP program, permitting an early assessment of processing schedule, utility and chemical requirements
- Define the required CIP Program to achieve the desired cleaning results in the CIP functional requirements
- Develop detailed functional (software) design specifications to ensure configurable cleaning recipes to permit circuit optimization during commissioning
- Define a commissioning plan that confirms CIP operations and enables CIP cycle optimization
- Define CIP functional testing that confirms acceptance criteria for each program phase

CIP operations planned in this manner will ensure that there are no surprises during qualification, with the effort being performed relatively hands-off by recording observations, doing swabbing and rinse sampling, etc., to confirm the successful optimized cycle.

REFERENCES

1. Howard T, Wienceck M. Biotech CIP cycle development. Pharm Eng 2004; 24(5).
2. Verghese G. Selection of cleaning agents and parameters for CGMP processes. In: Proceedings of Interphex Conference. Philadelphia, PA, March 17–19, 1998.
3. Seiberling DA. Design principles and operation practices affecting clean-in-place procedures of food processing and equipment, cleaning stainless steel. ASTM STP 538. Philadelphia, PA: American Society for Testing and Materials, 1973:196–209.
4. Stewart JC, Seiberling DA. Clean in place—the secret's out. Chem Eng 1996; 103(1):72–9.

6 CIP System Components and Configurations

Dale A. Seiberling
Electrol Specialties Company (ESC), South Beloit, Illinois, U.S.A.

INTRODUCTION

The end result of the design effort applied to clean-in-place (CIP) skid(s) for the biopharmaceutical industry is presently as varied as the number obtained by multiplying the number of engineering firms, consultants, and owners involved in the design process. However, perhaps as many as 85% to 90% of all of the CIP circuits validated during the past 15 years could be cleaned with CIP skids of not more than two different sizes (based on flow rate capacity), of several different configurations, determined by the return flow motivational force chosen for the project. The author has been associated with about 1200 to 1400 different CIP systems during a period of 45 plus years. These CIP systems have been applied in dairy, brewing, beverage, and food processing and, during the past 15 years, in biopharmaceutical applications and have been composed of combinations of almost every type of pump, valve, and instrument sensor of sanitary design commercially available.

CIP Skid

The combination of tanks, pumps, valves, and interconnecting piping designed to supply the flush, wash, rinse, and sanitizing solutions to a CIPable process, all generally mounted on a common frame to simplify installation, electrical wiring, and utility and drain connections, is generally identified as a CIP skid. A heat exchanger is included on most CIP skids and part or all of the required sensors and controls may also be included in a suitable enclosure mounted as an integral part of the skid. Chemical feed equipment and even chemical supply vessels may be included as part of the CIP skid, and one or more water supply tanks may also be incorporated. However, it is not adequately recognized that the "magic" of CIP does not reside in the CIP skid, but rather in the design of a CIPable process, to be described in a subsequent chapter. An overwhelming amount of time is spent in developing minimal CIP process instrumentation diagrams (P&IDs) supported by hundreds of pages of documentation, by individuals that have not had personal experience with the startup, operation, or commissioning of such equipment. The design and purchasing process adds considerable cost to the acquisition of what could be an "off-the-shelf" component for many projects.

This chapter will define the components, explore their function and operating requirements, and then assemble them in several different configurations, several of which would meet most application needs. The final choice may differ depending upon the function of the process to be CIP cleaned, i.e., a product development or research and development (R&D) facility may find a portable CIP skid will provide controlled cleaning in a flexible manner, and be easily adapted to changing process needs. The large single- or multiproduct manufacturing process will require higher recirculation rates, vastly different utility resources, a fixed CIP supply (CIPS) and

CIP return (CIPR)-piping to convey the flush, wash, and rinse solution to and from the process.

Definition of CIP Skid Capability

In some instances, the CIP skids are designed, specified, and ordered in the early stage of the project design, being considered to be long lead time items, this in turn being the result of the design, specification, and purchasing procedure of choice. A more effective procedure is to develop the overall process design first in a simple but complete schematic manner that allows the definition of all piping requirements based on planned operational procedures. This can also be the basis of defining every CIP circuit, which preferably will include a vessel and all interconnecting piping available for cleaning with the vessel. Then, by fine tuning the relationships of loads to the required CIP skid(s), the geography of the process equipment and the CIP skids can be optimized to accomplish validatable CIP with a minimum of water, chemicals, and time.

Table 1 presents the results of an analysis of water required to fill CIPS/R piping runs between 65 vessels and 5 CIP skids in a large biopharmaceutical facility of current vintage. The estimates of line length, diameter, and volume required to fill all of the product and CIP piping were developed via simultaneous reference to the schematic flow diagram, required process transfer rates, and CIP flow rates based on vessel diameter.

By judiciously locating the CIP skids with respect to the carefully estimated loads the range of CIPS/R piping volume for the process varied mainly from 25 gal for 200 L media prep vessels on a third floor to 99 gal for 15,000 L bioreactors with 700 ft of tubing of three diameters in the combined CIP and process piping runs, holding 99 gal. Two formulation tanks close to the skids at the first floor level required only 15 gal of water. The largest volume, 154 gal, was required to clean buffer hold and purification vessels in combination, in sizes of 2000, 4000, and 10,000 L. These data were extrapolated to estimate total water and time requirements and were found to be very much in line with the needs at commissioning, which in turn were facilitated by having such information available.

ENGINEERING CONSIDERATIONS

The selection and application of readily available CIP equipment, or the design of a special unit, the most common but unnecessary approach, is influenced by factors including (i) required delivery (gal/min), (ii) delivery pressure—psig, (iii) required sequence of treatment, (iv) number of tanks (based on water availability), (v) delivery temperature, and (vi) physical space available.

The number of tanks, pumps, and valves and the relative location on the support frame are of little engineering significance with exception of cost and space. More important are the variations in the use of these components to produce the required sequence of treatment.

Typical Fixed CIP System Skids

CIP systems are available in two substantially different forms as single-tank or multi-tank systems, this choice being driven by water supply rather than CIP cleaning needs. If the facility water supply at the CIP skid loop is not equal or greater than the required delivery rate, a water supply or surge tank must be

TABLE 1 Comparison of Water Requirements to Fill Clean-In-Place Supply and Return Piping and Process Piping Associated with Vessels of Various Sizes and Types in a Large Biopharmaceutical Facility

Vessel type	Volume (L)	CIPS (ft)	CIPS dia (in.)	CIPS (gal)	CIPR (ft)	CIPR dia (in.)	CIPR (gal)	Total line (gal)
Media prep	200	140	1.5	11	180	1.5	14	25
Media prep	4,000	200	1.5 and 2.0	31	210	2.0	30	61
Media prep	12,000	200	1.5 and 2.0	31	210	2.0	30	61
Bioreactor	120	280	1.0 and 1.5	21	200	1.5	15	36
Bioreactor	600	350	1.0 and 1.5	26	85	2.0	26	52
Bioreactor	3,000	450	1.0 and 1.5	35	180	2.0	26	61
Bioreactor	15,000	700	1.0, 1.5, and 2.0	83	140	2.5	16	99
Buffer prep	4,000	220	2.0 and 2.5	50	205	2.0	29	79
Buffer prep	13,000	220	2.0 and 2.5	50	205	2.0	29	79
Purification	4,000	190	1.5	16	90	2.0	13	29
Purification	10,000	210	1.5	18	160	2.0	26	44
Formulation	3,000	80	1.0 and 1.5	6	60	2.0	9	15

Abbreviations: CIPR, CIP return; CIPS, CIP supply.

provided, generally on the skid. Some designers provide a separate supply reservoir for each flush, wash, and rinse solution required by the CIP program.

Presented in Figure 1 is a composite design of all the components required to configure a one-tank or two-tank skid to operate in three different manners. The circled numbers 1–20 are referenced in parentheses in the following paragraphs that explain each component's function.

Major Components

A CIP recirc tank, sometimes used also as the only water tank, is a part of all CIP skids (1). The combination of a conical (or dish) bottom tank and outlet leg provides adequate net positive suction head (NPSH) for the CIP pump with a minimal volume of water in the tank. A tank outlet valve (TOV) (2) controls supply to the CIP pump (3). A casing drain on this pump is at the lowest point on the skid and will fully drain the complete skid at the end of the program. A shell and tube heat exchanger (4) is shown vertically mounted for space considerations and drainability is provided by installation of a tangentially drilled restrictor between the inlet and outlet tees. Heat exchangers may also be mounted horizontally, a preferred practice for large units. Steam supply and temperature control is discussed in Chapter 7. The CIPS piping on the skid includes the conductivity sensor, pump discharge pressure sensor, and CIPS temperature sensor, all near (5), and the flow continues to the flow element (generally a vortex meter, mass flow meter, or turbine meter) and a flow control valve (6). The choice of a throttling-type valve or a pump with a variable frequency drive for flow rate control in turn based on the meter analog output is described in Chapter 7. The two close coupled valves (7) are optional and will be discussed as part of three different configuration scenarios. The CIPS then continues to the process and the circuits.

A CIPR manifold on the skid is the general location for a return-flow probe (discrete), return-temperature probe, and resistivity probe, all (8). A return flow hold-back valve (HBV) (9) is optional. The means of establishing recirculation through the tank or around the tank will be discussed as part of the configuration scenarios.

Single- or Multi-Tank Variation

A single-tank skid would require the provision of the water supply to the CIP recirc tank. A single ambient water for injection (AWFI) source is shown via valve (14) to a spray device (12). Two waters of different quality, or temperature, are commonly used, and both are supplied through individual spray devices.

A multi-tank CIP skid may be fitted with one or more water tanks to accommodate the different water qualities or temperatures. The hot water for injection (HWFI) TANK (16) is shown as an American Society of Mechanical Engineers (ASME) vessel equipped with a rupture disc and is supplied with HWFI by valve (18) through a spray to flush the tank head and sidewalls, under control of a level probe. A vent filter is shown and a temperature sensor and sample valve are indicated (19). If this tank is intended to be SIP'd it would be insulated.

Chemical supply to the skid is shown by means of a chemical loop that originates at optional valve (20), with a restrictor adjacent downstream to control flow through the loop (see Chapter 8). The return connection for this chemical loop is optional and will vary with the operating scenario.

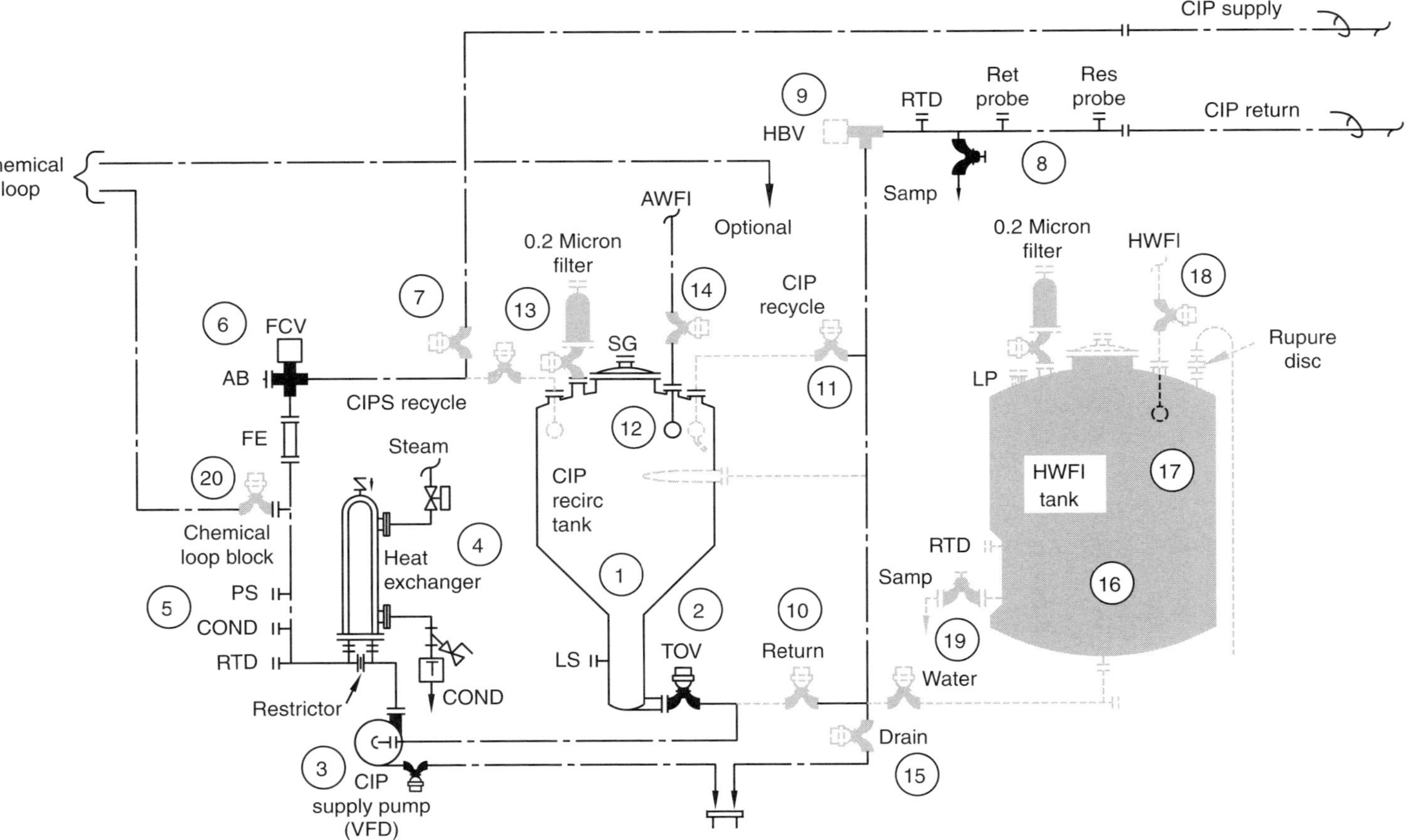

FIGURE 1 These major components can be assembled in a number of different configurations to create effective clean-in-place skids to serve any purpose.

Single-Tank CIP Skid Operational Concepts

Assuming the user facility has water loops of adequate capacity to support flow to the CIP skid at the use rate, attention will now be given to how best to configure the system. This in turn is determined on the basis of knowledge of the proposed CIP circuits and any special operating requirements.

Single-Tank Bypass Operation

Figure 2 illustrates the most simple and very reliable configuration of a single-tank CIP skid, capable of cleaning a vessel of any size with the minimum amount of water possible, by pumped or gravity return flow. Any piping in the circuit must discharge to the vessel being cleaned to enable air disengagement before return to the skid.

The skid mounted tank is a break tank between the water loops and the CIP pump and never contains a chemical solution. The valves shown permit the tank to be drained and rinsed to use waters of different quality for the pre-rinse and solution wash (lower quality) and final rinse (higher quality) purposes. All water is supplied through a spray device to flush the tank head and sidewalls with each addition of water. The large-scale heavy lined schematic in Figure 2 is for wash recycle and defines the flow path on the skid, through the chemical loop, and the CIPS/R to the process and circuits. The chemical loop return is to the CIPR manifold downstream of the CIPR sensor locations and permits the chemical loop to be flushed to drain during rinses, following which the chemical loop block valve may be closed to conserve water.

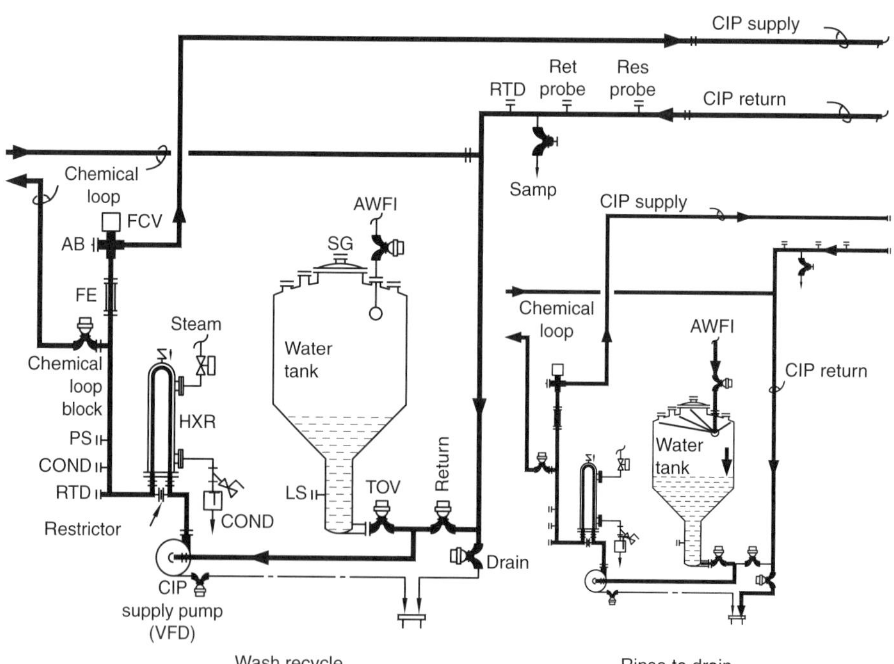

FIGURE 2 This schematic diagram illustrates *single-tank bypass operation*, the most reliable and simple method of cleaning vessels of any size.

The small heavy lined circuit in Figure 2 illustrates the rinse to drain flow path. All rinsing involve supply of water of the required quality to the break tank, delivery to and return from the connected circuit, and discharge to drain. If gravity return is applied, the static head must be sufficient to allow the CIP pump to operate at less than 8 to 10 in. vacuum during recirculation. If a return pump is utilized, it should be sized and controlled to stuff the CIP pump at 1 to 3 psig (7–20 kPa).

The wash recycle will begin with a system fill step with the drain and return valves closed, using the meter to control the delivery of sufficient water to fill the CIPS and CIPR piping and create a minimal puddle of 3 to 5 gal (12–20 L) in the vessel being cleaned. A flat plate vortex breaker will be required to operate at this desirable low level. Following the system fill, the TOV will close and the return valve will open. The unique operation of this system when washing tanks (or a tank in combination with a line) eliminates the problem of "balancing" flow produced by CIPS and return pumps, and also the problem of "air-locked" pumps that frequently occurs in the application of multi-tank systems, as an air bubble in the return pump will be drawn through by the supply pump, and an air bubble in the supply pump will be pushed through by the return pump. Recirculation at the set point flow and pressure is achieved with great reliability. Following confirmation of recycle by flow and pressure sensors, heating will be initiated and chemicals will be introduced. Cleaning cycle timing will begin following confirmation of return flow temperature and conductivity.

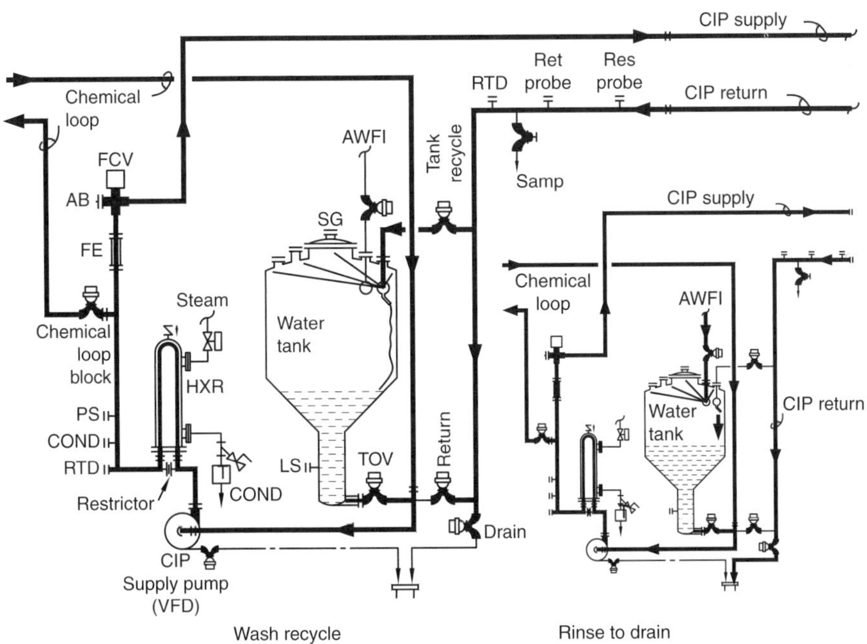

FIGURE 3 The addition of one valve to the CIPR line will allow *single-tank bypass* or *recycle operation for cleaning piping circuits.*

Single-Tank Bypass or Recycle Operation

If it is necessary to clean a line circuit, i.e., a circuit consisting of only piping or piping plus equipment such as a homogenizer, centrifuge, or filter housing(s), the addition of a CIPR recycle valve, as shown in Figure 3 enables the above-described system to operate with WASH RECYCLE through the CIP skid mounted tank.

The suggested design using a cone bottom tank and leg to create the desired NPSH with minimum volume in the tank is beneficial to the reduction of water, chemicals, and time. The return flow must be introduced to the tank through a spray to continuously flush the tank head and sidewalls and prevent the development of the traditional "bathtub ring" during the solution wash. The RINSE TO DRAIN is as described previously.

Single-Tank Total Recycle Operation

Some CIP applications may include circuits or operating conditions that are expected to produce substantial air entrainment in CIPR flow, examples being the use of a liquid ring return pump to overcome undesirable CIPR design conditions, or the utilization of injected air to the CIPS for spray CIP of large filter housings. The single-tank bypass skid performance would be degraded by the presence of massive air in the CIPR flow, but the many other advantages of the small, simple CIP skid concept may be achieved by configuring the CIP skid as shown in Figure 4. The supply side of the system is unchanged but CIPR flow is always to the CIP recycle tank via the CIPR recycle valve. Most of the minimal water, time, and chemical use advantages of the single-tank bypass CIP skid may be retained by the

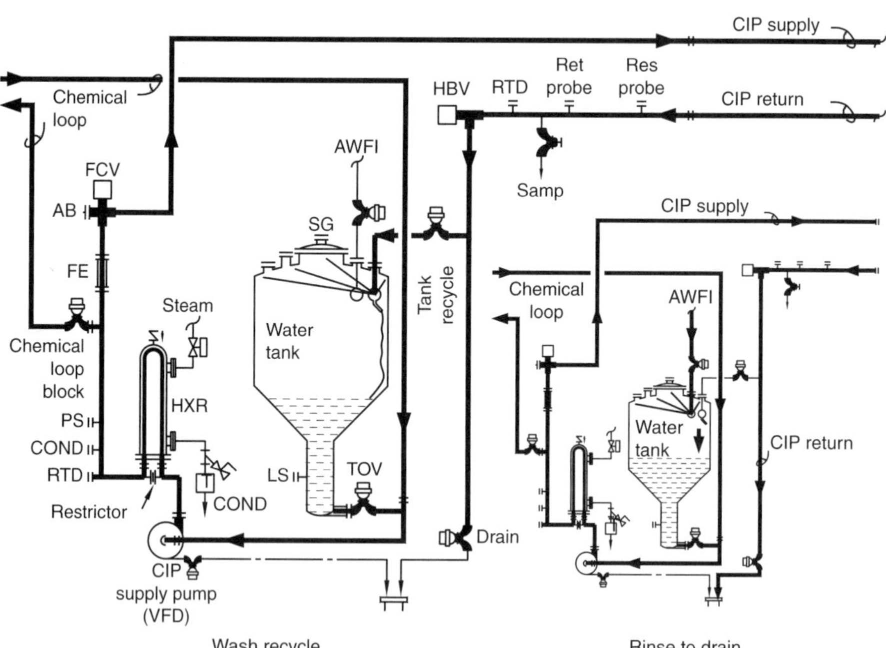

FIGURE 4 The *single-tank single-use recirculation operation* system is effective in handling return flow with substantial entrained air.

addition of a HBV to the return line. This valve, of throttling-type design, will be controlled by level in the CIP recycle tank during wash recycle, the level sensor (LS) being located low on the leg for maximum sensitivity to minimal changes in tank and static volume.

The large-scale heavy lined schematic in Figure 4 again depicts wash recycle and defines the flow path on the skid, the chemical loop, and the CIPS/R to the process and circuits. The small heavy lined circuit in Figure 4 illustrates the rinse to drain flow path, unchanged from the previous configuration. The wash recycle will begin with a system fill step with the drain valve closed and the CIPR recycle valve open, using the meter to control the delivery of sufficient water to fill the CIPS and CIPR piping, create a minimal puddle of 3 to 5 gal (12–20 L) in the vessel being cleaned and fill the CIP recycle tank (now a solution tank) to a level just above the tank leg. On completion of the system, the HBV will be placed under analog control based on the CIP recycle tank LS. The control of the system fill volume and CIP recycle tank level provides indirect control of the puddle in the vessel being cleaned and again, the unique operation of this system when washing tanks (or a tank in combination with a line), eliminates the problem of "balancing" flow produced by CIPS and return pumps. All other aspects of the operation of this system configuration are as described previously.

Note that CIPR flow to the CIP recycle tank is also by a spray device, in this instance one fitted with a reduced discharge tube to direct the major flow to the sidewall, to maximize air disengagement. The reduced discharge will create the required back pressure of 5 to 15 psig (35–100 kPa) to assure full spray coverage during reasonable variation in system flow rate.

Single-Tank Single-Use Recirculation Operation

Both of the above-described configurations make up the smallest possible volume of solution required, for every individual circuit, at the required concentration, use it once for a controlled period of recirculation, and then discharge the spent solution to the sewer or facility waste system at the end of each cycle. These single-tank systems are smaller in size, simpler in design, lower in initial investment, and more importantly, flexible and reliable in application. All chemicals are fed automatically in the proper proportions and sequence from shipping containers, day use tanks, or bulk storage tanks as described in Chapter 8.

The single-tank total recycle CIP skid shown in Figure 5 is capable of operation at up to 80 gal/min (300 L/m) and a CIPS pressures of 78 Psi (640 kPa). The CIPR flow enters the solution tank at two points, via a tangential inlet and spray to facilitate rapid air disengagement from return flow and enable operation with only 12 to 14 gal (45–50 L) in the CIP recycle tank, while assuring continuous flushing of all solution tank surfaces. The spray device with a tangential outlet leg in Figures 3 and 4 will accomplish the same function alone at lower cost. A requirement of all simple single-tank CIP skids is a facility water supply(s) equal or slightly exceeding the maximum CIP flow rate for the largest diameter line circuit, or biggest vessel to be spray cleaned. Any of the above CIP skid configurations could be of approximately the same size, with operating capability to 80 gal/min (300 L/m), using 2-in. valves and skid piping. The delivery rate could be increased to 100 gal/min (380 L/m) by increasing the CIP pump suction to 2.5 in., and increasing the pump and heat exchanger size, with nominal (if any) increase in overall size.

Single-Tank Single Pass

Some applications demand absolute assurance of freedom from any potential cross-contamination. An example might be the use of common mixing and filling equipment to contract package different products for different drug manufacturers. Even experienced users of CIP, with substantial understanding of the ability to design, operate, and validate a CIP skid and CIPable process in which all surfaces of all components are subjected to the same rigorous conditions of time, temperature, and control, may elect to consider still another option which avoids the need for recycle, hence the opportunity for cross-contamination. The single-tank CIP skid shown in Figure 6 would designed and controlled to operate as follows:

Pre-rinse—The solution makeup tank would be filled and then recycle through the heat exchanger would be initiated to adjust the rinse temperature per the solution makeup schematic in Figure 6. If only HWFI was available, and a lower temperature rinse is desired, a shell and tube cooler might be substituted for the heater, of for maximum flexibility both might be installed in series. The volume of water would be adjusted to provide a single continuous rinse of adequate duration per the rinse to destination flow path in Figure 6, or two rinse

FIGURE 5 This total recycle clean-in-place skid can operate at up to 80 gal/min with an adequate water supply, and occupies an area of only 36 in. by 84 in., including the control panel.

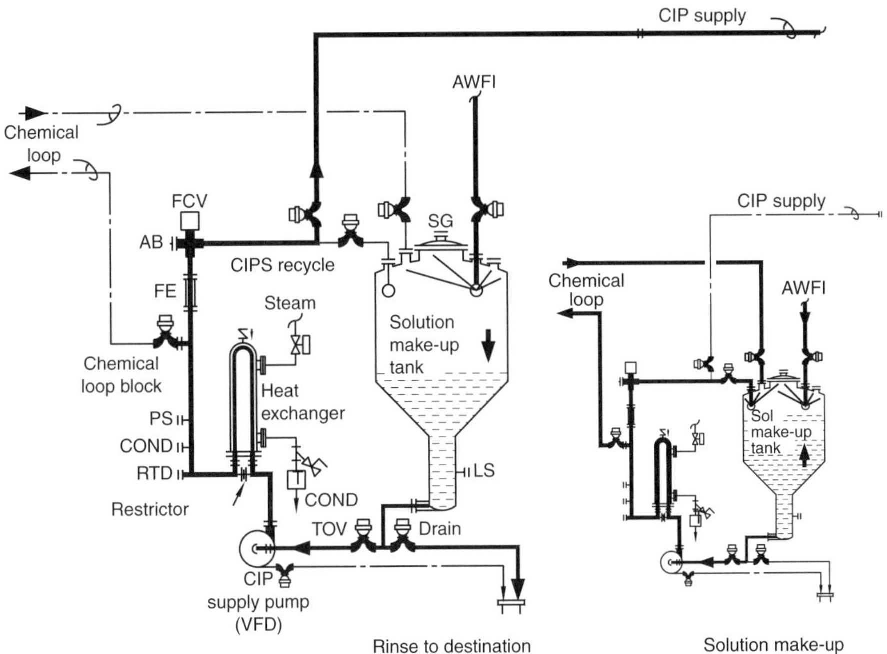

FIGURE 6 For processes which demand maximum protection against cross-contamination, the standard components can be configured to provide *single-tank single-pass* operation.

steps might be employed, followed by an air blow of the CIPS to and through the circuit, and then a circuit drain step.

Alkaline Solution Wash—Next, the tank would be filled with sufficient water to provide a batch of adequate size to meet the time requirement at the desired flow rate. For example, if a wash time of five minutes was required at 50 gal/min (190 L/m), the solution batch would be 250 gal (950 L) plus the amount required to fill the CIPS piping to the beginning of the circuit. Recycle from the tank, through the heat exchanger and back to the tank would be initiated as would chemical feed, and the solution would be adjusted to the desired temperature and conductivity. Then, it would be supplied to the circuit at the required flow rate for the specified time. An air blow of the CIPS to and through the circuit would follow.

Post and Final Rinse—The solution makeup tank would then be rinsed to drain and then filled with water of the required quality for these operations, conducted as described for the pre-rinse above.

Instrumentation for program documentation may be more costly and complex than for CIP skids which use recirculation, as the result of single pass discharge to multiple destinations.

Options include (1) multiple temperature, conductivity, and resistivity sensors at the circuit destinations, (2) validation that controlled CIPS conditions always produce the desired conditions at the end of the circuit, and (3) the installation of an eductor supported CIPR collection system to bring flush, wash, and rinse solutions back to a common sensing point, perhaps on the same skid on which the above-described components are located.

Multiple Tank CIP Skid Configurations

The major reason for the addition of one or more and generally larger tanks to a CIP skid is the lack of adequate water of the required quality for CIP rinse operations. If the facility high quality water supply is not equal to the maximum CIP delivery rate, a tank must be provided to accumulate water in advance of the start of the rinse operations. To make this point most vividly, consider the water requirements for rinsing the 15,000 L bioreactor for which the total circuit volume was estimated to be 99 gal (380 L). The rinsing of a line may be accomplished by pumping a volume of 1.5 times the line volume through the entire line. When the circuit contains multiple flow paths, each of those paths must be rinsed in sequence. The pre-rinse may be only two or three passes through but to meet final rinse resistivity criteria, the final rinse may require five to seven times the circuit volume, which, for example, would be 500 to 700 gal (1900–2700 L). The typical calculation of water tank size would be based on the deficiency of delivery gal/min less the supply gal/min, in this example 50 gal/min, and for a rinse duration of seven minutes, the tank would need to be 350 to 400 gal (1300–1500 L approximately). This tank adds to the weight, space, and cost of the CIP skid, and lacking elaborate software to control filling and emptying, may cause loss of time for draining and rinsing when water of different qualities is used for the pre-rinse, wash, and post-chemical wash rinses as compared to the final rinse. If the tank is used for water of product quality the tank may be of ASME construction and protected with a sterile vent filter. If the tank is to be SIP'd it will be insulated for personnel protection.

Multi-Tank CIP Configuration with CIP Recirc Tank Before CIPS to Circuit

If the CIP delivery rate exceeds the water supply flow rate, a separate water tank will be beneficial. The solution tank may vary in configuration and the two tanks may be combined in several different manners.

This first multi-tank CIP skid illustrated in Figure 7 is essentially the single-tank CIP skid reviewed in detail as Figure 4 with a separate HWFI tank added, HWFI supply to this tank, and a valve to supply the CIP pump from this tank in addition to supply from the CIP recirc tank.

The large-scale heavy lined schematic in Figure 7 is again for wash recycle and defines the flow path on the skid, the chemical loop, and the CIPS/R to the process and circuits. The small heavy lined circuit in Figure 7 illustrates the rinse to drain flow path, unchanged from the previous configuration. Both rinse and wash phases of the program will now draw the water from the HWFI tank. The wash recycle will begin with a system fill step with the DRAIN valve closed and the CIPR recycle valve open, using the meter to control the delivery of sufficient water to fill the CIPS and CIPR piping, create a minimal puddle of 3 to 5 gal (12–20 L) in the vessel being cleaned and fill the CIP recirc tank to a level just above the tank leg. On completion of the system fill, the HBV valve will be placed under analog control based on the CIP recirc tank LS. The control of system fill volume and CIP recirc tank level provides indirect control of the puddle in the vessel being cleaned and again, the unique operation of this system when washing tanks (or a tank in combination with a line), eliminates

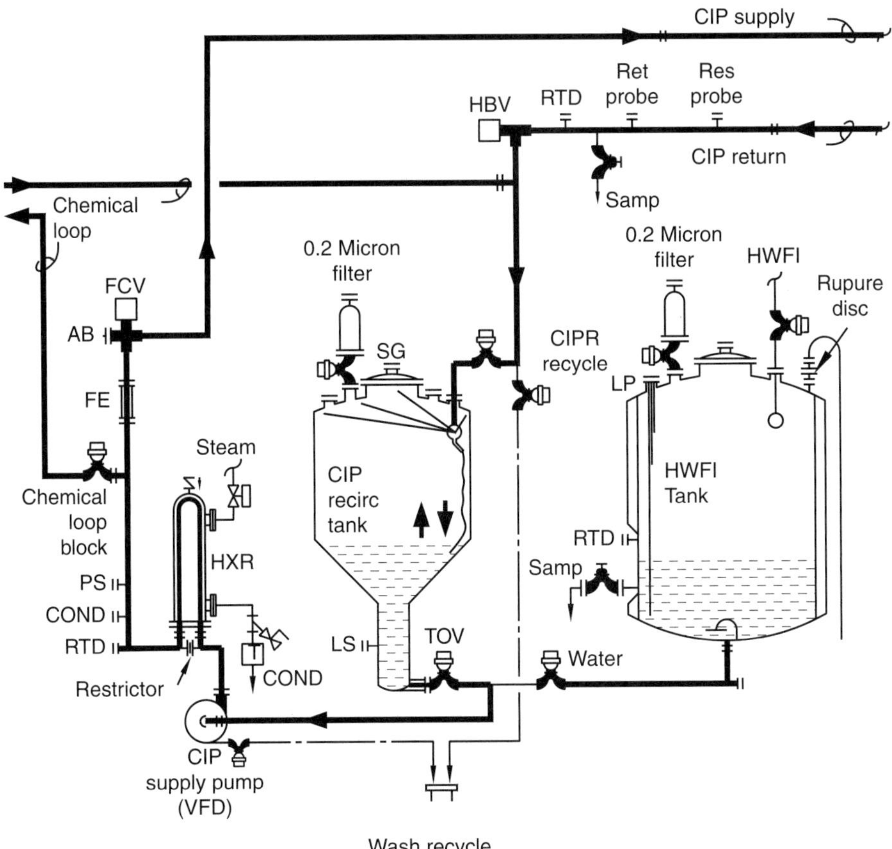

FIGURE 7 The heavy lines on this diagram depict wash recycle flow through a conventional two-tank clean-in-place skid.

the problem of "balancing" flow produced by CIPS and return pumps. Note that CIPR flow to the CIP recirc tank is by a spray device with a reduced discharge tube to direct the major flow to the sidewall, to maximize air disengagement, as discussed previously.

The addition of a HWFI tank to the CIP skid will also allow provision of a lower quality water for the pre-rinse, chemical solution washes, and post-chemical solution wash rinses directly to the CIP recirc tank (not shown). This eliminates the need for draining, rinsing, and draining the HWFI tank if a lower quality water is used for the initial phases of the CIP program. All other aspects of the operation of this system configuration are as described previously.

The rinse to drain illustrated by the heavy line portion of Figure 8 starts with the supply of HWFI to the HWFI tank. The water valve admits this water to the CIP pump, the TOV being closed. The rinse is delivered to the circuit by the CIPS piping and returns through the CIPR piping, passing the return sensors enroute to drain.

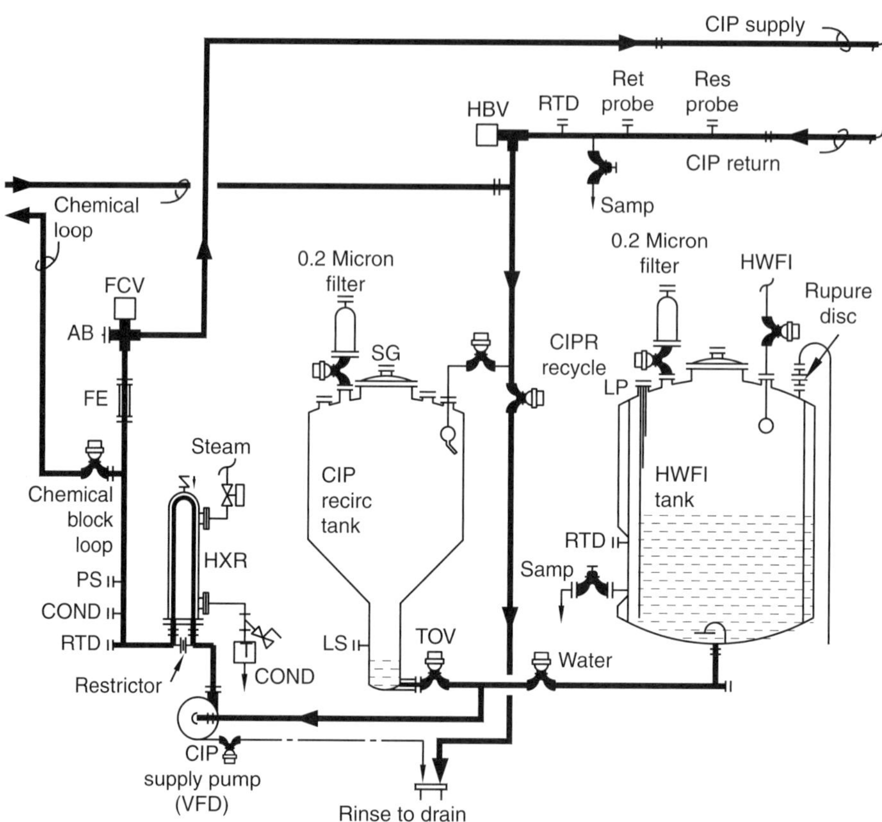

FIGURE 8 The heavy lines on this diagram depict the rinse to drain flow through a conventional two-tank clean-in-place skid.

Multi-Tank CIP Configuration with CIP RECIRC TANK after CIPR to Drain

This alternative to Figures 7 and 8 provides mix-proof valve separation of the CIP recirc tank and HWFI tank, assuring no possibility of either a lower grade water from the CIP recirc tank, or wash solution and soil contacting any part of the HWFI tank and supply piping. And, placing the CIP recirc tank on the end of the circuit assures that it is fully subjected to all of the time, concentration, and temperature criteria applied to cleaning the process circuit.

This second multi-tank CIP skid illustrated in Figures 9 and 10 is similar to the above but differs in two respects. A standard shutoff valve controls flow through the top port of a mix-proof valve (not through the valve passage) for flow from the CIP recirc tank to drain. The mix-proof valves on the HWFI tank and CIPR recycle tank outlet lines connect the vessels to the CIP pump with the equivalent of "double-block-and-bleed" separation of the two streams. The wash recycle circuit is illustrated in Figure 9 and the rinse to drain in Figure 10.

The two systems are comparable with respect to program control equipment including instrument sensors and I/O, and the multi-tank CIP skids also

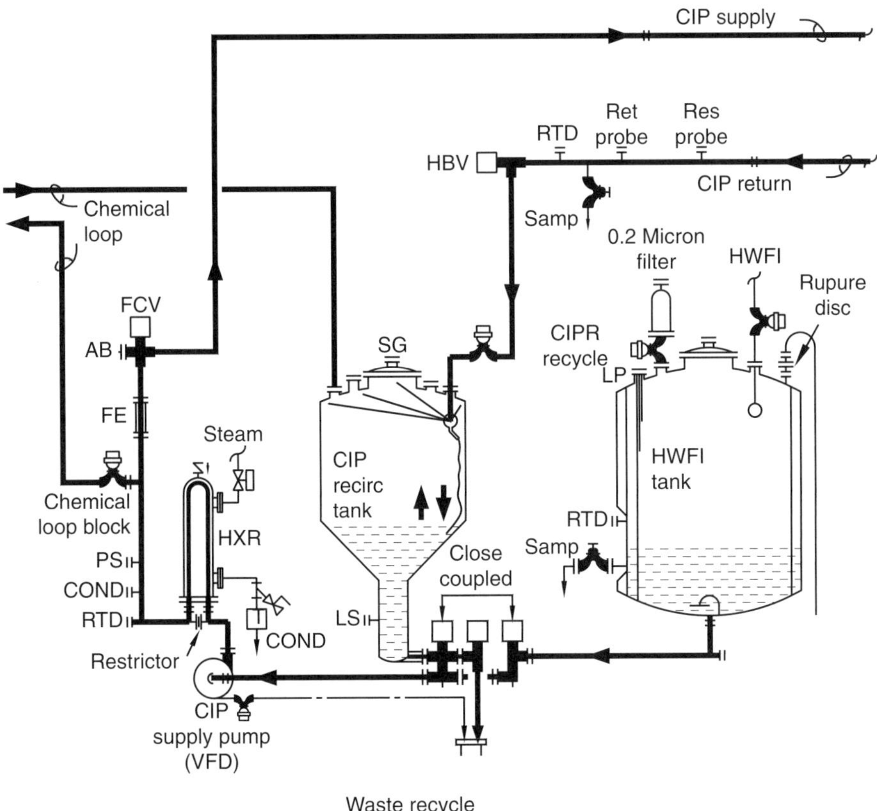

FIGURE 9 The heavy lines on this diagram depict the wash recycle flow through a two-tank clean-in-place skid with the CIP recycle tank on the return end of the circuit.

vary little in instrumentation and control components overall as compared to the single-tank configurations. A major variation is the additional level control system required for the HWFI tank. However, multi-tank systems require more space.

The two-tank CIP skid shown in Figure 11 is essentially the system depicted in Figures 9 and 10, with added features including

- The CIP recirc tank was used as a makeup tank for single-pass CIP, by installing the two valves shown (7) in Figure 1.
- The CIP recirc tank was increased to 150 gal (570 L) to provide a single-pass solution wash of five minutes at 25 gal/min (95 L/m) plus a small reserve for filling CIPS lines of varying lengths.
- Both the CIP recirc tank and HWFI tank were designed for SIP and insulated.

A visual comparison of Figures 5 and 11 should enable the reader to see that the multi-tank CIP skid was perhaps twice the physical size and weight of the smaller single-tank skid, the single-pass operating capability being the only added feature.

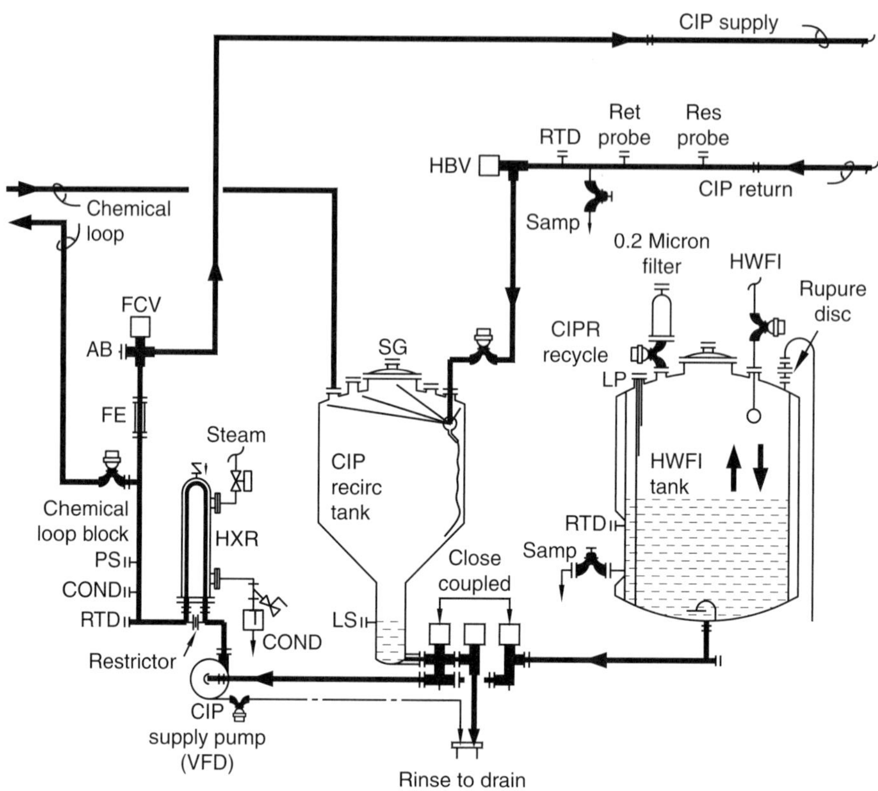

FIGURE 10 The heavy lines on this diagram depict the rinse to drain flow through a two-tank clean-in-place skid with the CIP recycle tank on the return end of the circuit.

Single-Use Eductor-Assisted CIP Skid

Any single-use CIP skid intended to function with minimal water requires a reliable CIPR system. The *single-use eductor-assisted* (SUEA) CIP unit shown schematically in Figure 12 fulfills this need as the ultimate, and highly preferred, solution to meeting criteria for minimum circuit volume, minimum vessel puddle, reliable recirculation, and rapid evacuation of all solutions from the circuit between CIP program steps.

This proprietary design uses an eductor as a pumping device, by injecting water through an orifice into an enclosed chamber, to create a vacuum of 16 to 18 in. under normal operating conditions, and approximately 12 in. at 80°C. The vacuum may be used to assist gravity return, or continuously prime return pumps which normally handle a 50/50% air/water mixture. The air separation/recirculation tank used with the eductor-assisted return system makes it possible to achieve recirculation of cleaning and sanitizing solutions at flow rates ranging from 50 to 120 gal/min (190–450 L/m) with as little as 12 to 15 gal (45–57 L) of solution in the air separation/recirculation tank. The vessel being spray cleaned, if fitted with a flat plate vortex breaker, will contain only a minimal puddle, if any. An air blow at

FIGURE 11 This clean-in-place skid combined the use of a CIP recycle tank on the return side of the circuit with the ability to operate as a single-pass CIP skid. The large water surge tank was required to accommodate a low flow rate water-for-injection loop.

the origin of the CIPS distribution piping system will clear all solution from the supply piping to the spray, or from the complete line circuit. The eductor-assisted return system will quickly draw all solution from a connected vessel and the CIPR piping.

A basic SUEA CIP unit, less water tanks(s), occupies a space of only 3 ft×6 ft (1 m×2 m). Supply tank(s) for water will increase the total size to that of the unit shown in Figure 13. Seiberling (1) described the prior dairy and food plant use of this system in a book written for pharmaceutical readers in 1987. The operation of the eductor-assisted CIP/sanitation unit and an alternative multi-tank eductor-based recirculating CIP system was discussed in a 1990 publication by Adams and Agaarwal (2) and in 1986, Seiberling (3) again described the use of this system which by then had been successfully applied as part of many biopharmaceutical projects. Perhaps a hundred or more of these CIP skids are operating in the United States today, in both large production applications and in more modest pilot plant and product development facilities.

Figure 14 is an installation photograph of a SUEA system fitted with a second CIP pump, heat exchanger, and chemical feed system supported by a single water

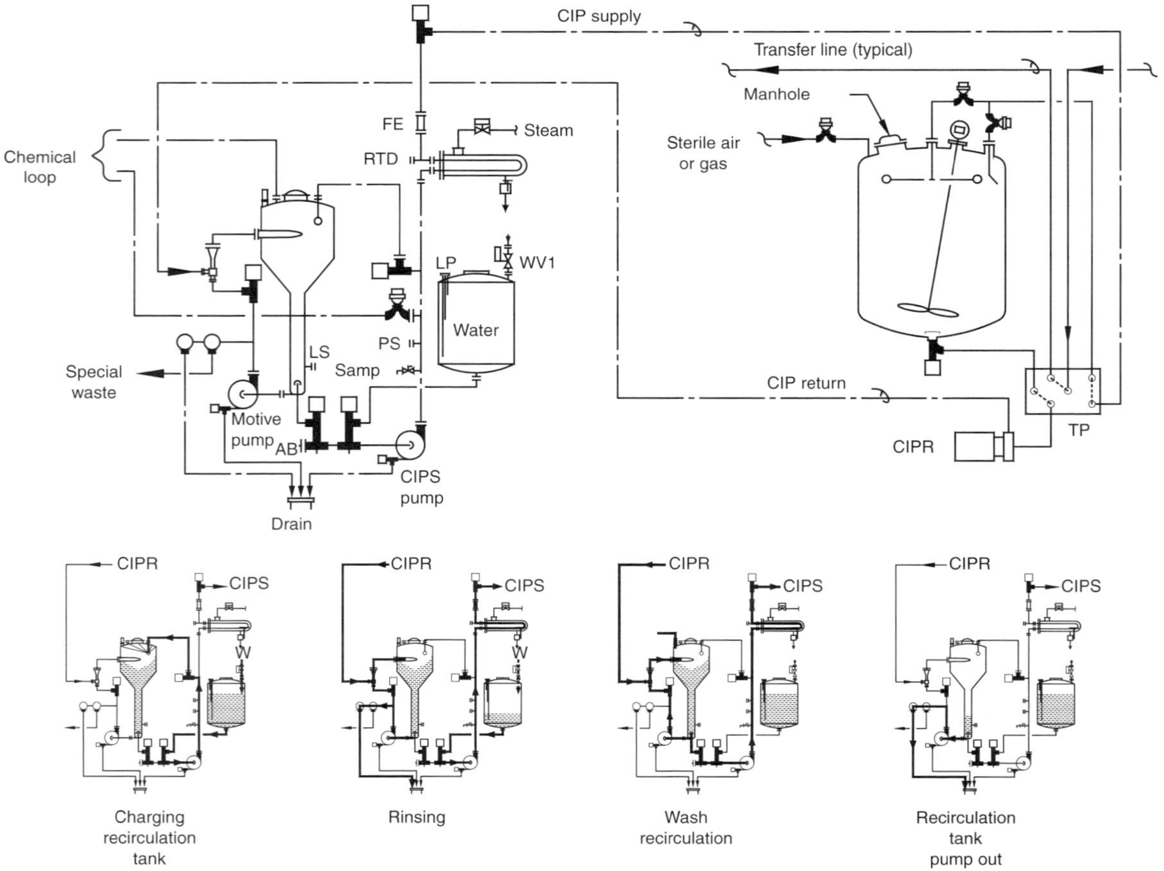

FIGURE 12 This schematic diagram of a single-use eductor-assisted clean-in-place system connected to a pharmaceutical tank includes diagrams of the four major operating conditions used for each phase of the program.

FIGURE 13 A single-use eductor-assisted CIP skid with one water tank. Note tangential inlet on air separation tank above eductor.

tank to provide automated CIP of two UF systems, by provision of water, heat, and chemicals to the ultrafiltration (UF) feed tank. The SUEA skid supplied water, heat, and chemicals to the UF feed tank and the UF system pump and valves were programmed to cycle these fluids through the remainder of the system and associated piping. Storage solution was injected from a portable tank connected to an automated valve on the suction side of the UF feed pump.

This part of this chapter will be concluded with Table 2, a comparison of the capabilities of the various CIP skids described above. Twelve *Selection* Criteria are listed and the nine different CIP skid configurations are rated Not Recommended (NR), Good, or Best for each criteria. Some explanation follows

- Some NR ratings are because the skid lacks the physical capability, i.e., a CIPR recycle valve to the water tank for the bypass only skid and portable skids. This option is normally not provided on portable skids because of the lack of water and drain capability through desirable small flexible hose connections.
- The single-tank skids and portable system are rated NR for application where the required water flow rate is less than the CIPS flow rate. A skid with a water tank is desirable.

FIGURE 14 This single-use eductor-assisted CIP skid has a second CIP pump and heat exchanger on the left-hand end as a system capable of supplying flush, wash, and rinse solutions to a ultrafiltration process skid, from a common water tank.

- A Good rating for any skid/criteria combination is based on the expectation of proper engineering design and installation.

Large High-Volume Dual CIP Skid

Whereas this chapter has focused on how a relatively standard or common CIP skid could meet most pharmaceutical and biotech needs, it is recognized that not all CIP circuits are "standard."

Figure 15 is a photograph of a CIP skid which included two *single-tank total recycle operation* CIP skids upgraded to 200 gal/min (760 L/m) by use of larger tanks, pumps, and heat exchangers. The two large tanks were installed on the skid were mounted and piped as surge tanks for the high-quality water required for the final rinse and available in limited supply, and for CIP waste neutralization.

CIP Skid with a Process Function

A Case History of a project that resulted in a CIP skid being much more, and hence more affordable, may be of interest to some readers. The original request was for a CIP skid to clean a small, but very expensive mixer, which was to be applied to introduce small quantities of readily soluble drugs into water being recirculated from a mixing tank. Further discussion led to the recognition that the four tanks and piping also needed to be cleaned.

The CIP skid illustrated in Figure 16 was designed and installed to perform the functions of drug mixing, CIP of the lines on completion of a mixing operation, CIP of the mixing tanks as they were emptied, and then CIP of the downstream process at the end of the production day. The CIP skid was purchased for a lesser cost than the

TABLE 2 Comparison of Nine CIP Skid Configurations on Basis of Selection Criteria

	Single-tank systems (for water supply equal to CIPS flow)				Multi-tank systems		Eductor-assisted systems		
Selection criteria	Bypass only	Bypass and recycle	Recycle only	Single pass	Basic	High purity	Basic	High volume	Portable skids
Tank circuits	BEST	BEST	GOOD	GOOD	GOOD	GOOD	BEST	NR	GOOD
Line Circuits	NR	BEST	GOOD	GOOD	GOOD	GOOD	GOOD	NR	NR
Water used per cycle	BEST	GOOD	GOOD	NR	GOOD	GOOD	BEST	BEST	BEST
CIPR hydraulics	GOOD	GOOD	GOOD	NA	GOOD	GOOD	BEST	BEST	GOOD
Low available water supply	NR	NR	NR	NR	GOOD	GOOD	BEST	NR	NR
Water-for-injection quality final rinse	GOOD	GOOD	GOOD	BEST	GOOD	BEST	BEST	BEST	NR
Zero cross-contamination	NR	NR	NR	BEST	NR	GOOD	GOOD	GOOD	GOOD
125–200 gal/min CIPS flow rate	NR	NR	NR	NR	GOOD	GOOD	NR	BEST	NR
Large air volume in CIPR	NR	NR	GOOD	NA	GOOD	GOOD	BEST	BEST	NR
Pilot plants	GOOD	GOOD	BEST	NR	NR	NR	NR	NR	BEST
Small-scale mnfg plants	GOOD	GOOD	GOOD	NR	GOOD	GOOD	BEST	NR	GOOD
Large-scale mnfg plants	GOOD	GOOD	GOOD	NR	GOOD	GOOD	BEST	NR	NR

Abbreviations: CIPR, CIP return; CIPS, CIP supply; mnfg, manufacturing; NA, not applicable; NR, not recommended.

FIGURE 15 This large CIP skid included two *single-use total-recycle* systems operating at up to 200 gal/min supported by large pure water and neutralization tanks.

original dedicated mixer, and the cost of all piping and valves in the project was attributed to the process equipment budget. Only the CIP skid, four sprays, and four valves to control flow to the sprays were actually attributable to the CIP cost.

Portable CIP Skids

The rapid development of many small biopharmaceutical R&D facilities has created a need for small portable CIP skids which can be moved to and connected to fixed equipment such as small bioreactors, small buffer prep tanks, and portable tanks often used for product transfer, holding, and even purification processes.

The mini-CIP skid in Figure 17 is perhaps the smallest ever fabricated. The full featured single-tank single-use bypass type system occupies only a 30 in. by 45 in. footprint, yet is capable of 45 gal/min flow rate. A Venturi chemical feed system draws chemicals (only a few ounces required for most programs) from self-contained reservoirs on the skid and heating is via an electric heat exchanger (HXR). Operation is from a panel view human–machine interface (HMI) with screens including a P&ID, program selection, process variable input and adjustment, and maintenance screens. This is used for teaching and training purposes includes a manual operations screen, which allows automated control of selected

FIGURE 16 This CIP skid is a drug mixing system that can clean four mixing tanks, and all associated piping to three downstream processes.

phases of a CIP program to be manually initiated. A 230-V AC power supply compressed air and one or two waters are required for operation.

Desirable Criteria for Pharmaceutical and Biotech CIP Systems

The desirable criteria for a pharmaceutical CIP unit were listed by Seiberling (3) in 1996 and are included here for easy reader access.

1. The unit should be constructed of components and materials which meet 3-A sanitary design requirements, or the evolving pharmaceutical industry equivalents as outlined by Clem in Chapter 12.
2. All solution contact surfaces should be polished to a Ra 20 to 25 or better finish. This is a minimum polish applicable to the CIP skid tanks, pumps, valves, and interconnecting piping. Electropolish is specified by some users.
3. The CIP unit should operate reliably with a minimal quantity of solution in the total system, to reduce water, chemical, and steam requirements, time for filling and draining for each program phase, and the cost of treating aqueous waste.

FIGURE 17 This mini-CIP skid is used for demonstrating fully automated CIP to an aseptic process course and a cleaning validation course. *Source*: Courtesy of the Parenteral Drug Association (PDA)-Training and Research Institute (TRI).

4. The CIP unit design must provide for isolation of soft water and pure water supplies from chemical solutions.
5. The CIP unit recirculation tank and all interconnecting piping, pumps, and valves should be fully "self-cleaning" and drainable, and at the end of any completed program the CIP skid solution contact surfaces should be as clean as the equipment to which it was connected.
6. The CIP skid and the associated control software should be of robust design and construction, to comply with Roebers suggestion (see Chapter 2) "If you cannot clean your process equipment and piping in a robust, validated way, do not even think about making a pharmaceutical or biological product!"

Location of CIP Skids

Though CIP has been employed in the biopharmaceutical industry for more than 15 years, not all processors, or support companies, are knowledgeable about the application to a new process or production facility. Space for CIP equipment is often not provided in the initial design, and when the concept is ultimately included, adequate space is not available in the proper location. The effective, and economical, installation and operation of a CIP System requires the CIP skid to be located as close as possible to the center of the tank (spray) CIP loads. The CIP recirculation unit(s) should preferably be located *below* the origin and termination of all CIP circuits to best utilize gravity as the means of removing all solutions from interconnecting piping. A location at the same level as the process is the next best choice, and acceptable, but requires the use of CIPR pumps. Alternatives include (1) grouping the CIP skids in a central area to facilitate provision of water, chemicals, and waste discharge and (2) providing substantial gray space areas below and around the process to enable the skids to be in close proximity to all equipment. The CIP skids must also be accessible by operating and maintenance personnel, and permit movement of chemical supplies to the CIP chemical feed equipment. The recirculating unit should be placed near a drain of adequate capacity to handle the maximum discharge flow rate during rinsing operations. Clear areas of 18 to 24 in. (45–60 cm) on the rear and 36 in. (90 cm) on each side are recommended for ease of maintenance. The floor area under the CIP skid(s) should be constructed of or covered with corrosion-resistant floor material. Area drainage via floors pitched ¼ in. per ft (21 mm/m) is desirable. The CIP Unit drain valve should discharge to a hub drain, rather than to the floor. The drain must be adequately sized to handle the maximum discharge plus 25% from each drain valve. Since most CIP recirculation tanks are vented to the atmosphere some vapor and/or chemical discharge may occur during operation. The system may occasionally be troubled by leaky valves, pump seals, and connections. The tanks and piping may be hot. Good ventilation is essential.

CONCLUSION

The cost of purchasing, operating, and maintaining CIP skids can be substantially reduced by the following procedure:

- Carefully analyze the CIP requirements of the project, giving consideration to vessel size, piping size, and the length and volume of product and CIP piping runs.
- Evaluate the capability of all water supply systems that must serve CIP and consider dedicated loops from the source tanks and water systems, as the means of minimizing the number and size of tanks on the CIP skid.
- Design the facility to maximize the use of gravity to drain vessels and piping to destination vessels for both production and CIP operations.
- Share the conceptual design criteria with knowledgeable and experienced CIP skid vendors and seek a recommendation for suitable CIP skids of established designs and configurations.
- Become a partner with the vendor and the project A&E to fine tune both the CIP skid design and the design of a CIPable process.

REFERENCES

1. Seiberling DA. Clean-In-Place/Sterilize-In-Place (CIP/SIP). In: Olson WP, Groves MJ, eds. Aseptic Pharmaceutical Manufacturing. 1st ed. Prairie View, IL: Interpharm Press Inc., 1987:247–314.
2. Adams DG, Agaarwal D. CIP system design and installation. Pharm Eng 1992; 10(6):9–15.
3. Seiberling DA, Ratz AJ. Engineering considerations for CIP/SIP. In: Avis KE, ed. Sterile Pharmaceutical Products–Process Engineering Applications. 1st ed. Buffalo Grove, IL: Interpharm Press Inc., 1995; 135–219.

7 CIP System Instrumentation and Controls

Barry J. Andersen
Seiberling Associates, Inc., Beloit, Wisconsin, U.S.A.

INTRODUCTION

Successful clean-in-place (CIP) operations require a controlled combination of time, temperature, chemical concentration, and mechanical action to provide satisfactory performance on a repeatable basis. This chapter discusses and reviews the instrumentation and control concepts required to ensure that above criteria are met.

This chapter will consider the following items:

- Common instrumentation utilized to control the CIP process and verify system performance
- Review performance requirements as they relate to the selection of programmable logic controller (PLC) systems and distributed control systems (DCSs)
- Software development concepts as they relate to CIP
- CIP System Quality Control Tools

A BASIC CIP UNIT MODEL

A basic recirculating CIP unit model is shown in Figure 1. The minimum instrumentation and control elements required to control time, temperature, concentration, and mechanical action are highlighted in this diagram. The intent of this model is not to illustrate the actual mechanical details of a CIP system (as in Chapter 6) but to identify the most basic instrumentation and control needs for any application.

A brief discussion of the various components utilized to achieve this control follows.

Time Control

In the very early days of CIP, the control systems used for these applications generally consisted of relays, cam timers, and stepping switches. The cam timer served as a reliable and accurate means of controlling the lengths of various steps of the CIP program, including the duration of the chemical wash steps. Today's PLC systems and DCSs provide the required timing functions in software that replaces all of the old hardwired devices. Time control today is merely a function of adjustable variables within the operating programs for the system. Although it is easy to take time control for granted with today's modern control systems, there are many important decisions to be made during development of the CIP programs that have a direct impact on how well the system performs for all of the required circuits. The programmer must take into consideration the circuit-specific timed functions, which require flexibility (and are generally recipe driven), versus equipment-parameter timing functions that are non–circuit specific. Errors are often made in the development of these programs that cause flexibility limitations

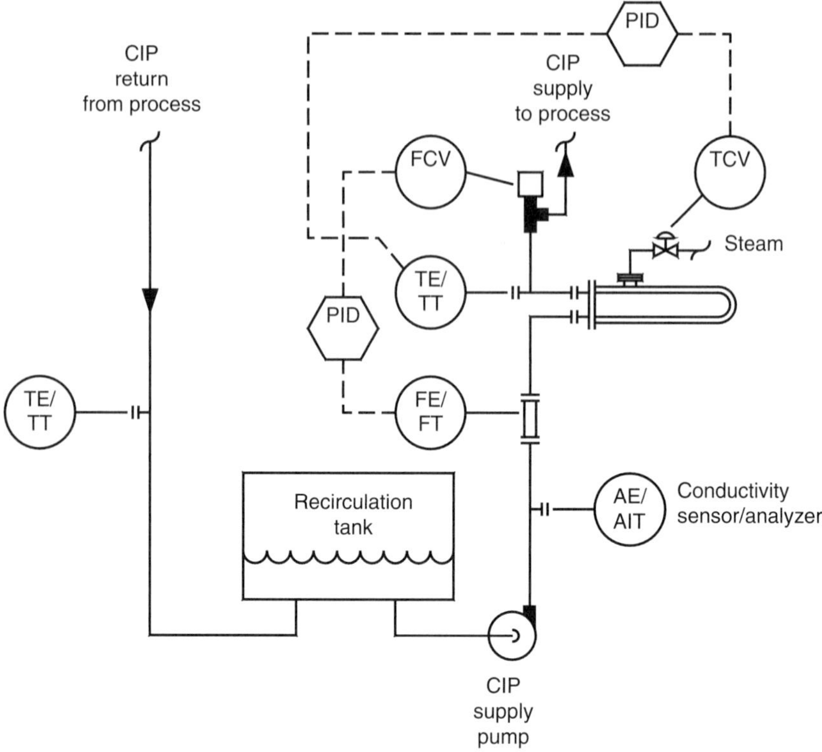

FIGURE 1 Simple CIP unit.

in the application and less than ideal system performance. Programming considerations will be reviewed later in the chapter in the interest of promoting software designs that are flexible and easy for the commissioning engineer and/or end user to configure for a variety of circuit applications.

Temperature Control

Our basic CIP Unit model highlights a supply side resistance temperature detector (RTD), shell and tube heat exchanger, and temperature control valve (TCV) to heat and control the CIP solution temperature. In addition, there is normally a return side RTD to monitor and confirm that appropriate return side temperatures are maintained. The majority of systems are equipped with some variation of this basic design. And, it is not uncommon for some pharmaceutical CIP systems to also be equipped with a heat exchanger for cooling, when the incoming purified water supply is too hot for satisfactory rinsing of proteinaceous soils which may be denatured. The RTDs can be identified in Figure 1 by the TE/TT designation. Similarly, the temperature control valve is designated by TCV.

RTD Temperature Measurement

Although the selection and specification of RTDs for CIP applications is not a challenging effort, there are a few considerations that should be made when choosing these devices. In particular, many pharmaceutical users prefer to specify RTDs with sanitary thermowells (rather than the sanitary direct insertion type) throughout the process system. The cited advantages include the ability to replace elements while the system is running and/or the ability to perform system calibration without opening the piping system. It is important to remember that CIP operations are not a continuous process; however, it is not likely that operating personnel will be operating the CIP system while an RTD is being replaced. One big disadvantage of the thermowell type RTD is the temperature lag that is introduced into the system. Not only will the control loop response be slower, but the additional time to attain wash temperatures should be anticipated. This in turn will add to the overall cycle time of the equipment being washed.

The Temperature Control Valve

Most pharmaceutical applications incorporate a modulating steam valve and proportional integral derivative (PID) loop to control supply side solution temperature as indicated in Figure 1. It is recommended that the steam valve be equipped with an electropneumatic or pneumatic positioner to ensure the best control loop performance by negating valve stem friction.

It is also possible to successfully control temperature using a simple on/off–type automatic ball valve if the deadband inherent with this type of control is acceptable. The advantages of this approach include simpler and less expensive control hardware and ability to bring the system up to temperature faster than with a PID configuration. This discretely controlled TCV has been used in many food/dairy and some pharmaceutical applications for many years. The use of a small valve and large valve in parallel permits on/off control to heat fast and maintain temperature easily.

Chemical Concentration Control

Chemical concentration control requires a chemical injection system and a separate conductivity monitoring system. It is important to note that the conductivity monitoring system is normally used to verify that the appropriate concentration of cleaning chemicals has been added to the system. The conductivity instrument is not typically used to control the chemical injection system directly, particularly in a recirculating CIP system. Since it takes a considerable period of time for the chemicals to blend in solution and be equally distributed throughout the system, direct conductivity control would generally result in inaccurate chemical concentration levels in the cleaning solution.

Chemical Injection Systems

The mechanical components and control strategy required to inject chemicals into the CIP system will vary depending on the type of CIP system to be used. More specifically, a single-pass CIP system generally requires more sophisticated injection equipment and controls than a recirculated system does. Various injection systems and controls will be discussed more thoroughly in chapter 8.

The Toriodal Conductivity Instrument

In most CIP applications, a sanitary toroidal-type conductivity instrument is used to verify chemical concentration. There are a various number of manufacturers of these instruments who utilize a microprocessor-based analyzer which incorporate built in chemical concentration curves and temperature compensation. In many applications today, the temperature compensation is deactivated and only the raw conductivity readings are utilized to verify chemical concentration. Since most systems employ more than one cleaning chemical, use of temperature compensation and/or chemical concentration curves yields erroneous readings for all but one of the chemicals in question. Utilization of raw conductivity readings will work well as long as there is minimal temperature variation of the cleaning solution when the measurements are taken. Typical chemical concentration curves for a sodium hydroxide (NaOH)–based cleaning solution are included in Figure 2 to will help illustrate this point.

There are some design considerations to keep in mind when utilizing the toroidal conductivity sensor. First, it is important to locate the sensor in a line where there is a minimal amount of air incorporation in the CIP solution. In many recirculating CIP systems, significant air incorporation is present on the return side of the system, and this is not a desirable location for the sensor as the accuracy of the readings can be significantly affected. Second, it is important to note that the soil loads present in the cleaning solution will affect the conductivity readings. As a result, this instrument will not infer absolute chemical concentration, but will do a very good job verifying that minimum concentration levels necessary are achieved.

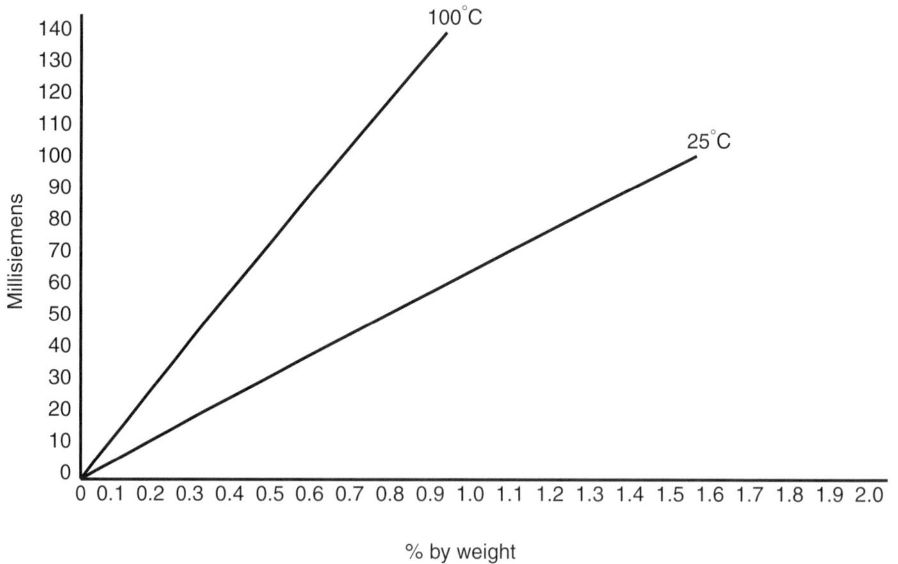

FIGURE 2 Conductivity curve for sodium hydroxide in solution.

Control of Mechanical Action

Control of mechanical action of cleaning solutions is generally accomplished by control of the CIP supply flow rate. This procedure ensures proper flow velocities through all CIP supply, return, and product piping and to all vessel spray devices. The objective is to assure full lines and spray coverage. Flow control is generally accommodated by utilization of a flowmeter, flow control valve, and associated PID loop. This concept is indicated in Figure 1.

The Vortex Shedding Flowmeter

In the past, a sanitary vortex shedding flowmeter was the instrument most commonly used to measure flow in CIP applications. This instrument has been generally reliable and very cost effective for this application. The biggest downside to these instruments is the fact that they are only available with either 2″ or 3″ flowtubes. Frequently, pharmaceutical users have CIP applications with very low flow ranges (less than 75 L/min) which are below a 2″ vortex meter can read. This, of course, necessitates a different type of instrument for the application. It should be noted that the CIP flow rate should never be less than that required for a velocity of 5 ft/sec in the largest portion of the CIP supply/return piping in the circuit.

A magnetic flowmeter might be considered for this application; however, most pharmaceutical applications utilize high purity water whose conductivity is so low that this type of instrument will not work.

The Mass Flowmeter

Mass flowmeters have gained a tremendous amount of popularity over the last few years and are often chosen for CIP applications also. These instruments, although much more costly, are available in various sizes that can measure low flow rates that the vortex shedding element cannot. Many earlier mass flow instruments were problematic with CIP applications; however, because of their very long recovery time when transitioning from an empty pipe to a filled system (also known as slug flow or two-phase flow). This resulted in unreliable flow readings during the initial burst rinses of a typical CIP tank cycle. Recent improvements in mass flow technology have eliminated this problem; however, and it is anticipated that these instruments will be the most popular for CIP applications in the future. It is recommended that the instrument manufacturer be consulted for the most appropriate flow tube design for a CIP application as the technology is being advanced all of the time. On a final note, one must be careful to specify a mass flow instrument that is both CIP cleanable and drainable. Many mass flow tubes require specific orientation to maintain drainability.

The Flow Control Valve and Variable Speed Drive

The vast majority of CIP applications use a modulating flow control valve to regulate the CIP supply. They are generally cost effective and provide the quickest response to system disturbances. In addition, the valve is sometimes used as a blocking valve during air blow steps to prevent air from blowing backwards through the CIP Unit.

Variable speed drives on the CIP supply pump provide an alternative method of regulating flow and this has become much more economical as drive costs have dropped while the reliability of this technology has improved considerably. Some energy savings can be expected with this approach, but the quick response of the flow control valve is lost, and a blocking valve may still be required as part of the air blow system. The flow control valve logic can incorporate a means of clamping the valve at a point near the normal operating position during the circuit and using a timer or counter to release the clamp when the fill is complete. This minimizes the time required to achieve stable flow.

It is also possible to utilize a flow control valve and variable speed pump combined. This is particularly useful on systems where a very wide range of flow is expected or when the CIP system may supply circuits over a wide range of vertical elevations. In such applications, it is sometimes difficult to specify a single modulating valve that will work for every circuit application. In general, the pump is run at a selected fixed speed during the selected CIP cycle while the flow control valve is actually used to regulate the CIP supply flow.

A MORE COMPLEX CIP UNIT MODEL

While the basic CIP unit model illustrates the most basic requirements for the proper control of cleaning solutions, it is not a practical model for real applications. Figure 3 illustrates a fully functional CIP system that requires additional instrumentation and controls to perform properly. These additional requirements are detailed in the following pages.

Level Control

A practical CIP system will require the application of level monitoring instrumentation to provide (*i*) control of the incoming water supply to the CIP skid water tank, (*ii*) monitoring and control of solution tank level during recirculation sequences,

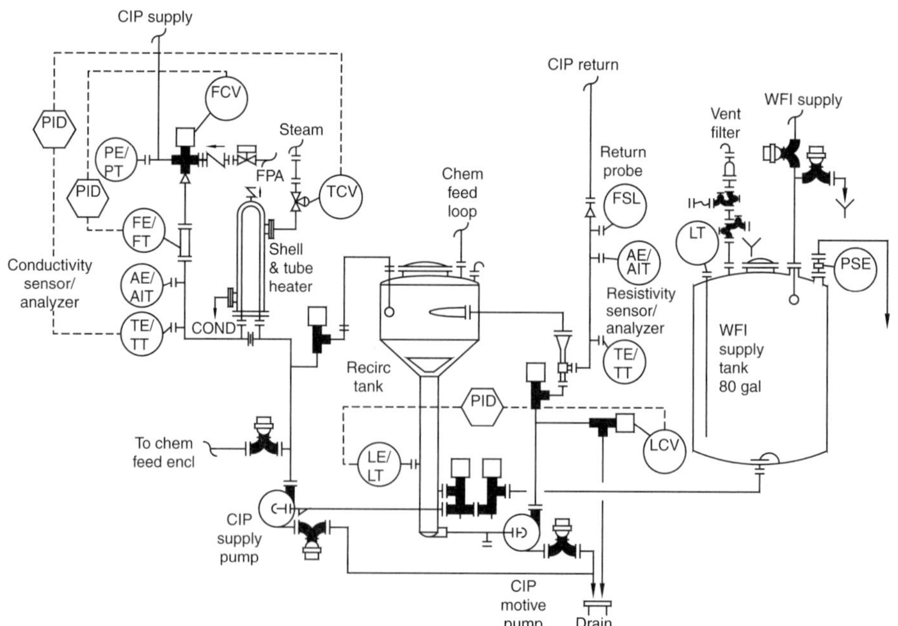

FIGURE 3 Pharmaceutical CIP model.

and (*iii*) monitoring and control during recirculation to detect water loss from the circuit. A properly designed system requires no level monitoring of a vessel being cleaned; however, as the general intent of CIP is to clean the target vessel with the smallest possible puddle of cleaning solution in the vessel. Control of level in the CIP skid solution tank is an excellent method of indirectly regulating level in the target vessel, when used in combination with meter-based system fill.

Figure 3 illustrates level instrumentation of both the purified water supply tank and the wash/recirculation tank. The types of instrumentation typically used in these tanks will be discussed as well as the pros/cons of some alternative types of instrumentation. It is important to note that most CIP applications are measuring 100″ w.c. or less which makes the level sensing technology to be used fairly critical.

The Magnetostrictive Level Transducer

Often, a magnetostrictive level transducer is recommended for level monitoring in the purified water supply tank. This type of sensor is most easily recognized as a long probe that extends to the bottom of the tank with a float that slides up and down on the probe. This instrument uses radar technology to measure the time it takes for microwaves to travel down the probe to the float and reflect back to the transmitter. The transmitter then simply calculates the level reading based on the measured time interval. This instrument is highly advantageous in this application since it is capable of performing accurately over wide temperature ranges, wide pressure ranges, and widely varying dielectric constants (of the purified water). Its primary disadvantage is the fact that one must have adequate clearance above the tank to remove and replace the probe in case of failure. Also, very tall tanks may exceed the maximum available length of these instruments, although this is usually not the case.

This technology may be considered for the recirculation/wash tank also, but this is not practical if the recirculation tank is of small volume, or if the CIP solution exhibits a high degree of turbulence in the tank, causing this type of instrumentation to perform poorly.

The Bubble Tube Level Transmitter

A bubble tube level transmitter has been successfully applied to many CIP system recirculation tanks over the years. Although this is very old level sensing technology, it outperforms most of the alternatives due to its stability over a wide range of temperatures and the fact that its electronics are isolated from the process. Figure 4 highlights the schematic details of one these systems.

This instrument uses a combination of pneumatic and electronic technology to sense the backpressure in a pneumatically purged line connected to the bottom of the recirculation tank. The backpressure measured is proportional to the static head of the water or CIP solution in the recirculation tank. The rate of the purging gas is controlled by a pneumatic flow controller which is critical for accurate level measurements.

As previously stated, this instrument will maintain accuracy over a wide temperature range. This is particularly important since CIP systems often operate over wide temperature ranges, and inaccurate level readings in the wash/recirculation tank can cause substantial performance problems.

There are several downsides to this type of instrumentation that the user should be aware of. First, instrument technicians often are not experienced with these devices, and there is often a lot of confusion about how to properly calibrate

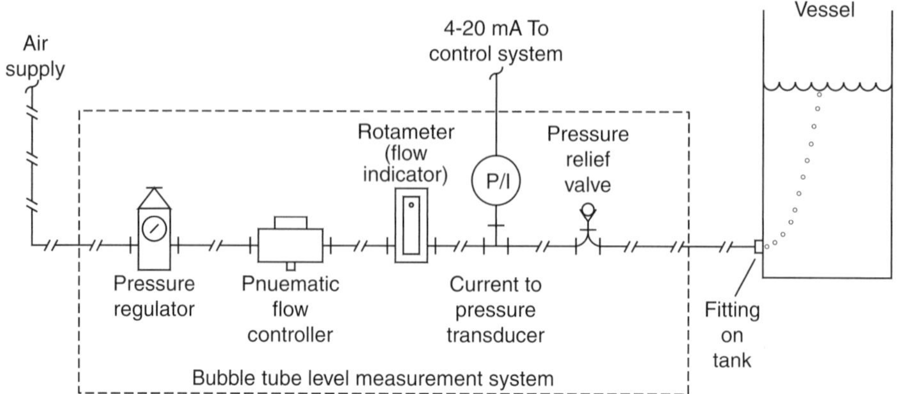

FIGURE 4 Bubble tube level transmitter.

one of these instruments. Second, these instruments often have some amount of zero offset which cannot be fully calibrated out of the system. The zero offset is caused by the residual backpressure in the pneumatic purge line when the recirculation tank is empty. (The purge line backpressure cannot be taken into account by the instrument manufacturer since it is custom fabricated for each application.) Although this offset really does not hurt anything, it causes a lot of consternation because everyone would like the instrument to read zero at empty, and the cause of the problem is not readily understood. Finally, the bubble tube level sensor will not work on a non-vented recirculation tank.

Other Level Sensing Technologies

There are many other types of level sensing technologies available to the user; however, each technology has some disadvantages which makes their use in CIP applications generally impractical. Several typical types of instruments will be reviewed.

Hydrostatic (diaphragm) level sensing technology has improved considerably over the years; however, this type of instrumentation is still plagued with excessive zero drift caused by the thermal shock that typical CIP systems are exposed to. There have been some recent advances with ceramic elements which are much more temperature stable, but these materials are generally not recognized as acceptable in pharmaceutical applications.

Capacitance probe–type sensors are not applicable in CIP systems since they are sensitive to changes in dielectric constant of high purity water at varying temperature causing unacceptable accuracy errors.

Ultrasonic level measurement technology will not provide accurate readings in a wash/recirculation tank. Any foaming of the CIP solution will cause substantial instrument inaccuracy. This author has not tried applying this technology on a purified water supply tank and will defer comment over the acceptability in this application.

Typical non-contact radar systems are generally affected by any moisture or condensation that forms or sprays on the instrument horn, causing substantial accuracy problems. In addition, other factors such as low dielectric constant of the water and the overall design of the vessel can interfere with proper performance of these devices.

Supply Pressure Monitoring and Control

Unlike their level sensing counterparts, the use of modern sanitary diaphragm type pressure transmitters to measure CIP supply pressure will work very well as the effects due to temperature variation are almost negligible. Also the role of the pressure transmitter is much less critical in nature and is generally used as more of a monitoring device than regulating device in CIP applications.

Many applications interlock the chemical feed sequences and the heating controls so that these systems cannot be operated if a minimum supply pressure is not maintained. This is implemented as a safety measure to prevent steam and chemicals to be applied to the system when the supply pump is not discharging water or CIP solution. This interlock function could also be readily applied utilizing a minimum flow setpoint instead and is frequently the means utilized in pharmaceutical applications.

Return Flow Check

For years the food and dairy industry has incorporated instrumentation in the CIP return line to check and verify that the first pre-rinse of the CIP program has successfully returned back to the CIP unit. Failure of the return flow check would result in an aborted CIP program along with an appropriate alarm message to the operator. This functionality was primarily provided to ensure that all manual swing connections on the CIP supply side and CIP return side were configured properly. Even with the advent of fully automated hard piped mixproof valve systems, it is still considered good practice to include this functionality in all applications, including pharmaceutical.

The Conductivity Sensing Element

The instrument of choice for many years has been an inexpensive sanitary stainless steel probe mounted in an insulating material and clamped in place within a shallow tee fitting. The probe is, in turn, wired to a relay designed to trigger based upon the difference in conductivity between air and water. These conductivity probes were originally designed to be used for discrete liquid level controls, but have also served reliably in this application many times over. One must be careful to specify an appropriate relay with a conductivity range which will work with the water to be used in the application, however. Water such as water-for-injection will require a very sensitive relay that can detect very low conductivity levels.

Recently, some pharmaceutical users have rejected the use of these devices as the manufacturers of them have been slow to provide proper material certifications and/or performance specifications for their products. Hopefully, this situation will improve over time such that product acceptance will not be a problem.

The "Tuning Fork"

A more recent alternative to the conductivity-based device is another instrument that was designed originally with level sensing in mind. The "tuning fork" level switch utilizes a pair of paddles that vibrate at a known resonant frequency. When the forks are immersed in a liquid, the frequency changes and the associated electronics detect the change. These tuning fork devices have also been used successfully in return flow check applications and are currently providing the appropriate material certifications and performance specifications. The biggest downside to these devices is that they are simply much more expensive than the conductivity-based predecessor.

Other Detection Devices

This author has seen other devices specified for the return flow check application, usually with unsatisfactory results. In particular, many users are tempted to specify thermal flow switches in their designs. Experience has shown that these switches will not function reliably in most CIP applications. Although the reasons for this are not fully known, it is suspected that the substantial amount of air incorporation present in the water on the return side of most systems prevents repeatable thermal conductivity and resultant flow detection.

Final Rinse Conductivity Check

In many pharmaceutical applications, it is a requirement that the final rinse water be tested to confirm that the final rinse is not contaminated with residual cleaning chemicals or soil, providing additional assurance that the process equipment in question has been cleaned properly. This test involves measuring the conductivity (some users prefer resistivity) of the rinse water using a sanitary electrode type conductivity instrument to measure the very low conductivity ranges encountered in this application. The toroidal instrument discussed previously is not capable of measuring conductivity at levels this low and is unsuitable for this duty.

Out of necessity this final rinse instrument must be placed in the CIP return line as this rinse program will run as a single pass unrecirculated process. Like the toroidal sensor, the electrode type unit must remain fully flooded to provide accurate readings. As discussed previously, the water on the return side often exhibits substantial air incorporation, and the electrode sensor is often installed in a drainable instrument well to keep it flooded and retain accuracy. Figure 5 illustrates an instrument well that has been designed for this application.

Also, the electrode type instrument like the toroidal type comes equipped with built in temperature compensation. Since one is only interested in measuring the final rinse water conductivity, it makes more sense in this application to utilize this instrument with the temperature compensation activated.

Valve Limit Switches

Most pharmaceutical applications today employ the use of valve limit switches or proximity switches as feedback devices on all automatic valves to verify that the valve in question is in its expected position. This would include process, CIP distribution, and CIP skid valves. There are exceptions; however, and it is possible to successfully perform CIP and be assured of cleaning effectiveness without this type of device monitoring in the controls scheme. In particular, small processes or pilot processes may benefit by reducing system costs in this manner. Historical CIP trends and alarm logs can be a very useful quality control tool to verify proper cleaning performance in the absence of these feedback devices. The owner should perform a risk analysis to determine if these devices are required for the application in question.

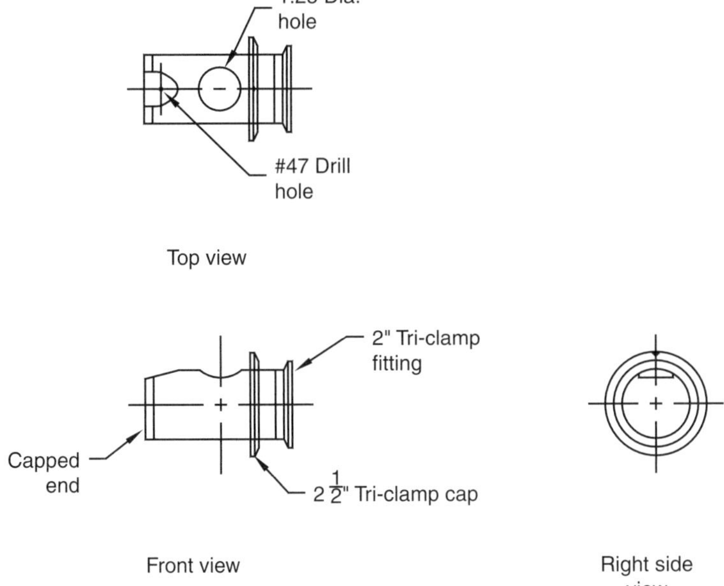

FIGURE 5 Resistivity instrument well.

"Off Skid" or Remote Instrumentation

In the case of most recirculating CIP systems, the instrumentation required to monitor and control CIP operations is located on the CIP skid itself. There are generally no requirements for field mounted instrumentation to successfully perform CIP. In some cases, however, certain mechanical components such as tanks or heat exchangers may be located remotely from the skid which may necessitate remote installation of some instruments.

Single pass CIP systems normally will require off skid instrumentation since there is no CIP return back to the unit. Return temperature measurement, final rinse conductivity, and return flow checks would all require remote field instrumentation at some point in the process drain piping of the system.

CONTROL SYSTEM SELECTION AND PERFORMANCE REQUIREMENTS

Most pharmaceutical users probably do not consider the performance requirements or programming considerations associated with their CIP applications early on in a project when the control system architecture is being defined. In particular, one should consider control system response time requirements, software coding requirements, and communication loading between system controllers. CIP applications are more demanding in these three areas than many typical process operations. There are many control systems that do not perform well for the CIP portions of the process because of basic conceptual or architectural decisions that were made very early in the project.

Typical performance specifications for a CIP application would include the following:

- Fast I/O solving response. Ability to sequence device outputs with less than one-second response time. This response time includes logic solving and system/controller intercommunications delays.
- Ability to support many complex discrete sequences. Sequences may accommodate many discrete devices and may include 50 or more operational steps.
- Ability to solve PID loops
- Fast, robust communications abilities between all controllers in the process related to the CIP application. System intercommunications should not have a noticeable impact on control system response.
- Ability to develop sequences that can be modified easily during commissioning.
- Flexible online programming change ability during commissioning.

PLC Control

In general, most modern higher level PLC systems have the capabilities required to meet the performance specifications listed above. PLC systems are generally a good fit for CIP applications since they are very capable of handling complex discrete sequences. When considering the overall system architecture, it is important to note that it is to one's advantage to implement larger capacity controllers and minimize the controller count in the overall system. This has the advantage of minimizing the communications between controllers during device sequencing operations, which can improve system performance and simplify logic coding. This approach is probably contrary to how most pharmaceutical systems are implemented, but would yield the most efficient performance. Figure 6 illustrates this "ideal" system architecture.

It is important to note that the utility systems in the plant are a good place to draw boundaries and implement on separate controllers since intercommunications with these systems during CIP is much more limited.

DCS Control

DCSs are typically much more challenging to implement in a process application requiring CIP operations for a variety of reasons.

Most DCSs were originally designed for continuous processes with a large number of PID loops and not a lot of discrete sequencing. Oftentimes, these systems ran processes that were started up and shutdown in a more or less manual fashion, so the primary function was to provide loop controls. Since these systems were not originally intended for CIP control applications, it is challenging to develop responsive logic sequences, although improvements in some recent DCSs make this easier to accomplish. Also, because of the required code complexity, it is usually found that it is much more time consuming, challenging, and costly to make field software modifications during the commissioning process.

Another problem with DCSs is the fact that these systems are typically made up of many distributed process controllers. This, by nature, means that there will have to be a lot of intersystem communications required to coordinate all of the device sequencing operations related to a particular CIP circuit. This communications requirement could easily span across four or five controllers for just a single

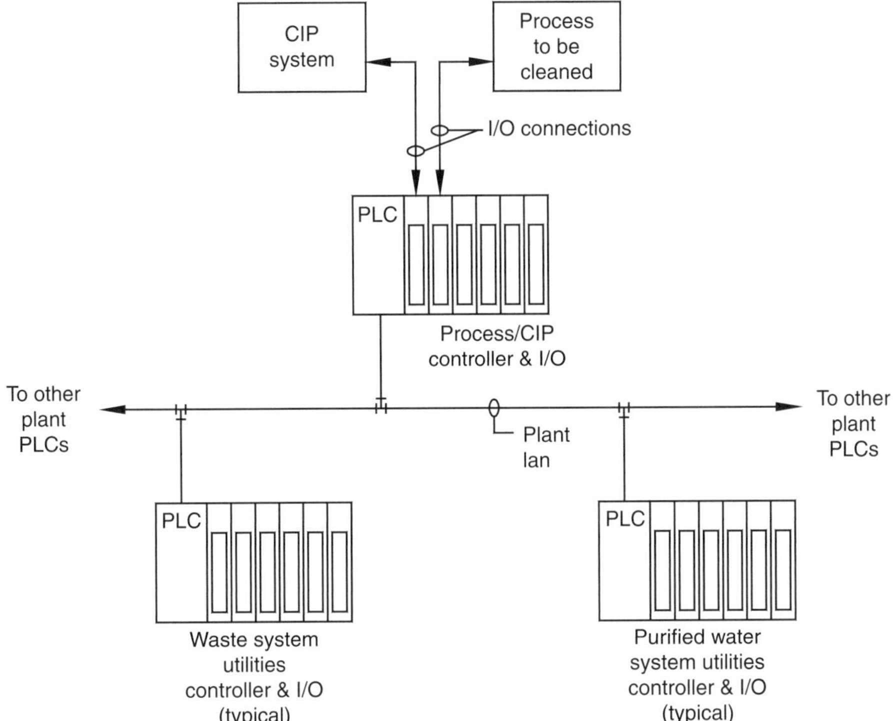

FIGURE 6 Ideal control system architecture.

circuit. A heavily loaded communications network can add additional complication causing device sequencing delays, subsequent process hydraulic imbalance, and resultant poor performance during CIP operations. Of course, increased software complexity can also be expected with these additional communications requirements.

Finally, DCSs in general are not known for high processing speed. Again, this would not be a requirement for a continuous process. It is important to be able to sequence devices quickly, in order to maintain proper hydraulic balance, and many DCSs struggle with these fast processing speed requirements.

This author is not stating that a DCS cannot be made to work in a CIP application. Nor is this author claiming that all DCSs are plagued with the above-mentioned problems. On the contrary, successful installations do exist. However, one should usually anticipate a substantially more challenging and costly programming and commissioning effort.

Hybrid Systems

There have been a number of instances in the past where the user has chosen to use a DCS for the process controls and a PLC-based system for the CIP system controls. All of the complex CIP unit sequences and valve sequencing programs are then developed in the PLC system and the valve sequence commands are simply transmitted over to the DCS. The concept with this approach was to eliminate the complex sequencing code on the DCS side which has been quite successful in a number of applications. The primary consideration to be made is whether the user is willing to have two separate control technologies in place in the process control system. In reality, there are often other "specialty" PLCs in the process anyway, where DCS control simply is not practical due to processing speed reasons.

Obviously, careful planning will be required to develop the software interface between the two control systems, but this approach can be a good alternative way to overcome DCS limitations that may be difficult to deal with.

PROGRAMMING CONCEPTS

Most pharmaceutical users today employ good automated manufacturing practices (GAMP4) guidelines in the development of their system documentation for automation-related projects. This practice can be successfully applied to CIP applications but is not the complete solution to properly document these types of systems.

General Software Documentation Package

During the design process, a user requirements document, functional requirements document, and system design document will usually be assembled to outline the basic system requirements as well as the fine details pertaining to a particular CIP application. It is not the intent of this text to focus on the content requirements for these documents since this is well outlined by GAMP4. Of greater interest is the supplemental documentation that one needs to develop to adequately define the system.

Development of Matrix Charts for Sequence Definition

CIP Unit Matrix

Since CIP sequences are typically fairly complex operations, it is much easier to document the various program steps and their functionality using a CIP unit matrix. The function of typical CIP systems in the food and dairy industry have been documented this way for many years and this approach works very well in pharmaceutical applications also. This CIP unit matrix will serve as the major tool for defining the operational sequences and the system design document will fill in all of the gaps that the matrix chart does not adequately define. A typical CIP unit matrix is shown in Figure 7.

As can be seen in Figure 7, a description of each program step (or phase) is listed across the top of the chart. The steps will execute from left to right. The left-hand column lists the events that must occur in order for the system to advance to the next step as well as the devices that must be sequenced at each program step. This chart provides an excellent summary of the overall function of the CIP unit program. The overall sequence can become complicated rather quickly, and the chart format provides a better way to present this functionality, rather than attempting to document entirely by narrative in a system design document.

CIP "Recipe" Definition

Along with the matrix charts, the individual CIP circuit setpoints are documented in a recipe format that provides easy interpretation by the CIP system user as well as

			Step description	Charge recirculation tank	1st rinse	Return flow check	1st drain	2nd rinse	2nd drain	3rd rinse	3rd drain	Pump out	Charge recirculation tank	System fill	Heat and chemical feed	Delay to temperature	Recirculate
	CIP1	CIP1		1	2	3	4	5	6	7	8	9	10	11	12	13	14
Bit #	Sequence #	Tag #	Sequence description														
8000	N51:0/15		Counter (Gallons)		X			X		X				X			X
4000	N51:0/14		Timer (Sac)				X		X		X	X					
2000	N51:0/13		Return flow req'd			X											
1000	N51:0/12		Recirc level req'd	X									X				
800	N51:0/11		Temperature req'd													X	
400	N51:0/10		Caustic feed												X		
200	N51:0/09		Acid feed														
100	N51:0/08		Sanitizer feed														
80	N51:0/07	FV2000	Water supply valve from FWI	X	X					X			X	X			
40	N51:0/06	FV2001	Recv'd raw sol supply Vlv from SR2														
20	N51:0/05	FV2002	Recv'd past. sol. supply Vlv from SR1					X									
10	N51:0/04	FV2003	Recirculation supply Vlv												X	X	X
8	N51:0/03	FV2004	Spray supply valve, X=open	X									X				
4	N51:0/02	FV2005	Flow control valve (Throttling)	X	X			X		X			X	X	X	X	X
2	N51:0/01	FV2006	Eductor supply valve									X					
1	N51:0/00	FV2007	Recv'd sol. disch. Vlv to RC4 (Throttling)														
			Step	1	2	3	4	5	6	7	8	9	10	11	12	13	14
8000	N51:1/15	FV2008	Recv'd sol. disch. Vlv to SRI (Throttling)														
4000	N51:1/14	FV2009	Drain valve (Throttling)		X	X	X	X	X	X	X	X					
2000	N51:1/13																
1000	N51:1/12	CSP1	Supply pump	X	X			X		X			X	X	X	X	X
800	N51:1/11	MP1	Motive pump		X	X	X	X	X	X	X	X		X	X	X	X
400	N51:1/10	TCV1223	Heat (caustic)												X	X	X
200	N51:1/09	FV2010	Air blow														
100	N51:1/08	TCV1223	Heat (acid)														
80	N51:1/07		Return pump		X	X	X	X	X	X	X	X		X	X	X	X
40	N51:1/06		ToV pulse				X		X		X						
20	N51:1/05		Seat lifter pulse														X
10	N51:1/04		Check conductivity														X
8	N51:1/03		Pulse CIP supply valves								X						
4	N51:1/02																
2	N51:1/01		CIP complete														
1	N51:1/00		Sanitizer complete														
			Step	1	2	3	4	5	6	7	8	9	10	11	12	13	14
			Hex bit pattern / Hex bit pattern register	N41:001 = 108C	N41:002 = 8084	N41:003 = 2000	N41:004 = 4000	N41:005 = 8024	N41:006 = 4000	N41:007 = 8084	N41:008 = 4000	N41:009 = 4002	N41:010 = 108C	N41:011 = 8084	N41:012 = 0414	N41:013 = 0814	N41:014 = 8014
			Hex bit pattern / Hex bit pattern register	N41:051 = 1000	N41:052 = 5880	N41:053 = 4880	N41:054 = 48C0	N41:055 = 5880	N41:056 = 48C0	N41:057 = 5880	N41:058 = 48C8	N41:059 = 4880	N41:060 = 1000	N41:061 = 1880	N41:062 = 1C80	N41:063 = 1C80	N41:064 = 1CB0
			Step	1	2	3	4	5	6	7	8	9	10	11	12	13	14

FIGURE 7 CIP unit matrix chart (tank circuit, rinse phase).

the software developer. In particular, this documentation provides guidance to the programmer about the needed flexibility to adjust these setpoints easily and quickly at commissioning time. An example is shown in Figure 8.

Device Sequence Matrix

The device sequence matrix (also known as a valve sequence matrix) is developed to identify the devices in a particular CIP circuit that must be sequenced in order to

Step	Step description	CIP 1 SR1-2 Step register and length seconds/liters	CIP 1 RT1-3 Step register and length seconds/liters	CIP 1 Single Tanker Step register and length seconds/liters
1	Charge recirculation tank	N40:0001 = 0	N40:0061 = 0	N40:0121 = 0
2	1st rinse	N40:0002 = 80	N40:0062 = 150	N40:0122 = 500
3	Return flow check	N40:0003 = 0	N40:0063 = 0	N40:0123 = 0
4	1st drain	N40:0004 = 20	N40:0064 = 30	N40:0124 = 20
5	2nd rinse	N40:0005 = 90	N40:0065 = 180	N40:0125 = 300
6	2nd drain	N40:0006 = 20	N40:0066 = 30	N40:0126 = 20
7	3rd rinse	N40:0007 = 45	N40:0067 = 90	N40:0127 = 75
8	3rd drain	N40:0008 = 30	N40:0068 = 30	N40:0128 = 20
9	Pump out	N40:0009 = 60	N40:0069 = 100	N40:0129 = 15
10	Charge recirculation tank	N40:0010 = 0	N40:0070 = 0	N40:0130 = 0
11	System fill	N40:0011 = 90	N40:0071 = 80	N40:0131 = 25
12	Heat and chemical feed	N40:0012 = 0	N40:0072 = 0	N40:0132 = 0
13	Delay to temperature	N40:0013 = 0	N40:0073 = 0	N40:0133 = 0
14	Recirculate	N40:0014 = 600	N40:0074 = 1350	N40:0134 = 1125
15	Air blow	N40:0015 = 30	N40:0075 = 30	N40:0135 = 30
16	Wash drain	N40:0016 = 30	N40:0076 = 90	N40:0136 = 45
17	1st post rinse	N40:0017 = 40	N40:0077 = 230	N40:0137 = 150
18	1st drain	N40:0018 = 20	N40:0078 = 60	N40:0138 = 20
19	2nd post rinse	N40:0019 = 80	N40:0079 = 150	N40:0139 = 75
20	2nd drain	N40:0020 = 30	N40:0080 = 30	N40:0140 = 20
21	Pump out	N40:0021 = 60	N40:0081 = 120	N40:0141 = 15
22	Charge recirculation tank	N40:0022 = 0	N40:0082 = 0	N40:0142 = 0
23	System fill	N40:0023 = 90	N40:0083 = 80	N40:0143 = 25
24	Acid feed	N40:0024 = 130	N40:0084 = 200	N40:0144 = 0
25	Recirculate	N40:0025 = 300	N40:0085 = 600	N40:0145 = 500
26	Air blow	N40:0026 = 30	N40:0086 = 30	N40:0146 = 30
27	Acid wash drain	N40:0027 = 60	N40:0087 = 300	N40:0147 = 45
28	Post rinse	N40:0028 = 0	N40:0088 = 0	N40:0148 = 0
29	Drain	N40:0029 = 30	N40:0089 = 30	N40:0149 = 45
30	Pump out	N40:0030 = 15	N40:0090 = 15	N40:0150 = 15
31	Program complete	N40:0031 = 0	N40:0091 = 0	N40:0151 = 0
32	Charge recirculation tank	N40:0032 = 0	N40:0092 = 0	N40:0152 = 0
33	Rinse	N40:0033 = 120	N40:0093 = 180	N40:0153 = 150
34	Return flow check	N40:0034 = 0	N40:0094 = 0	N40:0154 = 0
35	Drain	N40:0035 = 60	N40:0095 = 60	N40:0155 = 30
36	System fill	N40:0036 = 90	N40:0096 = 80	N40:0156 = 25
37	Feed sanitizer	N40:0037 = 0	N40:0097 = 0	N40:0157 = 0
38	Recirculate	N40:0038 = 200	N40:0098 = 400	N40:0158 = 375
39	Air blow	N40:0039 = 30	N40:0099 = 30	N40:0159 = 30
40	Drain	N40:0040 = 30	N40:0100 = 90	N40:0160 = 45
41	Post rinse	N40:0041 = 0	N40:0101 = 0	N40:0161 = 0
42	Drain	N40:0042 = 0	N40:0102 = 0	N40:0162 = 0
43	Pump out	N40:0043 = 15	N40:0103 = 15	N40:0163 = 15
44	Program complete	N40:0044 = 0	N40:0104 = 0	N40:0164 = 0
45		N40:0045 = 0	N40:0105 = 0	N40:0165 = 0
50		N40:0046 = 0	N40:0106 = 0	N40:0166 = 0
		N40:0047 = 0	N40:0107 = 0	N40:0167 = 0
Spare		N40:0048 = 0	N40:0108 = 0	N40:0168 = 0
Caustic (strokes)		N40:0049 = 10	N40:0109 = 15	N40:0169 = 25
Acid (strokes)		N40:0050 = 5	N40:0110 = 7	N40:0170 = 14
Spare		N40:0051 = 0	N40:0111 = 0	N40:0171 = 0
Spare		N40:0052 = 0	N40:0112 = 0	N40:0172 = 0
Conductivity (mS)		N40:0053 = 25	N40:0113 = 25	N40:0173 = 30
Temperature (XX°F) wash		N40:0054 = 140	N40:0114 = 140	N40:0174 = 150
Temperature (XX°F) acid		N40:0055 = 0	N40:0115 = 0	N40:0175 = 0
Flow rate (XXX Gpm)		N40:0056 = 65	N40:0116 = 90	N40:0176 = 80

FIGURE 8 CIP recipe table (tank circuits, partial for two different vessels).

Bit #	Sequence #		1	2	3	4	5	6	7	8	9	10	11	12	13	14
		Step decription	Path 1	Transition	Path 2	Transition	Path 3	Transition	Path 4	Transition						
		Step length	15	5	15	5	15	5	15	5	0	0	0	0	0	0
		Step length register	N42:001 =	N42:002 =	N42:003 =	N42:004 =	N42:005 =	N42:006 =	N42:007 =	N42:008 =	N42:009 =	N42:010 =	N42:011 =	N42:012 =	N42:013 =	N42:014 =
			1	2	3	4	5	6	7	8	9	10	11	12	13	14
Bit #	Sequence #	Sequence discription														
8000	N42:54/15	FV1414	X	X						X						
4000	N42:54/14	FV1415		X	X	X										
2000	N42:54/13	FV1416				X	X	X								
1000	N42:54/12	FV1482						X	X	X						
800	N42:54/11															
400	N42:54/10															
200	N42:54/09															
100	N42:54/08															
80	N42:54/07															
40	N42:54/06															
20	N42:54/05															
10	N42:54/04															
8	N42:54/03															
4	N42:54/02															
2	N42:54/01															
1	N42:54/00	Reset								X						
		Step	1	2	3	4	5	6	7	8	9	10	11	12	13	14
		Hex bit pattern	8000	C000	4000	6000	2000	3000	1000	9001	0000	0000	0000	0000	0000	0000
		Hex bit pattern register	N43:001 =	N43:002 =	N43:003 =	N43:004 =	N43:005 =	N43:006 =	N43:007 =	N43:008 =	N43:009 =	N43:010 =	N43:011 =	N43:012 =	N43:013 =	N43:014 =
		Step	1	2	3	4	5	6	7	8	9	10	11	12	13	14

FIGURE 9 Device sequence matrix chart (bioreactor vessel and legs).

clean all product contact surfaces (of each device) or to maintain proper hydraulic balance among various circuit flowpaths. The device sequence matrix functions much like that of the CIP unit matrix except that it may cycle through a particular sequence many times during a single CIP Unit Program sequence. Again, this documentation provides guidance to the programmer about the needed flexibility to adjust setpoints as well as sequence steps easily and quickly at commissioning time. An example is shown in Figure 9.

Avoiding Some Typical Programming Pitfalls

As with any technology, there have been many different approaches to software design by different parties over the years. Some approaches have been more successful than others, and this author has attempted to identify a number of design methods or concepts that (in my opinion) have met with undesirable complications or problems. Most certainly, there are valuable lessons learned from understanding what has been previously attempted in the past.

Understanding the CIP Process

Although this may seem very obvious, this author feels the need to point out the fact that there is no substitute for utilizing software developers who have previous design and commissioning experience with CIP. Oftentimes, the development of the CIP code will be more complex than that of the process code and previous experience will make a substantial difference in the quality of the software application. Many pharmaceutical users choose their system integrator with the process controls in mind and the CIP portion of the programming effort is just an afterthought. One should not underestimate the design effort required for a CIP application. The pharmaceutical user may have to spend considerable time in locating a system integrator with this type of experience as many do not have it.

Overly Complex Software

Sometimes the end user or the system integrator utilizes software concepts that are far more complex than necessary. This increases project costs and time during design, commissioning, and validation. Some examples that this author has seen are noted in the following paragraphs. It is always recommended during the design process to perform some reality checks to ensure that the design is not unnecessarily complicated.

Acquiring the Flow Path

One must provide appropriate software interlocks such that a CIP process cannot be activated for a particular circuit if part of that circuit is currently in production use (or is in some other state when CIP would be undesirable). In other words, the flowpath to be cleaned must be acquired for CIP in order to clean it. On the other hand, there have been applications developed in the past where every segment of pipe between any automation devices or major pieces of equipment is treated as a separate entity and must be acquired individually in software before CIP can be activated. This level of complexity is unnecessary and burdensome to the production operators who must use the system.

The Process Controls in the CIP System

Most CIP applications involve a single CIP unit that supports the cleaning of several process areas or "cells." Oftentimes the user has separate controllers for these process cells, particularly if the control system is DCS based. There have been some approaches where parts of the higher level supervisory code for the CIP unit are designed as part of the process cell and located in that respective controller. This can lead to redundant code that must be developed, tested, commissioned, and validated in each process cell controller. It is this author's recommendation that all of the control code for the CIP unit be located in the CIP controller in order to minimize complexity.

The device sequencing (or valve sequencing) code (for the circuit being cleaned), may, on the other hand may be better located within the individual process cell controller. This last decision is really application specific and needs to be analyzed on a case by case basis.

The CIP Queue

Some CIP users have requested that queue logic be incorporated in their CIP applications in order to "queue up" cleaning circuits ahead of time. In reality, there is usually no benefit to this. Oftentimes the cleaning priorities change during the production day so that a predetermined schedule of cleaning operations is not practical. This is a case where humans are needed to make these operational decisions and developing software for this simply adds additional complexity and cost.

Unnecessary Interlocks During Valve Sequencing

This author was involved in a recent audit of a CIP system that was not performing well. One of the problems discovered was the fact that the valve sequencing (device sequencing) programs were written so that the sequences would not step advance until all of the valve position switches for all of the valves in the circuit transitioned to the expected position for that step. This not only added a lot of complexity to the software, but also the DCS encountered difficulty in processing all of the additional code during valve sequence execution. This resulted in significant unexpected delays in the valve sequence programs which, in turn, caused the overall CIP programs to run much longer than normally necessary. Normally, all that is needed is standard alarm module logic that compares valve output status to position input status for each and every valve. The alarm code runs independently as stand alone modules.

Flexibility Problems

Sometimes software developers take shortcuts and hard code sequences or setpoints that should not be hardcoded. This lack of flexibility will be paid for at commissioning time, because there are many parameters that must be fine tuned by trial and error and need to be easily adjustable. Also, other problems can be encountered such that the CIP programs can not be optimized because a hardcoded setpoint for all of the circuits must be set for a worst-case condition. This usually results in increased cleaning circuit times and operational costs.

Sometimes, device sequences must be recoded in the field to correct for unanticipated hydraulic imbalance or to eliminate hydraulic shock. The software design needs to provide proper functionality so that the appropriate setpoints and sequences can be easily modified in the field.

Properly defined matrix charts, recipe parameter charts, and software design documentation will direct the programmer as to what should not be hardcoded. Periodic software reviews will help ensure that these requirements are being followed.

Finally, proper control system selection is important here. If sequencers must be hardcoded because that is the only means with the selected system architecture, then you have the wrong system to begin with!

Fixing Mechanical Problems with Software

This author's employer has been involved in the audit of many CIP systems over the years. Time and time again, we have observed systems that did not function properly because of mechanical problems or mechanical design issues. Often, attempts are made to correct for these problems with control system software modifications. These fixes rarely correct the problem, and the user is usually left with a poorly performing system. Sometimes the software band-aid ends up causing other unanticipated issues. The best rule of thumb is a simple one "… do not expect to fix mechanical problems with software!"

Use of Batch Control Software

In the last few years, new software technology has appeared on the scene in an attempt to make the software development process simpler and more flexible for processes with complex batch sequences. Known as batch engines, these software packages are offered by numerous vendors and work in conjunction with the controllers in the system, providing overall supervisory sequential control while the nuts and bolts of the control software are still handled at the controller level.

Some users have utilized this batch control software in CIP applications with mixed results and caution should be exercised if one is considering this for their application. Probably the biggest concern is the fact that the execution speed for these batch engines is not always fast and repeatable. Some steps of the CIP program may only run for a few seconds whereas some batch engines simply cannot sequence that fast, leading to program steps that run significantly longer than expected. Some systems with these batch engines have experienced problems with varying CIP circuit cycle times due to different system loads at various times of the day. Remember that the proper control of time is one of the most basic CIP requirements and the batch engine needs to be able to keep up with the process!

Designing Software for Maintenance and Troubleshooting

Much attention has been paid in the last few years to developing software for CIP applications that does a proper job of ensuring that the CIP process is tightly controlled such that the cleaning effectiveness is maintained consistently, day after day. As a result, newer systems are now equipped with many more critical alarms then their predecessors. These alarms are usually programmed to abort the CIP process if problems are encountered.

One thing that is often overlooked; however, is the fact that these alarms can become a real nuisance during commissioning or when troubleshooting a problem. The alarm itself is not the problem, but the resultant aborted sequence can be. It is often desirable to continue on with the sequence to make observations of the root cause of an alarm when commissioning and troubleshooting. It is recommended that provisions be made to be able to ignore these abort conditions if desired. (Of course, certain abort conditions must not be disabled such that personnel safety is compromised or equipment is possibly damaged.) Also, the appropriate security provisions will need to be in place so that this functionality cannot be used inappropriately.

CIP QUALITY CONTROL TOOLS

Alarm Functions

Today's modern control systems allow easy development of alarms to monitor and cause appropriate action for the required critical process parameters related to CIP operations. Obviously, alarm monitoring of flow, conductivity, and temperature are critical to this process, but most systems have alarm functions for many other process parameters as well. The various system alarms and desired actions need to be determined on a case by case basis for each application. The user will most certainly have a substantial interest in the development of these system requirements so that CIP performance is closely monitored and controlled to help ensure quality control during cleaning operations.

CIP Report

CIP1 Report

Typical CIP report

Start Date	Start Time	Circuit	Mode	Elapsed (Minutes)	Water Used (Liters)	Caustic (Strokes)	Sanitizer (Strokes)
Thursday, July 26, 2001	4:29:21 AM	MT1	CIP	38.15	1500	18	8
Thursday, July 26, 2001	8:02:00 AM	SISTEMA LOTEO	CIP	68.05	3616	40	25
Thursday, July 26, 2001	9:34:30 AM	RC1	CIP	37.78	600	14	4
Thursday, July 26, 2001	11:03:13 AM	MT2	CIP	38.50	1499	18	8
Thursday, July 26, 2001	12:37:56 AM	AS1	CIP	100.97	8972	48	18
Thursday, July 26, 2001	2:37:51 AM	TBA19 LLENADORA	CIP	62.02	3263	30	8
Thursday, July 26, 2001	3:42:02 AM	MT1-2 TO UHT	CIP	34.02	1413	22	7
Thursday, July 26, 2001	6:00:31 AM	RC1-2 TO MT1-2	CIP	30.77	1214	23	8
Thursday, July 26, 2001	6:41:02 AM	RC2	CIP	36.43	600	14	4
			Daily Totals	446.68	22677	227	90

Record: 1 of 9

FIGURE 10 Typical CIP report.

CIP Reports

Modern CIP systems generally incorporate reporting functions that detail the circuits that were run during a typical time period. These reports serve as useful quality control tools to verify that CIP circuits have been successfully cleaned. The reports usually contain some useful information to confirm system performance such as circuit time duration, and water and chemical usage. Significant variance in any of these factors indicates that the user should thoroughly observe system operation in order to identify and correct potential problems. An example of a typical CIP report is shown in Figure 10.

Historical Datalogging

As discussed previously, modern CIP systems are designed to tightly control the CIP process to ensure consistent cleaning operations, day after day. Oftentimes, however, mechanical components can degrade or other problems can occur which may begin to impair CIP performance, but not to the point where the system is not being properly cleaned. The various system software alarms may not be able to detect a problem until it is far worse and to the point where CIP operations cannot be successfully completed. As a result of this, software tools have been developed to allow the user to spot and correct performance problems, often before they cause a major failure.

The historical trend capabilities of modern control systems provide the basis for one of the most powerful quality control tools available. An example of a typical CIP data logging trend is shown in Figure 11. This figure, and the also

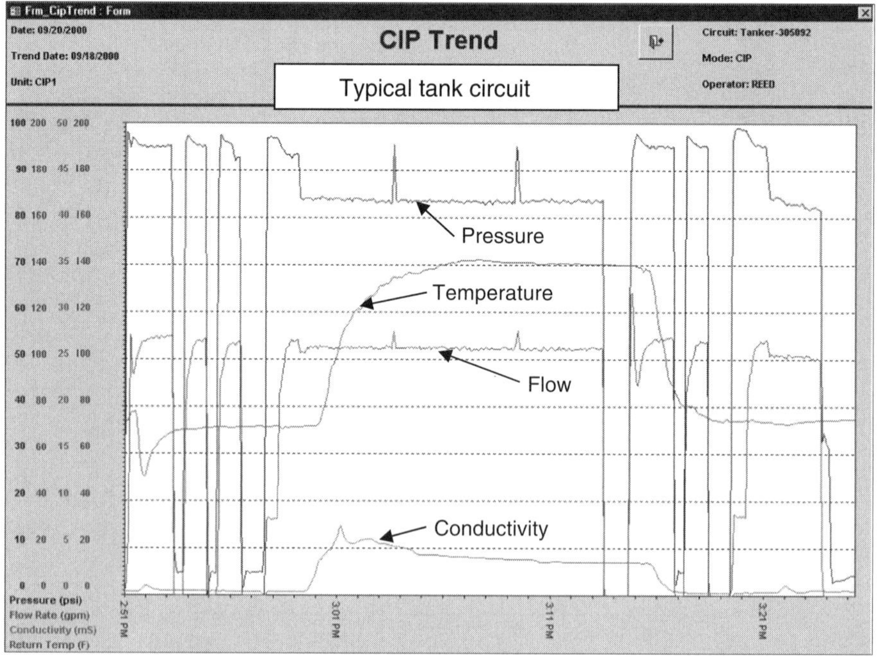

FIGURE 11 Typical CIP historical trend.

Figure 12, were created by manipulating hardware and software for a CIP system installed on a comestibles process subject to less stringent control than the typical pharmaceutical or biopharmaceutical installation. The lines are identified on the actual chart records of the study program.

Figure 11 is a textbook example of a well-performing tank cleaning program. The various steps of this CIP program are identified as follows and include three pre-rinse/drain sequences, followed by a recirculated wash sequence, next two post-rinse/drain sequences, and one recirculated sanitize sequence, a typical program for CIP in many comestibles facilities.

This particular trend indicates two pressure/flow spikes during the recirculated wash. This particular circuit exhibited a small amount of water loss due to system leaks associated with the vessel being cleaned. The pressure/flow spikes occur when the CIP unit senses that the circuit is running low on water and automatically adds makeup water to compensate. In this case, cleaning performance is not impaired as the water addition was small, but excessive system leakage and resultant water makeup would be a situation that would need immediate attention.

Figure 12 is a good example of a CIP circuit with some serious operational difficulties.

This particular CIP circuit included two vessels being cleaned in parallel, though sprayed intermittently. During the first part of the recirculated wash, program flow was blocked on the return side of one of the vessels by closing the outlet valve. CIP solution therefore accumulated in one vessel, and the CIP unit

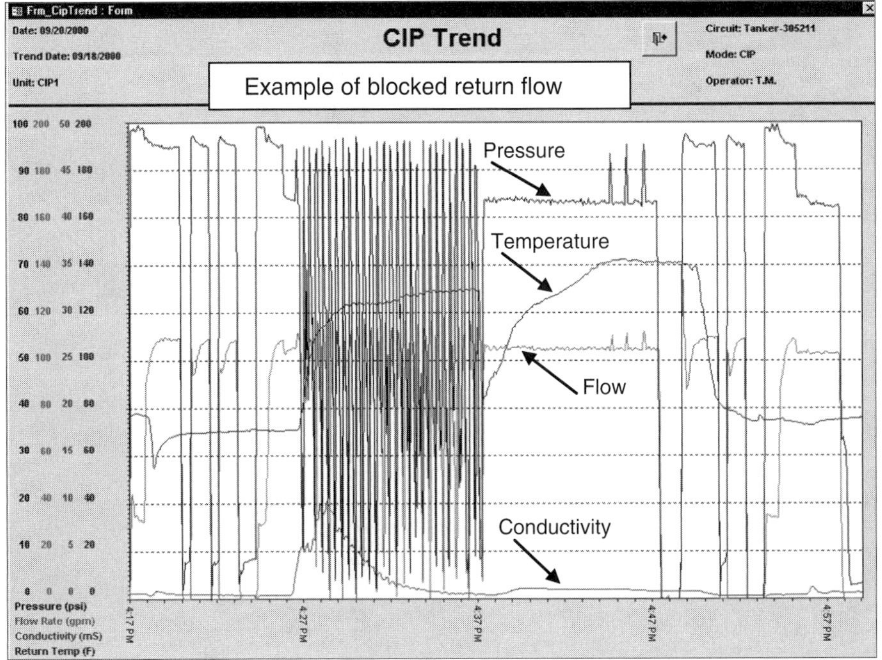

FIGURE 12 Historical trend with blocked return flow.

added makeup water to compensate for loss of return flow. In addition, the CIP solution conductivity dropped due to excessive dilution. (A number of system alarms were disabled to allow the CIP program to continue to completion. Normally, the CIP system would abort the program either due to valve position, excessive water addition, or low conductivity alarms. This circuit was run for demonstration purposes.) After a period of time to create this record, the outlet valve was again opened and the program performed normally after dumping the excessive water to drain (no record of this) following which the accumulated cold water was again brought to temperature.

The resultant pressure/flow trends, as recorded on paper, or displayed on screens, revealed the serious performance problem that would normally be recognized and corrected by operating personnel.

A problem occasionally found with nearly all CIP systems is an occasional partially plugged spray ball (Fig. 13). This will cause supply pressure to be somewhat higher, but this will often not be noticed as the normal location of a pump discharge pressure sensor is immediately downstream of the pump and will

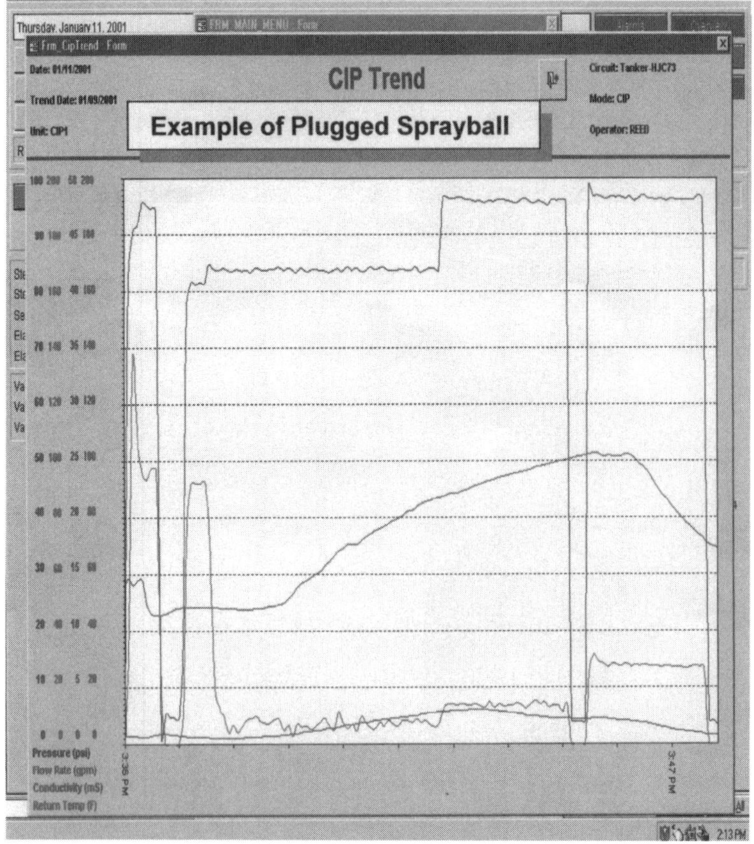

FIGURE 13 Historical trend with partially plugged sprayball.

display a pressure high on the pump performance curve. Supply flowrate, however, will be low, as will the rate of chemical mixing.

Many developing operational problems can be defined by the continuous review of these historical trends. As each circuit will have its own CIP "signature," a user can compare daily trends to a good known trend for each particular CIP circuit. It is possible to spot performance issues before they begin to inhibit proper cleaning of the circuit in question.

8 Cleaning Agent Injection Systems

Samuel F. Lebowitz
Electrol Specialties Company (ESC), South Beloit, Illinois, U.S.A.

INTRODUCTION

The clean-in-place (CIP) process is based on the cleaning kinetics of time, temperature, and chemical concentration. Fundamental to any automated CIP system is the capability to efficiently add cleaning chemical agents in an automated and highly reproducible manner.

Chemical cleaning agents are the major consumable materials of the CIP process, other than utility functions such as fresh water, steam, and compressed air. Handling of bulk cleaning agents and maintenance of chemical feed systems often becomes the primary CIP-related source of labor for a pharmaceutical process facility. Early in the architectural layout for the facility consideration must be given to the material flow of cleaning agent concentrated raw materials through the facility to the CIP area. Lack of proper planning will lead to inconvenience for the plant operators and possibly unsafe work conditions.

The process engineer developing CIP system design must select a robust dependable means of introducing cleaning agents into the CIP flow stream. Historically, chemical feed systems are a design challenge with respect to assuring for a chemical delivery rate that meets the particular CIP system concentration requirements, reproducibility, maintainability, and safety within a reasonable budget.

SAFE CHEMICAL STORAGE AND HANDLING PRACTICES

In the early stages of planning to incorporate an automated CIP system in a pharmaceutical facility, the architectural space allocations and relationships must be defined. Layout considerations include the following:

- Hazardous material storage and handling
- Chemical area location and "traffic pattern"
- Chemical pump location
- Double containment and curbing
- Safety equipment

At some location in the pharmaceutical facility concentrated cleaning agents will be received. For facilities with modest cleaning agent consumption, concentrated CIP chemicals are delivered in drums, on order of 30 to 55 gallon volume. The next step up in consumption is supplied by bulk totes, on order of 300 to 500 gallons, which are designed for forklift handling. In large facilities with multiple CIP systems delivery may be via tanker truck with transfer to a storage tank and distribution within the facility by a chemical feed header.

For CIP chemical concentrate delivered in drums or totes, an easily accessible path must be available to the location of the cleaning agent feed system. The access should be considered from the aspect of both chemical delivery and also the potential need for clean up of an accidental spill.

Double Containment

If the cleaning chemicals are received in barrels or totes, as a contingency for the possibility of accidental damage to a concentrated chemical container, double containment of the contents should be provided in some manner. One effective means of containment is construction of a curbed area. Design factors for the construction of a curbed area include structural integrity, containment capacity, chemical resistance, chemical container handling, and spill clean up method. Curbed areas provide both concentrated chemical double containment and a mechanical barrier to prevent impact by lift trucks, dollies, etc.

Figure 1 shows a commercially available spill pallet containment system installed as part of a prefabricated modular CIP system.

Safety Showers and Eye Wash Equipment

The need for emergency shower and eye wash equipment is mandated by regulations of the Occupational Safety and Health Administration (OSHA), "where the eyes or body of any person may be exposed to injurious corrosive material." The "Emergency Eye Wash and Shower Equipment" standard of the American National Standards Institute provides direction for selection and installation of emergency equipment to meet OSHA requirements. The required safety equipment must be provided at all locations where the full strength chemicals may

FIGURE 1 A commercially available spill pallet containment system installed as part of a prefabricated modular CIP system.

be encountered, specifically, in the receiving and storage area, and in the area where the chemical feed equipment is located.

The Minimum Requirement

Figure 2 illustrates a common layout for a CIP skid, CIP chemical containers, and a chemical feed pump enclosure. These components may all be located in a corner or along a wall on raised pad pitched to a floor drain that also receives the discharge of the CIP system to waste.

If not already located nearby, safety shower and eye wash fountain can be mounted along a wall at the end of the chemical resistant pad.

Chemical Pump Location and Mounting

CIP chemical feed pumps are usually located on a wall or base-mounted stand or in a cabinet at a height of about 2 ft above the concentrated chemical containers or day use tank. This orientation is preferred to minimize the lift and friction head encountered by the chemical feed pump, and to localize the concentrated chemical feed pumps for personnel protection. The chemical resistant pad pitched to drain serves the function of a curbed area but provides convenient access to the chemical

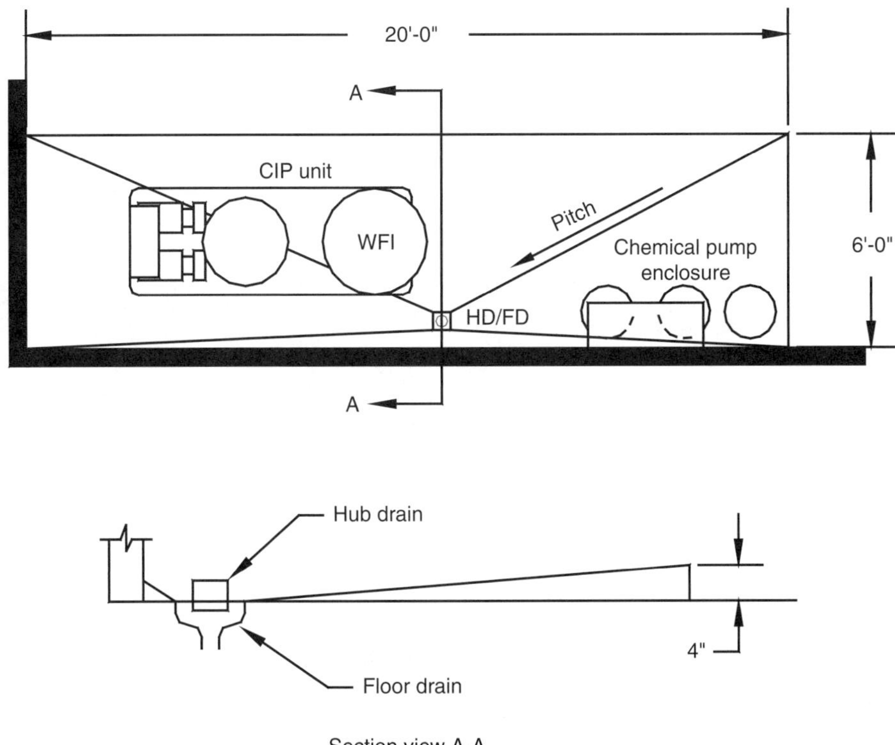

FIGURE 2 Typical layout for a CIP skid, CIP chemical containers and chemical feed pump enclosure.

barrels and the chemical feed pump enclosure. Leakage from any component may be washed down easily.

When the chemical dosing pumps are pre-mounted on the skid of a modular factory-assembled CIP system, similar contingency for the possibility of accidental spills must be addressed.

Tote-Based Chemical Feed Systems

A day-use tank system is a preferred method of handling chemicals delivered in bulk totes or tanker truck transfer to storage tanks. The intermediate tank provides capabilities including:

- Isolation of the CIP skid from direct connection to a large storage container, thus eliminating the accidental transfer of large volumes of chemicals as the result of a control valve, pump, or system failure.
- When used in the appropriate manner, the day use container may facilitate control of chemical usage.

In this case, the concentrated chemical is transferred by pump to a day-use reservoir typically of 15 to 55 gal capacity. Figure 3 shows a bulk caustic detergent container (a tote) on the left, from which transfer to the day-use container on the right is by a low level pump (not shown). Chemical feed pumps for multiple CIP skids are located in the stainless cabinet between the two containers. Weigh cells under the tote, as shown in the photo, provide a means of determining when an empty tote must be replaced, by monitoring loss of weight.

FIGURE 3 A bulk detergent tote on the left, day-use tank container on the right, and a stainless steel chemical feed pump enclosure supports multiple CIP skids.

CHEMICAL FEED METHODS

Central to the creation of CIP chemical cleaning solutions is the equipment that provides for chemical addition (sometimes referred to as "dosing"). Both under- and overfeed of chemicals are detrimental to the CIP cleaning program. Underfeed can lead to an ineffective cleaning cycle. Overfeed is wasteful, and places an unnecessary load on downstream waste treatment. Instrumental methods using solution conductivity are generally used to indirectly measure and control chemical agent concentration.

The following chemical feed methods are commonly used in conjunction with automated pharmaceutical CIP systems:

- Air-operated diaphragm pump
- Electromagnetic-operated diaphragm pump
- Motor-operated diaphragm pump
- Peristaltic pump
- Venturi feed
- Direct injection from pressurized header

Diaphragm Pump

Construction and operating principle of dual-chamber diaphragm pumps are extremely simple. In the left and right pump chamber, elastic diaphragms are clamped at the circumference. A rod connects the movable centers of the diaphragms. The opposite side of the diaphragm is the hydraulic chamber. A driving force (air, electromagnetic, or motor) moves the connecting rod in a reciprocating motion. This produces a periodic change in volume of the two hydraulic pump chambers. In combination with inlet and exhaust check valves, this change in volume produces a pumping action.

Air-Operated Diaphragm Pump

Air-operated diaphragm pumps are low cost, simple, durable, and relatively easy to maintain. An air distribution block shuttle valve located between the chambers alternately pressurizes one chamber while venting the other. The pumping flow rate is varied to an extent by increasing and decreasing supply air pressure. Air-operated diaphragm pumps are sized for the anticipated margin of required flow rate at full speed operation. Delivery quantity is then set by time of operation.

An alternative mode of operation is to replace the off-the-shelf air distribution block shuttle valve with a replacement distribution block that individually ports the air chambers. Using a solenoid valve arrangement, while one air chamber is pressurized, the other is vented, and in this manner the pump can be controlled stroke-by-stroke. The replacement distribution block arrangement has two benefits. Slow stroke-by-stroke operation ensures that the chambers have opportunity to completely fill to a known volume with each cycle. Precise control of the number of cycles permits the best degree of reproducibility for air-operated dual-chamber diaphragm pump dosing. Air-operated diaphragm valves require a plant air pressure on order of 100 psi. Air-operated diaphragm pumps usually do not require a pressure relief valve since maximum discharge pressure is limited by supply air pressure.

Electromagnetic-Operated Diaphragm Pump

An alternative driving method utilizes a solid-state electronic pulser to power the connecting rod cycling action between the pump chambers. Electromagnetic-operated pumps provide dosing output on order of a maximum of ½ gpm. To regulate dosing flow rate, the pumps are available with both analog speed and stroke adjustment. These pumps are especially beneficial for low-volume dosing applications, such as the fortification of caustic detergent solution by addition of sodium hypochlorite. Electromagnetic-operated pumps can operate on 110 volts alternating currents (VAC) at less than 1 A. The discharge pressure is self-limiting due to slippage limitation of the electromagnetic mechanism, so a pressure relief valve is usually not required.

Motor-Operated Diaphragm Pump

An electric motor may be used to turn a drive that is translated into a reciprocating motion to power the connecting rod cycling action between the pump chambers. Motor-operated pumps provide dosing output from one to several gallons per minute. Analog speed and stroke adjustment are available to vary dosing flow rate. The motor-operated pumps are much larger and heavier than both air- and electromagnetic-operated pumps. Motors typically operate at 240 or 480 VAC at ½ horse power (HP). The maximum discharge pressure of motor-operated diaphragm pumps is often quite high, so the need for a pressure relief valve should be considered.

Figure 4 shows an installation of both electromagnetic-operated diaphragm pumps (three smaller pumps on left) and a motor-operated diaphragm pump (larger pump on right). In this case, the electromagnetic-operated diaphragm pumps were used for metering CIP cleaning chemicals and the motor-operated diaphragm pump was used for a higher volume saline solution pre-rinse to provide cell lysis to aid the cleaning process of some cycles. This photo also illustrates the use of portable secondary-containment dollies as the means of moving barrels of concentrated chemicals to the location of the feed pumps.

Peristaltic-Type Chemical Pumps

Peristaltic pumps are used for cleaning chemical delivery to a lesser extent than diaphragm pumps. The peristaltic pump works similarly to the diaphragm pump, but substitute a flexible flow tube for a diaphragm and pinch rollers for check valves. The flexible tube is routed between a rotor that has two or more pinch rollers on it, and a channel that holds and routes the tubing around the rotor. As the rotor rotates within the pump housing, one of the pinch rollers compresses the tubing and pushes fluid through the outlet. Suction is created by the movement of fluid, and draws new fluid through the inlet, behind the moving pinched section of tubing. Whereas a peristaltic pump is quite satisfactory for low volume delivery of additives to production process, in which instance the tube is replaced for each subsequent batch, the risk of an undetected flow tube failure, and the resulting chemical leak, are factors limiting application of peristaltic pumps for CIP chemical delivery on a long term basis.

Venturi Chemical Feed Systems

Venturi-based chemical feed systems have gained in usage due to the simplicity and dependability afforded by a concept with no moving parts. Venturi-based systems cost significantly less than pump-based systems, and require little maintenance.

FIGURE 4 Electromagnetic-operated diaphragm pumps (three smaller pumps on left), a motor-operated diaphragm pump (larger pump on right), and portable secondary containment dollies for chemical drum handling.

To produce suction, motive flow is pumped under pressure through an entry section with a limiting orifice into the venturi section. As the motive flow exits the orifice, it expands, increasing in velocity just before the venturi section entrance. Based on Bernoulli's principle, vacuum (negative pressure) develops at the suction inlet port, which is located between the orifice and venturi section. Unlike the previously described pumping methods, venturi feed continuously blends the chemical dosing stream into the motive stream water rather than a pulsed injection, thus delivering a more even distribution. Venturi-based systems provide dosing output on order of maximum 1 gpm. Figure 5 is a photo of a venturi-based chemical system. The venturi is located at the stainless steel cross. The pressurized motive CIP solution slip stream flows through the ½ in. tubing up from the bottom and out through the top. A valve on each side of the venturi permits caustic or acid to be drawn into the slipstream as they are respectively required during the CIP cycle.

Venturi-based dosing systems are inherently self-priming due to the continuous suction. Although diaphram and peristaltic pumps are designed to allow the pump to re-prime itself under typical lift conditions, the pump will take several cycles (to perhaps many cycles) to completely prime depending on the application variables such as discharge pressure, so it is important to ensure the pump is

FIGURE 5 The venturi of this venturi-based chemical feed system is located at the stainless steel cross and a valve on each side controls delivery of acid and alkaline CIP chemicals.

primed before the CIP cycle starts, to ensure that the desired quantity is actually delivered during make up of the CIP solution. A "foot valve" located in the concentrated chemical storage container at the end of the suction line wand will keep the suction line full and prevent loss of pump prime.

A note of caution must be addressed regarding the possibility of accidental chemical release by siphon action. The check ball arrangement of diaphragm pumps prevent backward flow through the pump but *not* forward flow. Thus, it is possible to establish a scenario where concentrated chemicals are siphoned from the concentrated chemical container through the diaphragm pump *when the pump is not operating*. By this means a large-scale accidental release of concentrated chemical could occur. A combination anti-siphon/check valve located at the location point where the concentrated chemical discharge line connects to the CIP system piping can address the possibility of both accidental flow of water toward the chemical pump and accidental siphoning of chemicals into the CIP system.

For all of the above-listed chemical dosing methods, the suction line from the chemical supply container can be either a polyvinyl chloride (PVC) (or other compatible material) pipe wand or hoses slipped through a length of PVC pipe which in turn is located in the supply barrel. This piping is merely to keep the tubing straight and make certain that it projects to the bottom of the barrel. The end of the suction wand should be cut at a 45° angle to prevent blockage through contact with the bottom of the barrel.

All chemical-tubing connections must be clamped tight. An "air-locked" chemical dosing pump will seldom be self-priming. Likewise, a venturi will suck air rather than chemicals. Failure to deliver chemicals when required will result in an automatic low-conductivity alarm on systems equipped with chemical concentration feedback logic.

Meter-Based Chemical Delivery

Direct chemical injection from a pressurized header is another simple, reliable means of cleaning agent dosing. This method is suitable for pharmaceutical facilities that utilize bulk tank storage of concentrated chemicals with pumped flow through distribution headers. Concentrated chemical discharge is directed to the CIP system recirculation tank by an automatic shutoff or flow control valve. Double-protection valving, instrumental monitoring, and/or software safeguards must be utilized to ensure that a single-valve failure on the pressurized supply cannot lead to a large-volume accidental discharge from the bulk storage tank. A schematic diagram of this method will be discussed later in this chapter.

CLEANING-AGENT SOLUTION DILUTION

Utilizing the selected chemical delivery method, a known quantity of concentrated chemical is added to a known quantity of water to formulate the dilute cleaning solution of required strength. Concentrated chemical discharged by the selected delivery method should be piped the shortest possible distance prior to dilution, both from the perspective of personnel safety and system performance.

The greater the distance of concentrated chemical piping the greater the opportunity for spillage as result of accidental damage by forklift, carts, leaning ladders, or other means. Frequently the chemical piping is of polymeric material, such as PVC or Teflon, in which case added piping support and double containment are considerations. Also, long lengths of concentrated chemical piping provide the opportunity to contain large quantities of air, which must be displaced when the system is primed. So with greater priming volume comes less certainty of the amount of concentrated chemical actually dosed.

Chemical Injection via a Slipstream (or Loop)

A preferred method to minimize concentrated chemical piping is to bring the dilution water flow from the CIP system to the chemical pump, rather than pipe concentrated chemicals to the CIP system. This can be accomplished by the use of slipstream loop, which departs from the primary flow under pressure beyond the CIP supply pump discharge, and returns either at lower pressure on the suction side of the CIP supply pump or through a nozzle on the head of the CIP recirculation tank.

Figure 6 illustrates a small diameter chemical loop (1 in. diameter or less) slipstream, which starts at a branch tee in the CIP supply pump discharge line preceding the flow meter (FE) and reenters at a tee on the CIP return line manifold. Note that a restrictor orifice is shown at the supply line branch location. By this design, the chemical loop receives a continuous limited flow on order of 2 to 3 gpm, for chemical injection during solution wash steps (Drain valve closed and Return valve open) and rinsing (Drain valve open and Return valve closed) the loop with fresh water during the remainder of the CIP cycle. The pressure drop per foot will be consistent for the entire length of the loop and the pressure against which the chemical pumps must deliver is approximately one-half of the total, and is consistent for all CIP circuits cleaned from the CIP skid. In this configuration it is important that the chemical loop branch location precede the CIP flow meter element (FE), to ensure

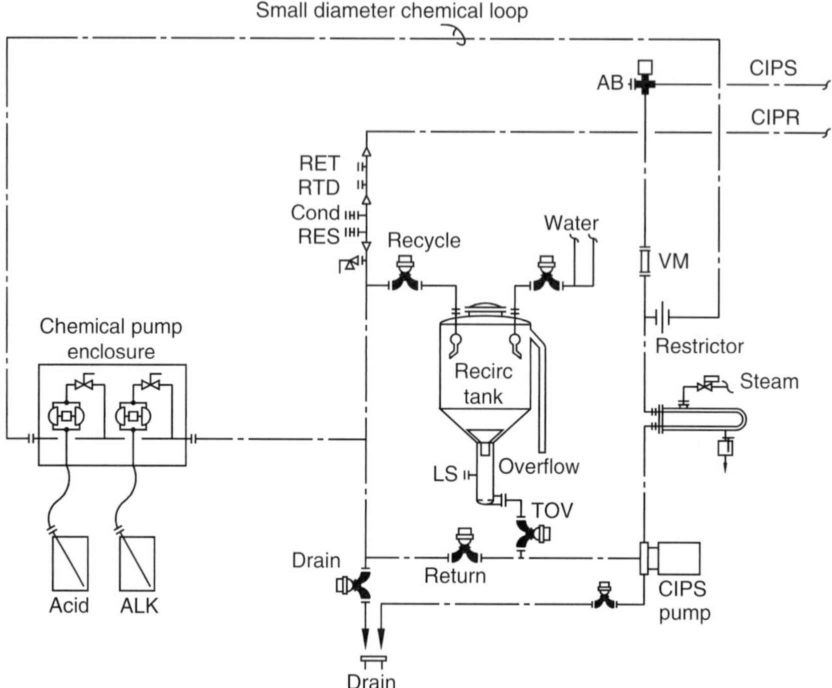

FIGURE 6 Illustration of a small diameter chemical loop (1 in. diameter or less) slipstream with return to CIP return line.

that the control software input is the "true" amount being delivered to the circuit, without regard for the small quantity being routed through the chemical loop.

Figure 7 illustrates an alternate configuration to that of Figure 6 for the small diameter chemical loop. In this case, the CIP system is of an eductor-assisted two-tank design, with separate fresh water tank and motive recirculation tank, which receives both rinse and recirculated wash water from the CIP return. Accordingly, the slipstream starts at a branch tee in the CIP supply pump discharge line preceding the FE and reenters at a nozzle on the head of the motive recirculation tank.

Figure 8 illustrates direct chemical injection from a pressurized header, without the use of a chemical loop. Concentrated chemical is directed through a small-diameter ($\frac{1}{2}$ in.) discharge manifold to a nozzle at the CIP system recirculation tank from the facility alkali and acid header. The chemical discharge manifold begins with an automatic shutoff valve (V1-A & B). A manual needle-type control valve (CV-A & B) is pre-set during commissioning to provide the required restriction for a slow feed of the needed chemical quantity over several minutes. A manual shutoff valve (HV-1A & B) blocks can block forward flow. This shutoff valve used in conjunction with a pair of calibration taps (HV-2A & B and HV-3A & B) is used for off-line calibration of the FE.

Two different FE types are illustrated on Figure 8. The alkali header supplies a magnetic FE, which is suitable for cleaning chemicals that are conductive. For nonconductive concentrated CIP chemicals, a mass FE can be used and this is illustrated on the acid header.

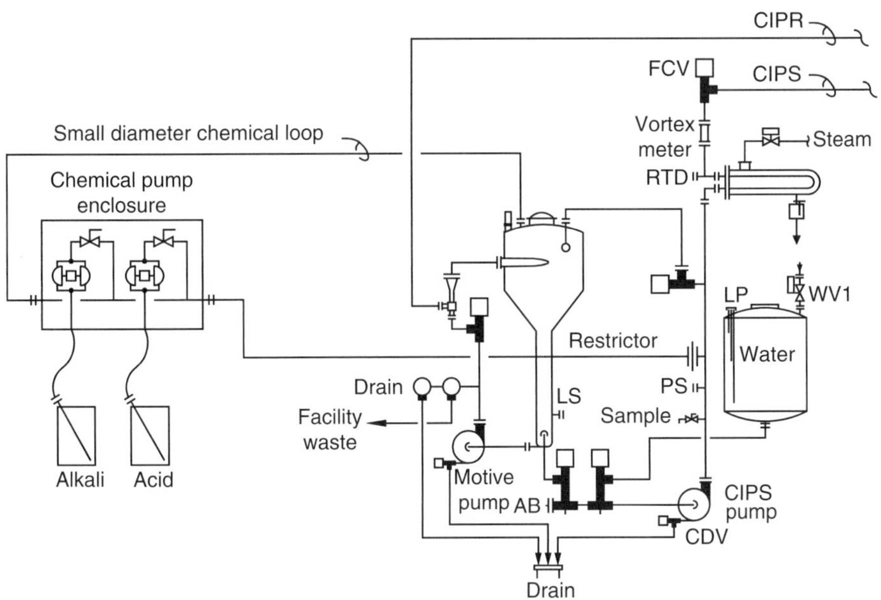

FIGURE 7 Illustration of a small diameter chemical loop with discharge through a nozzle on the head of the recirculation tank.

Mixing and Control of Application of Cleaning Agents

From the perspective of CIP cycle optimization, the most efficient and economical method of batching CIP chemicals is by direct injection into the recirculating wash solution either in the CIP return line manifold or the recirculation tank. Following the initial pre-wash rinsing to drain at the start of the CIP program, a "system fill" volume of water is pumped into the CIP circuit adequate to fill the CIP supply/return piping, accumulate a small "puddle" in the process tank being cleaned (if a tank cleaning program is in progress), and establish a minimum working volume in the CIP recirculation tank (if a recirculation tank is being utilized).

After the system fill, stable recirculation is established between the CIP system and process equipment, tanks, and piping that comprise the CIP circuit. The volume recirculating should now be the absolute minimum necessary to ensure that chemical dosing volume, heating energy, and subsequent effluent volume are all optimized to be the least possible. Heating is generally initiated at this time.

Chemical cleaning agents are then injected into the recirculating system fill water volume by any of the established delivery and dilution methods to obtain the desired concentration. In this manner, when the solution is at strength and well-mixed, the solution will likewise be at set point temperature for the wash. It is desirable to inject the concentrated chemical into the recirculating solution over a long period of time, preferably during five to seven passes through the circuit, to maximize uniform distribution. If excessive feed rates are used, i.e., oversized chemical feed pumps, the only opportunity for mixing to occur is in the small puddle of a vessel being cleaned and in the CIP skid solution tank.

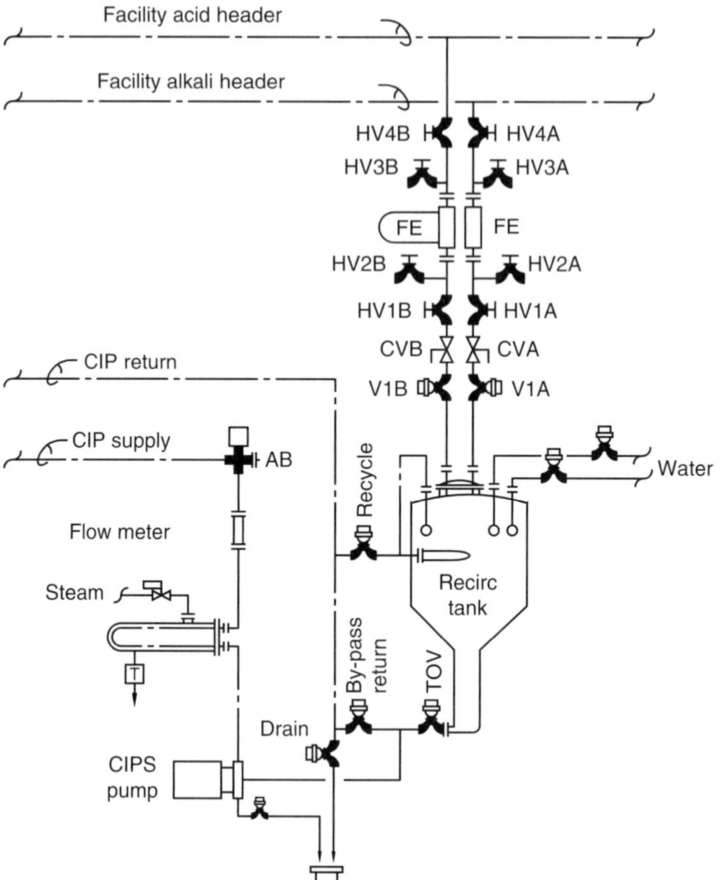

FIGURE 8 Illustration of direct chemical injection from a pressurized header, including requisite valves and instrumentation.

In-Line Chemical Injection for Single-Pass Cleaning

Direct in-line chemical injection for single-pass delivery is used for selected CIP applications, where a decision is made to provide single-pass chemical contact on a once-through basis rather than detergent wash recirculation. Examples of single-pass CIP are pre-CIP decontamination rinses, cleaning of a transfer line separating different classes of areas, and cleaning multi-product filling lines. The most common method of direct injection utilizes a motor-operated diaphragm pump with discharge connection to the CIP supply pump suction, either directly or through a slip-stream loop. The flow rate of a motor-operated diaphragm pump is near-linear to the speed of rotation of the drive motor. This type of pump is especially suitable when a large dosing volume is required in proportion to an analog FE output, as in the case for single-pass chemical delivery.

An alternate method for single-pass delivery is to pre-batch cleaning agent solution in a CIP recirculation tank that is located on the CIP system. The CIP recirculation tank is first filled with water. Then, local pumped recirculation is established on the CIP skid. Cleaning agent is then dosed into the tank by any of the established delivery and dilution methods to obtain the desired concentration. If desired, solution can be heated during the chemical dosing and recirculation mixing. After set point concentration is established, the solution is ready to be pumped out for single-pass delivery.

Verification of Cleaning Chemical Concentration and Residue Removal

The previously described chemical dosing methods all have a margin of error. That error in the dosing volume itself might not be of great consequence, except for the possibility (likelihood) that inevitably at some time a drum will go dry, hose will be kinked, valve will be closed, or some other malfunction causing less, or perhaps more, to be dosed than intended. This possibility of error, coupled with the need for validation data and batch electronic records, leads to the use of instrumental methods to ensure and document that proper cleaning chemical concentration has been established.

With recent improvements in solid-state strain gauges, analog scales have gained attraction for size, dependability, and cost. Moderately priced scales enclosed in stainless steel housing are available for weight measurements on the magnitude of that for a 10 to 55 gal chemical container. While chemical metering systems are "reproducible," weigh scales provide an absolute measure of the chemical dosage to within 1%. Any chemical feed system can benefit from the absolute dosing accuracy provided by weigh scales under day-use tanks. This is especially true for venturi-based dosing, which coupled with weigh scales provides a highly accurate, dependable, and cost-competitive chemical dosing option.

In-line conductivity measurement of cleaning agent, whether during recirculation or single-pass cleaning, is a frequently used instrumental method to verify that solution concentration is within threshold boundaries. During system commissioning and cycle development, the relationship between solution concentration and conductivity is established for the cleaning agent that will be used. During the CIP cycle, at a point in the program where cleaning agent solution should be fully dosed and mixed the conductivity test occurs. The most effective way to ensure complete chemical mixing is to add the concentrate slowly, over six to eight full recycled passes of the solution volume. By this means, a minimal mixing time will be needed before the system controller looks for a reading within a predetermined range of the set point conductivity.

If the requisite set point is not achieved an added dose of chemical will be added, and the conductivity check repeated. This process will be allowed to repeat a predetermined number of iterations, at which point required conductivity must have been achieved or an operator alarm will occur, alerting of a chemical addition malfunction.

Solution conductivity is customarily expressed in milli-Siemens (mS). For example, at 25°C:

0.3% active NaOH (3000 ppm) ≈ 20 mS
1% active NaOH (10,000 ppm) ≈ 53 mS
3.0% active NaOH (30,000 ppm) ≈ 145 mS

The use of a pH sensor to verify cleaning agent dilution concentration may initially seem appropriate but really is not of great practical value. Measured pH will shift rapidly in large increments with little change in solution concentration.

That is to say a weak acid solution can read pH 3 or a weak caustic solution pH 12. However, conductivity corresponds directly to solution strength. Also, for best accuracy pH instruments must be frequently calibrated and stored in buffer solution between uses. The latter condition is difficult to maintain for a CIP system located in a utility area remote from the process.

9 CIP Spray Device Design and Application

John W. Franks and Dale A. Seiberling
Electrol Specialties Company (ESC), South Beloit, Illinois, U.S.A.

INTRODUCTION

The in place cleaning of liquid handling processes began with a focus on the piping systems used to transport product and fluids used to prepare, produce, or package the end product. The cleaning regimen was developed to assure that the flush, wash, and rinse fluids fully contacted all surfaces, and control of pressure and flow was determined to be basic requirement. The control of flow and pressure resulted in control of the cleaning-solution velocity within the pipe. The early investigators determined that a flow velocity of 5 ft/sec (1.5m/sec) would assure a full pipe in horizontal runs and move insoluble soil through and out of the system. As attention was expanded to the inclusion of process equipment that was not practical to be filled and pressure washed at controlled high surface velocities, the need for an alternate approach became obvious. For this purpose, spray devices were developed to deliver solution to all product contact surfaces, and any surface that might drip, drain, or, otherwise transfer fluids to product contact surfaces. It was quickly recognized that a tank could be cleaned by spraying only the upper area at a rate which assured that the flush, wash, and rinse solutions passed over all other areas en route to the vessel outlet.

Highly refined spray devices are now utilized in biopharmaceutical and pharmaceutical processes and for active pharmaceutical ingredients (API) processes as well as in other sanitary industries to precisely and consistently deliver cleaning solutions or solvents and flushes or rinses to surfaces of process vessels including mixing tanks, hold tanks, and reactors, as well as "tank like" equipment such as charge chutes, vapor ducts, dryers, evaporators, dry ingredient mixers, and other equipment used in a specific production process.

It is critical that sprays be designed and installed to assure that all surfaces which require cleaning are fully covered and receive continuous replenishment of the required sequence of fluids during the cleaning and rinsing processes. The cleaning process is generally a chemical process whereby soils are suspended and continuously rinsed away. Properly designed spray distribution devices provide for complete solution coverage of all the intended surfaces but are not generally required to provide physical impact of solution on all surfaces.

This chapter will define the requirements for successful spray cleaning as well as describing the general types of sprays currently utilized in this industry and also focus on unique applications and design guidelines.

REQUIREMENTS FOR SUCCESSFUL SPRAY CLEANING

To successfully clean a process in a biopharmaceutical or pharmaceutical facility, a number of factors must be addressed, starting with the CIP philosophy, a CIPable design and a CIP sequence of program.

CIP Philosophy

The production process must be designed to incorporate a selected cleaning philosophy from its inception. Design criteria may include separation of CIP circuits by functional process areas, common CIP supply and return piping for a variety of vessel sizes, and integration of process piping for conveying CIP fluids.

CIPable Design

All equipment in this process must be designed to be CIPable or easily removed for manual cleaning. The sprays must be designed giving consideration not only to just minimum coverage requirements, but also maintenance of adequate velocities in all vessel CIP-related piping. The vessel outlet valves and outlet piping must be given consideration not only just for process flow rate, but also the generally higher CIP flow requirement. And, the individual equipment components in the process must be cleanable by spray methods. Vessels are the major item of equipment in most processes and this chapter will discuss vessel CIP. Figures 1 and 2 in Chapter 1 illustrate a typical vessel and define the CIP issues of concern.

CIP Programs

The cleaning flush, wash, and rinse solutions and the cleaning regime, i.e., combination of time, solution temperature, and chemical concentration must be capable of loosening and suspending the soil for subsequent removal.

Other chapters of this book provide detailed information on the above and related subjects.

TYPES OF SPRAY DEVICES

There are two basic types of spray devices utilized in this industry, those being fixed type and rotating type. Rotating sprays, especially those designed to provide physical impact on all surfaces, are less suitable for permanent installation in these types of applications, but are often applied in other industrial and commercial applications. Fixed sprays devices are by far the more prominent in both biopharmaceutical and pharmaceutical processes. The advantages and disadvantages of each are reviewed below.

Rotating Spray Devices

Rotating or "dynamic" spray devices are available in two types, single axis and dual axis. The single-axis type is sometimes described as "spinners." Most rotating spray devices of either type claim the use of impact or impingement as the means of accelerating the cleaning process, and using less water and solution. Rotating sprays by definition have moving components and bearing surfaces. Their inherent design makes them susceptible to wear (a validation concern) and potential internal cleanability issues. Foreign particles in recycled wash solutions may make them susceptible to failure due to lack of rotation. Rotation sensors are sometimes employed to assure a rotating spray is indeed rotating but that adds additional complexity to the entire process. Rotating sprays can be constructed of stainless steel, hastelloy, or other alloys, or polytetrafluoroethylene and often include some nonmetallic type bearing surface in the design. Most single-axis rotating sprays provide 360° coverage and therefore the quantity of solutions applied to the vessel sidewall is considerably less compared with a fixed spray approach where virtually all the solution is applied to the top head area and allowed to flow down the entire sidewall. Dual-axis rotating sprays provide coverage of only a small part of the total area at any instant and may require significant time for one complete rotation. Rotating sprays often require higher operating pressures than fixed sprays and that pressure requirement is additive to the pressure drop of the circuit which can sometimes dictate the upsizing of the supply pump.

Because of their usually higher operating pressure, rotating sprays may provide an advantage of greater impact velocity and the physical energy can supplement the chemical cleaning process at those points of impingement. If "standard" rotating spray designs can be utilized, they may be cost-effective from an initial procurement perspective. However, total operating costs and complexity should be considered, keeping in mind most rotating sprays cannot be left installed if in the product zone. Simple fixed sprays are more likely to be accepted in the product zone than rotating sprays and fixed sprays can also be designed to be above the product zone in most cases. In either case, the fixed spray is more likely to avoid the ongoing operating costs of the reinstallation and removal of the spray.

Fixed Spray Devices

Fixed or "static" spray devices by definition have no moving parts and are most always custom designed for each specific application in this industry. Fixed spray devices can be constructed totally of stainless steel, Hastelloy, or other alloys. Fixed spray devices are fabricated in many shapes and sizes to address each need. The design of the spray pattern and the spray position in the vessel are critical.

Fixed sprays are generally very robust and can tolerate variations in supply pressure of plus or minus 20% and still function satisfactorily for the intended purpose. Fixed sprays are generally designed to direct streams to specific equipment features, i.e., nozzles, manway, and agitator collars, as well as to the tank head, thereby allowing surface tension and gravity to continuously move the solution across the dish head and down the vertical sidewall surfaces. Continuous replenishment of the flush, wash, and rinse solution across the entire surface provides a chemical cleaning process of maximum effectiveness, because the boundary layer between the soil and cleaning solutions is being constantly refreshed.

When properly designed, installed, tested, and validated, a fixed spray can reliably provide the intended service for extensive periods of time, which is an extremely valuable characteristic for validated processes. This is why fixed sprays are predominantly selected for application in this industry. Therefore, the remainder of this chapter is primarily focused on fixed spray design and application, first for vessels, and then for filter housings cleaned with transfer piping.

DESIGN PRINCIPLES

Historical

The application of fixed spray devices was developed simultaneously with the development of automated CIP cleaning systems, initially in the dairy industry, dating back to the late 1950s. Initial designs were simple spheres but soon a more favorable pumpkin shape was developed and patented. The flattened top of the pumpkin-shaped upper half-sphere provided more upward coverage and reduced the number of sprays required for proper coverage of simple vessels. As spray cleaning was applied to dryers and evaporators, other configurations including tubes with bubbles and small half-spheres were developed. These have been further refined to provide the pinpoint coverage to address the numerous issues found in biopharmaceutical and pharmaceutical process equipment. This chapter can address only the most basic aspects of design and application.

Coverage Criteria

The early users of tank CIP quickly learned that it was easier to pump water into a tank than to remove it, reliably, under automatic control. Field experience suggested that horizontal, cylindrical, and rectangular tanks could be effectively cleaned at flow rates of 0.2 to 0.3 gpm/sq ft (30–35 Lpm) of the upper third of the surface for vessels in general and for all other applications of permanent spray devices. Other criteria include (1) run agitators and (2) spray both sides of all baffles. The advent of the large dairy silo in the early 1960s resulted in findings that 2.0 gpm/ft (25 Lpm/m) of circumference for vertical vessels, applied to the dish head only, would clean vessels of nearly any size, if the tank head contained no nozzles. When spray CIP was applied to biopharmaceutical and pharmaceutical industry vessels in the late 70s and early 80s, it was quickly discovered that standard spray designs used successfully in dairy, beverage, brewing, and food facility vessels would not work, primarily because of the number of head nozzles and other interruptions to a continuous surface. Directionally drilled sprays became the solution, by addition of streams to specifically target nozzles, manway and agitator collars, dip tubes, etc. Successful experience suggests the need for 1.0 to 1.5 gpm (3.8–5.7 Lpm) being required for each head nozzle and that the 2.0 gpm/ft (25 Lpm/m) of circumference based on diameter is needed for manway and agitator collars in addition to the 2.0 gpm/ft (25 Lpm/m) of vessel circumference. The total flow rate for vessels using this approach typically averages close to 3.0 gpm/ft (38 Lpm/m) of circumference.

Fixed spray devices have been found to perform very effectively at 25 psi providing good spray ricochet and coverage while avoiding excessively high system pressure requirements.

Flow Rate and Turbulence Considerations

When cleaning pipelines, the recommended velocity of 5 ft/sec (1.5 m/sec) is to assure a full line in horizontal runs and this velocity provides Reynolds numbers (*Re*) well above that required for Turbulent flow ($Re > 2100$). The velocity of a film passing down the sidewall of a vessel is governed only by gravity, but *Re* is affected by viscosity and temperature. Hyde (1) calculated the *Re* of turbulent falling films at 2.5 gpm/ft (38 Lpm/m) of circumference as 2060 for ambient water (20°C), 4360 for a caustic wash at 60°C, and 5668 for a water-for-injection (WFI) rinse at 80°C. Greene (2) previously cited a reference to *Principles of Chemical Engineering* (3) as the means of determining the *Re* for a film running down the sidewall of a tank and noted that temperature, via it's impact on viscosity, affected *Re* significantly. He suggested that for a 7.5-in. (2.3 m) diameter tank, a flow rate of 75 gpm (17 m^3/hr) would be required for a cold rinse as compared to only 25 gpm (5.5 m^3/hr) for a hot (80°C or 180°F) wash to achieve the *Re* of >2100 considered adequate for turbulent flow. The higher flow in a vessel of this diameter would equate to approximately 3.0 gpm/ft (38 Lpm/m) of circumference suggested above as the average design flow rate, and this combined with recognition that most flush, wash, and rinse solutions used in biopharmaceutical CIP applications are well above ambient temperature by virtue of the available pure water (PW) or water for injectables (WFI) supply to the CIP Skid would assure effective cleaning. Providing adequate solution to the surfaces is important to facilitate effective cleaning, however, the designer must keep in mind that all solutions entering the vessel must be quickly and constantly removed to avoid large solution accumulations in the vessel and thereby compromising the main objective.

Configurations and Applications of Typical Fixed Spray Devices

Typical Fixed Ball-Type Spray Device

Figure 1 illustrates a full featured fixed spray configuration which consists of a 2.5 in. (64 mm) pumpkin-shaped spray head, directionally drilled, affixed to a 1.5 in. (38 mm) spray supply tube via a slip joint adapter, and retained by a removable *Spray Retainer Clip. An Index Pin* welded to the tube assures proper location of the spray on the tube which is also properly positioned in the 3 in. (76 mm) clamp-type tank head nozzle by a *Spray Tube Index Rod* which projects through a tab welded to the nozzle by the tank fabricator. The spray supplier and tank supplier must carefully coordinate the locations of all indexing tabs (nozzles).

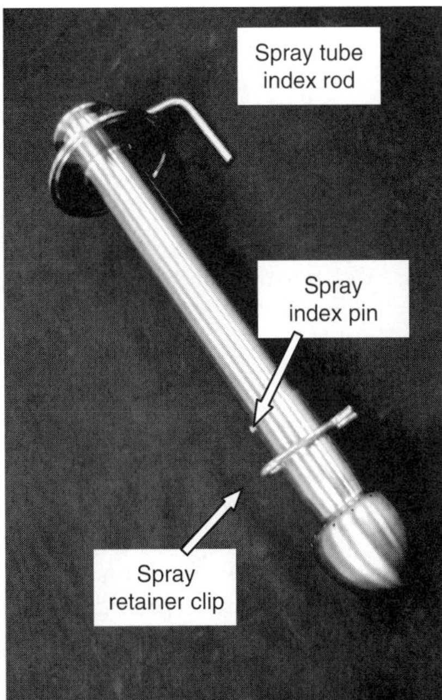

FIGURE 1 The above photos illustrate how a fixed directionally drilled spray device may be precisely located in a tank nozzle to assure coverage of all critical areas. The spray tube and spray head would be removed only for inspection or manual removal of debris often accumulated during commissioning.

Similar spray devices based on the use of 1-in. (25 mm) tube and a 1.75-in. (45 mm) ball capable of passing through a 2-in. (51 mm) nozzle are available.

Spray Positioning

The number of sprays and their optimum position will be highly dependent on the allowable depth in the vessel to which the sprays are to be located. The spray streams directed at nozzles must have an upward vector component of 30° or greater above horizontal to allow the spray to ricochet upward after hitting the target nozzle. Generally reductions in head space for the sprays will require a greater number of sprays to achieve adequate spray vectors.

The desirable approach is to establish the vessel head layout in the following sequence.

- Locate manway
- Locate agitator (180° opposite manway in smaller vessels and to minimize spray devices and nozzles)
- Locate baffles if required
- Locate sprays to optimize coverage with a minimal number of spray devices and nozzles
- Locate all other nozzles within the spray coverage circle based on the depth of installation

Manway collars, agitator collars or nozzles, and no foam inlets or dip tubes usually require spray streams from two directions to achieve a cross-chop and good ricochet coverage of the entire nozzle. Baffles require spray streams be directed to the top of each side of each continuous baffle.

The vessel depicted in Figure 2 was based on the above criteria and the decision to use three baffles which immediately required three spray devices to assure coverage of both sides of each baffle. S1 and S2 were placed to provide coverage of both the manway and agitator collars in addition to four of the six sides of the baffles. S3 provided coverage of the other sides of the baffles. Nozzles A–C were covered by streams directed from within the normal circle of coverage whereas D was targeted by streams at a slightly lower angle added to sprays S2 and S3. In this example, the wide angled agitator blade was expected to turn slowly through the ricochet from the multiple spray streams, a solution generally successful.

The space for nozzles in which to install sprays is often limited and the application of a spray "ball with an arm and a bubble" shown in Figure 3 can provide coverage of a manway or agitator collar from a spray otherwise not ideally located, thus providing two or even more sources of spray streams through a single nozzle. Such assemblies will pass through a short 3-in. (76.2 mm) nozzle and can be installed and removed from outside the vessel if care is taken to provide space above the tank head in which to move the spray tube as required for installation purposes. An alternative described as a "bent tube spray" may be comprised only of multiple bubbles at required locations on a single 1.5-in. (38 mm) tube. However, it is more difficult to assure full removal of debris from such sprays as compared to removable balls.

The example illustrated in Figure 4 began as a tank head with an agitator, a manway and nine nozzles on a single arc. This predefined head layout was fitted

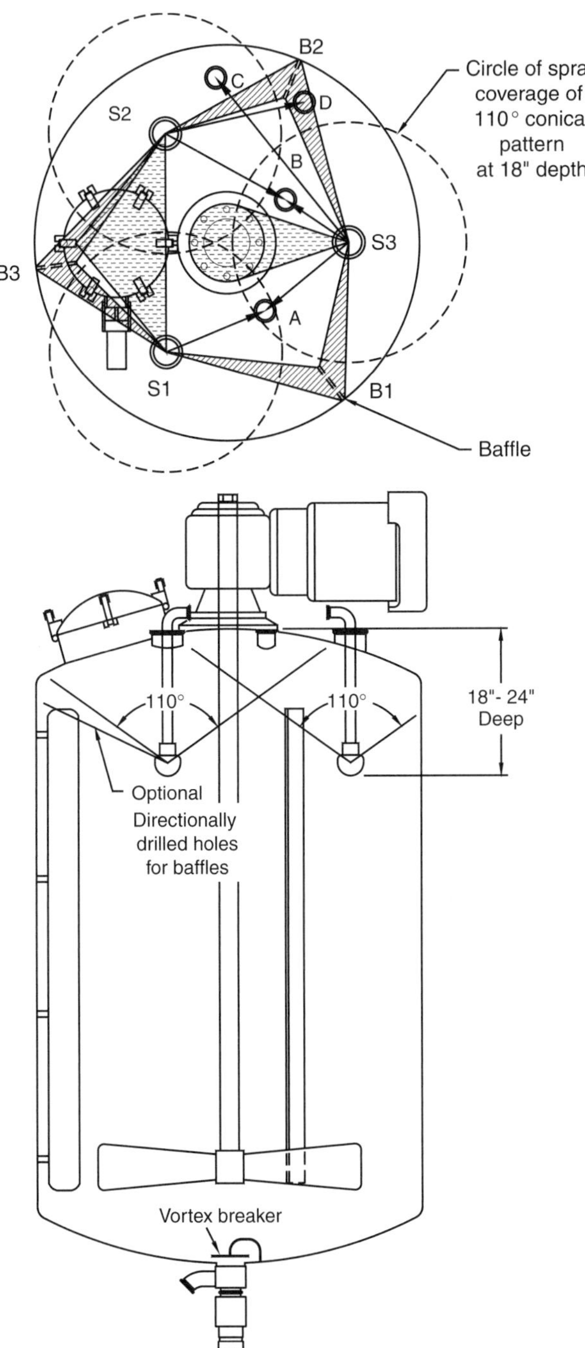

FIGURE 2 This line drawing of a large agitated vessel with baffles illustrates the preferred method for locating spray devices with reference to agitator and manway collars and baffles. Other nozzles are covered by streams from at least two sprays.

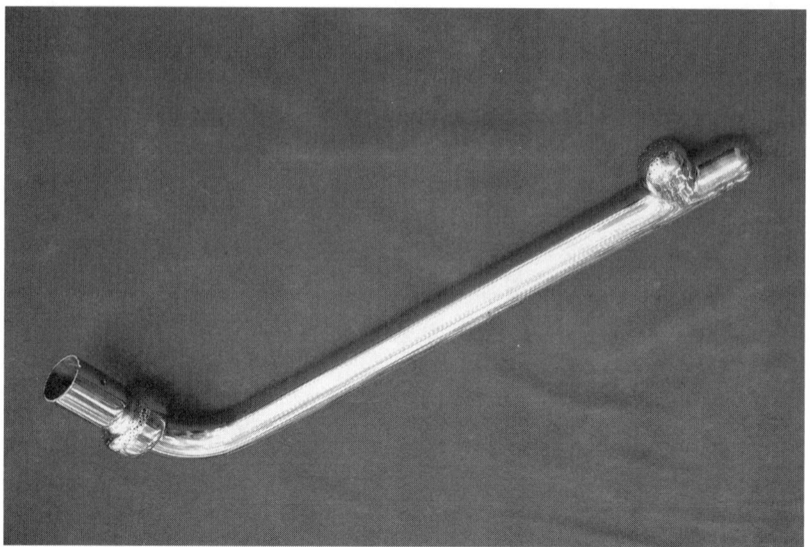

FIGURE 3 A single spray device consisting of a ball with an extension arm and a bubble permits the enhanced use of available nozzles in tank heads.

with three sprays to cover the nozzles and S3 was of the "ball with a bubble arm" to provide manway collar coverage from an ideal position.

Spray Coverage of Baffles, Agitator Blades, and Inlet Tubes

If a vessel contains baffles, each side of each continuous baffle must have spray streams directed at the top of the baffle at a rate of 3.0 gpm/ft of width per side, or a minimum of two streams per side. Agitators require similar special attention with spray streams directed specifically at the mounting nozzle and upper shaft as well as the agitator blades. The agitator should be rotated slowly during the cleaning cycle to allow the blades to pass through either normal falling ricochet or a line of spray streams directed at the blades outward from the centerline as shown in Figure 5. Agitators that have a flat continuous bottom surface or other designs which are shadowed from the top will require a spray device positioned at some point below the agitator. This can be best accomplished using a spray extending down from the top head. A J-hook welded to the sidewall may be required in heavily agitated tanks to support the long supply tube. Sidewall entry for permanent sprays is often not (acceptable) preferred. Depending on the application sometimes sparge tubes can be designed to spray clean the underside of agitators. Alternatively if lower sidewall sprays become the only choice, they may need to be removed between cleaning campaigns. Lower sidewall ports can be covered by spray streams from an upper spray or a lower positioned agitator spray if applicable. In some cases, lower ports can be cleaned by just the normal solution flow down the vessel sidewall if the sidewall port extension is minimal, the radius at the sidewall interface is generous and the upward slope angle is less than 5°. Special fittings, e.g., NA Connect® Millpore Corporation, Billerica, Massachusetts, U.S. (4), are frequently used in these applications as they minimize nozzle length.

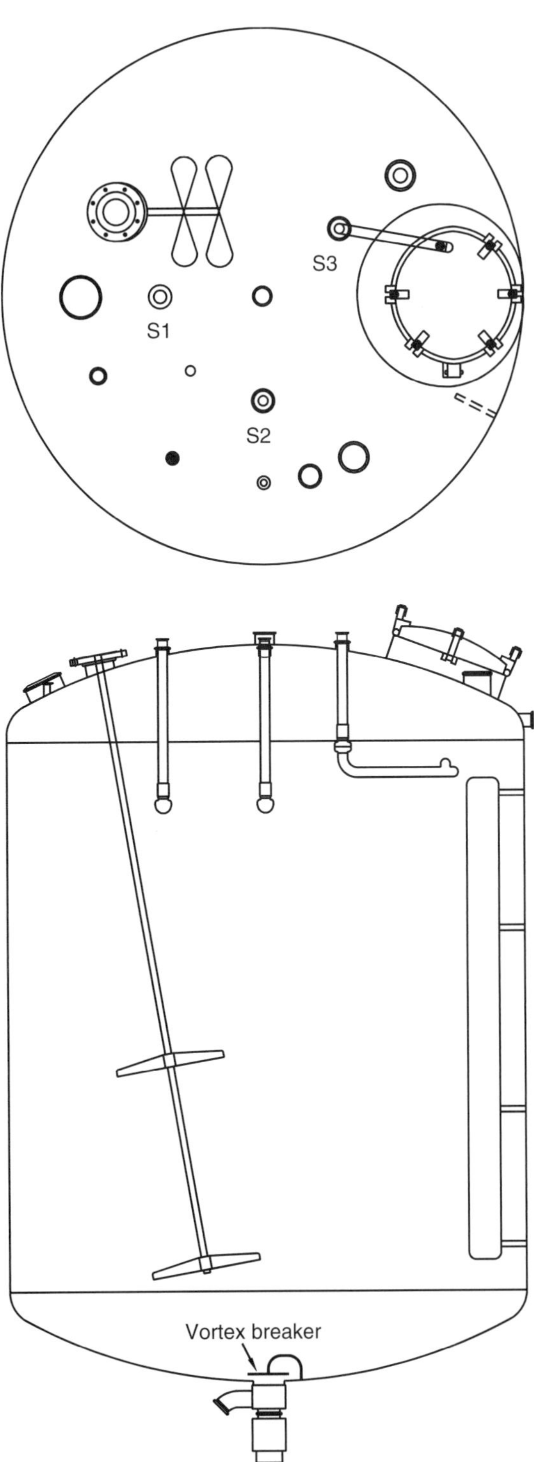

FIGURE 4 This tank head had been configured for all process nozzles and opening. Spray nozzles S1–S3 were added and located to meet clean-in-place coverage criteria.

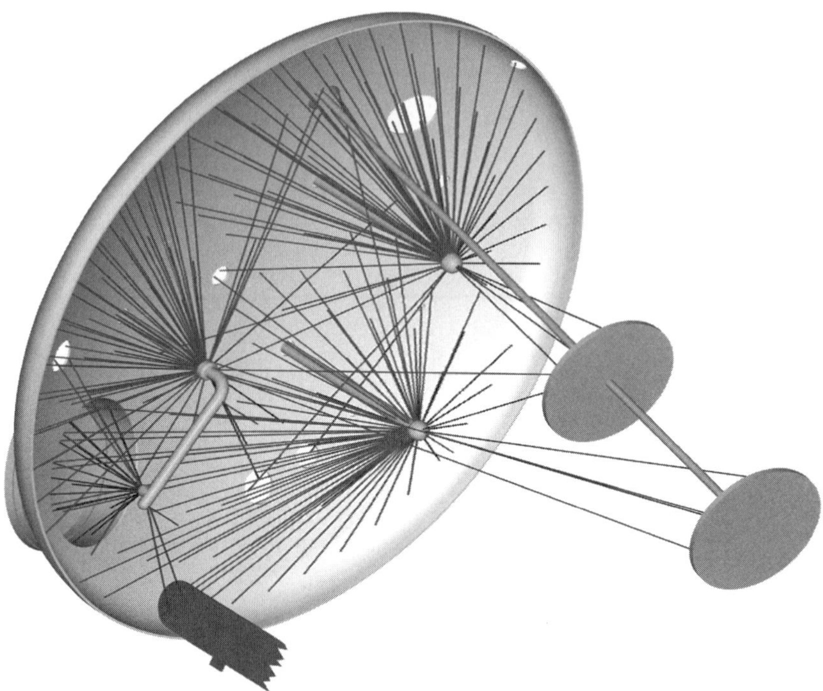

FIGURE 5 Computer-assisted three-dimensional spray design permits the precise location of all holes in multiple spray heads and viewing of the stream trajectories from multiple views.

Spray Flow Rate Criteria

The spray devices must be designed to meet the minimum flow rate requirement for the vessel or equipment item being cleaned. However, the final spray flow rate must give consideration to the requirements of the specific CIP circuit and the total CIP system. If the equipment is being cleaned in conjunction with other equipment or piping then the item requiring the highest flow rate will determine the flow rate for circuit. This will often be a substantially higher flow rate than is required and the sprays must be modified to use the additional solution to enhance spray coverage efficacy to optimize the CIP system application. Vessel outlets also need to be sized to meet the higher flow rate required for the proper function of the total CIP system, i.e., multiple circuits cleaned from a single CIP skid, CIP supply and return piping, and process piping. It is important to balance the overall CIP process within a facility and this is discussed by other chapters of this book.

Vortex Control

Directly related to spray coverage is the need for maintenance of velocities of 5 fps in all piping in the circuit. As the flow rate is raised the tendency for solutions to accumulate in the vessel will increase and if the spray rate exceeds the normal drainage rate, the buildup of solution in the tank bottom can result in what is referred to as "bath tub ring" at the solution/air interface, and also sedimentation of undissolved soil. A flat plate vortex breaker three times the outlet diameter permanently installed between 0.5 and 1 in. (13–26 mm) above the outlet will facilitate removal of solution by gravity, return pumps, an eductor, or any combination thereof, by eliminating the vortex and its inherent restriction of the outlet. Whereas sequencing flow through inlet legs in parallel with the sprays for a portion of the cycle is desirable, pulsing sprays to accommodate an undersized tank outlet is a very last resort.

SPECIAL SPRAY APPLICATIONS

Bubble Sprays

The half-sphere "bubble" shown on the arm of the spray in Figure 3 may also be welded to the end of a 1.5-in. (38.1 mm) supply tube and directionally drilled as shown in Figure 6. This design is applied for spraying horizontal and vertical ductwork. Custom drilling and spacing is required for horizontal duct sprays to achieve proper swirl in the duct and complete coverage. This spray can also be configured with flange connections for API applications in Teflon lined large diameter pipe or vent/vapor ducts as shown in Figure 7 which also includes a spray ball with a flange connection for installation in nozzles of glass lined vessels. Spray devices used in the API process are generally fabricated of hastelloy. Condensers on API process reactors are usually required to be part of the cleaning regimen. Bubble sprays or tubes with bubbles may be installed to spray the top and bottom heads of the condenser. The top spray(s) will provide upwards coverage of sufficient volume to flood the tube sheet and provide enough flow to equal 2.0 gpm/ft (25 Lpm/m) of total circumference of the tubes. The bottom spray is positioned to spray the lower tube sheet as the fluid falling down the tubes will not adequately cover the surface of the lower tube sheet.

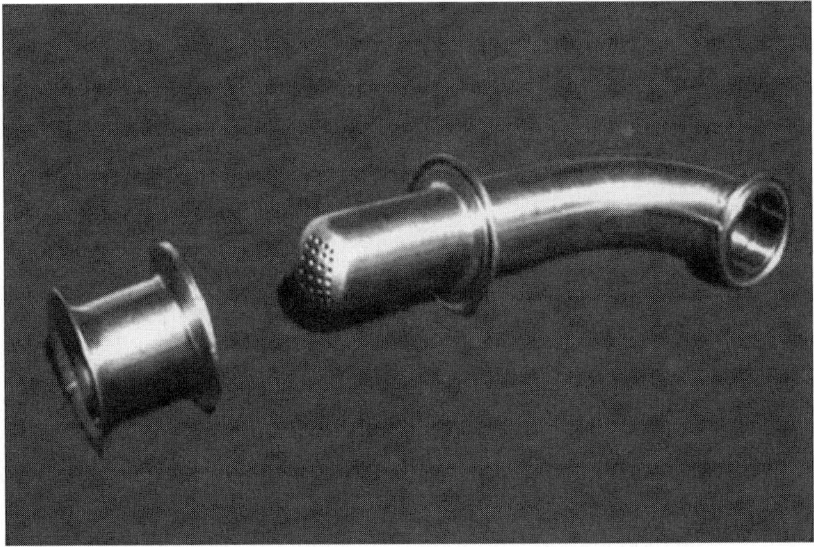

FIGURE 6 This specific "bubble spray" and installation sleeve were designed for installation in the sidewall of a screw conveyor. The bubble projects only slightly through the installation ferrule and, if necessary, can be made flush.

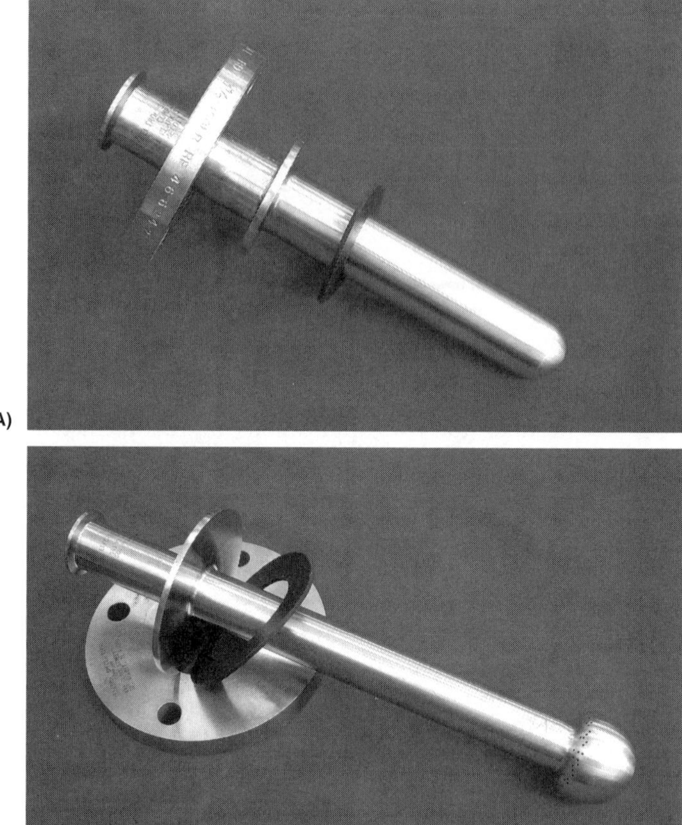

FIGURE 7 These ball and bubble sprays were designed for installation in American Society of Mechanical Engineers flanges typically used in glass-lined vessels and polytetrafluoroethylene line piping and ductwork of active pharmaceutical ingredients processes. (A) is a tube spray; (B) is a welded ball-type spray.

Filter Housing Sprays

Filter housings have been traditionally removed from the process piping and cleaned off line. The off-line method is time-consuming, labor intensive, and contributes to physical damage to the components due to handling. Some efforts to clean by simple flooding have been only marginally successful because of little fluid movement at the inside perimeter surface of larger diameter filter housings. Air pockets may form if flow is not bottom to top and cause no solution contact to occur in some areas.

During 2001, several major projects were headed by design teams that wanted to include filter housings in the normal CIP circuits, often in series with a vessel to be cleaned in combination with a transfer line. This opportunity contributed to the development of a modified CIP process and spray devices which effectively and efficiently clean filter housings of any diameter and height in line as part of the normal CIP circuit. Two different style low profile spray devices were developed for permanent installation in the housing. To spray clean a filter housing requires a means of keeping the housing essentially empty, which is accomplished by injecting air into the CIP supply line at the beginning of each phase of the CIP program, and during periods of recirculation. This air injection approach results in the development of a pocket of air in the housing, and allows the sprays to function as intended. Experimentation has demonstrated the ability to achieve flow rates of 50 gpm (190 Lpm) through a filter housing, sufficient to provide a velocity of 5 fpm (1.5 m/sec) through 2-in. piping in the circuit, and upwards of 80 gpm (300 Lpm), when required, to simultaneously clean a vessel downstream of the filter housing, while accumulating only 1 to 3 gal (3.8–11 L) in the housing.

A spray-type CIP device as shown in Figure 8 is required when the filter housing has nozzle connections on the dome in addition to the spray nozzle connection. This spray provides the necessary upward vector to assure coverage in the additional dome nozzles. A larger hole in the bottom of the machined element is sized to pass solution not required to clean the housing, and maintain operation at a low-pressure drop.

Figure 9 illustrates a disk distributor type CIP device applicable to any filter housing free of nozzles in the top dome. The width of the slot between the disc and the dome is based on flow rate and desired pressure drop through the housing. Figure 10 is a photo of an experimental setup in which a test dome fitted with a disk distributor sized for 50 gpm and 5 psi was rotated to discharge horizontally. The full film of water cascading off the edge of the dome projected several feet as a hollow tube of water.

The flow rates and operating pressure drop for these sprays are determined by other elements of the CIP circuit when the filter housing is being cleaned with piping and a downstream vessel or other equipment. Deliberate incorporation of air in the CIP stream requires special consideration for handling the CIP return flow.

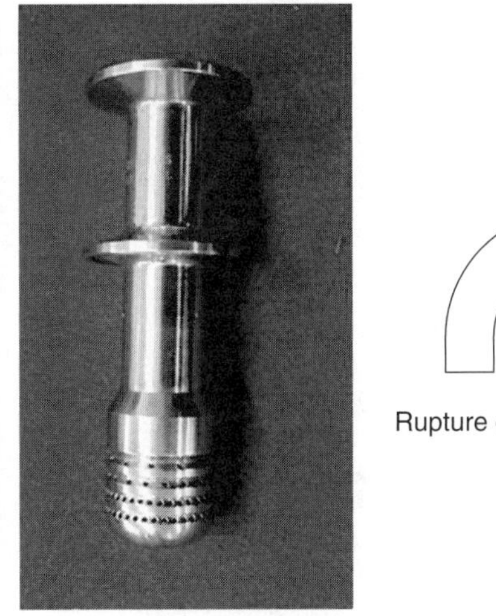

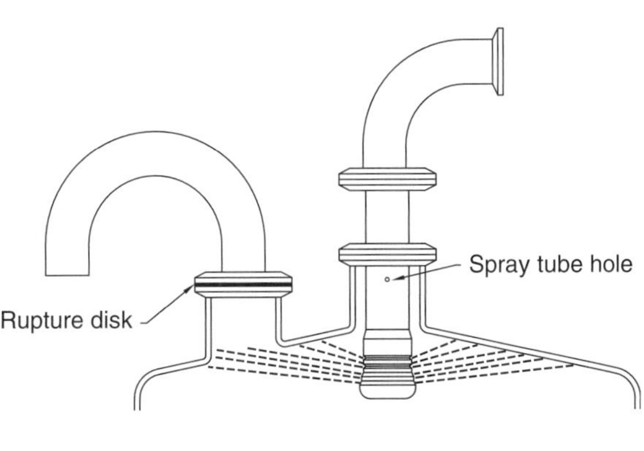

FIGURE 8 If a filter housing is fitted with nozzles in the top dome area, a spray-type device of short length and capable of being drilled for a wide variety of flow and pressure requirements is used to direct streams to the multiple nozzles, as when cleaning larger vessels.

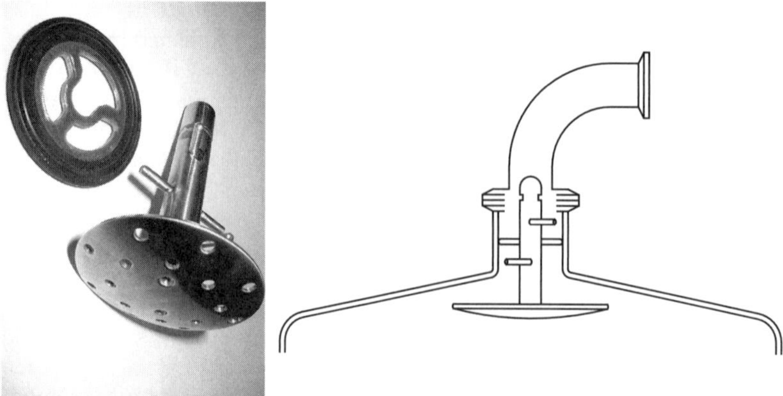

FIGURE 9 Filter housings absent nozzles in the dome may be cleaned with a simple disc distributor. This device may be left in place for all use of the filter and will distribute flush, wash, and rinse solutions to the housing dome and sidewall at any range of flow and pressure required to meet other CIP design criteria.

A vessel downstream of the filter housing will provide a means of disengaging that air before it reaches a standard centrifugal return pump. An eductor-assisted return has worked successfully in these applications and a liquid ring return pump may also be an alternative choice for this application.

FIGURE 10 This photo made during development testing of the disc distributor CIP device shows a hollow tube of water leaving the test dome when located in a horizontal position.

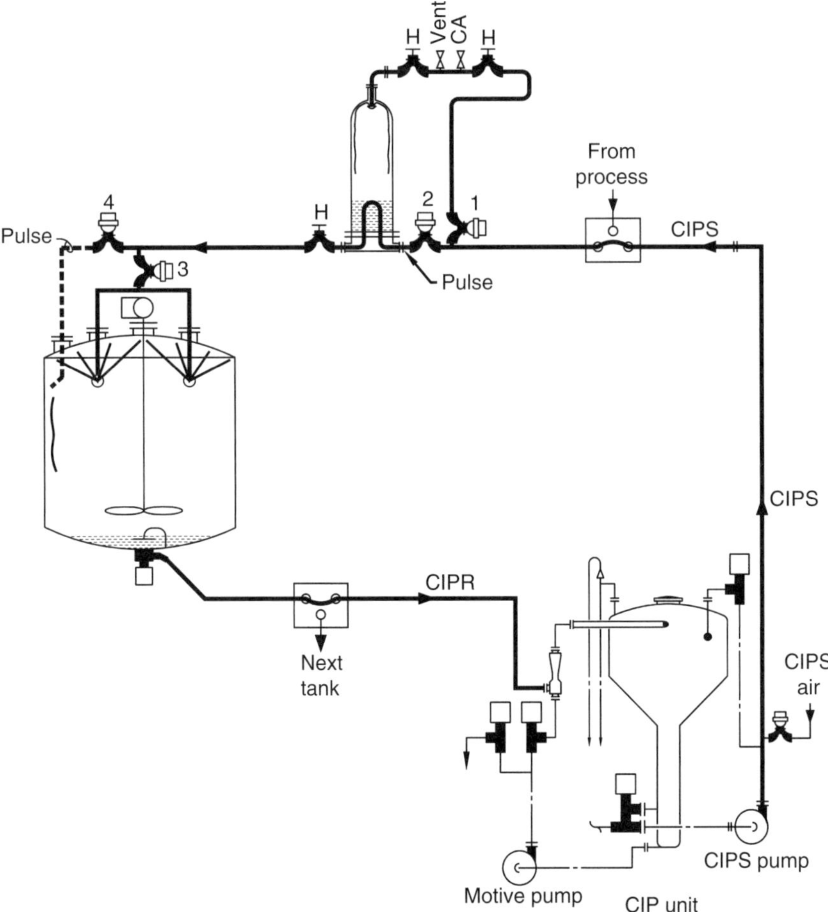

FIGURE 11 This schematic illustrates how a filter housing may be incorporated in the transfer line used to convey process and CIP fluids to the vessel and cleaned with the vessel and piping in a single circuit. The spray and disc distributor devices shown in Figures 8 and 9 may be used in this manner.

Figure 11 illustrates the application of a filter housing in a transfer line to a vessel used also as the CIP supply for the vessel sprays. Proper engineering design of the process piping at the filter housing inlet can provide for CIP of all of the fittings, valves, and gages typically removed and manually cleaned.

SPRAY COVERAGE VERIFICATION

In the mid-1990s, this industry began implementation of a test to assure spray coverage of all product contact surfaces of any vessel. No universally accepted standard has been established for this test, commonly referred to as the "riboflavin coverage test" and it has been based on end user preferences. The American Society of Mechanical Engineers (ASME) Bio Process Equipment (BPE) committee has set forth a recommended practice in its 2002 edition.

The variables that impact the results include the following:

- Riboflavin concentration
- Rinse times
- Drain times
- Rinse cycle repeats
- Rinse cycle lengths
- Rinse temperature
- Vessel cleanliness
- Method of riboflavin application
- Riboflavin drying
- Interfering white light
- Intensity of the black light
- Interpretation of the results

It is essential to understand that the riboflavin test is a coverage test and in itself only provides some assurance of coverage, and not ability to remove soil, which is a chemical process as discussed earlier in this chapter.

As of this writing, the current most commonly used coverage test is coating the surfaces with riboflavin at 200 ppm and running three ambient rinses of two minutes duration, separated by one minute drains the surfaces are then inspected with a black light to detect residual riboflavin. Occasional false positives (failures) of a riboflavin test occur because of gasket feedback related to general technique as well as poor pretest cleaning of the surfaces. And false negatives (passes) of a riboflavin test may occur as the result of excessive white light, inadequate black light, and interpretation of the results. Though much riboflavin coverage testing is accomplished in the tank vendor's facility, the preferred final test is generally done at the time of commissioning, under supervision of validation personnel, and using the installed CIP skid to supply water to the spray devices.

SUMMARY

CIP spray devices are precision fabricated components that play a critical role in the cleaning process applied to vessels and other equipment of all shapes and sizes. They must be designed and applied as an integral part of the CIP process, to assure proper coverage at a specific flow rate and pressure and be robust enough in design to tolerate some fluctuation of those variables. The design applications presented in this chapter when used in creative combinations can be applied to clean the vessels in any system that is otherwise of CIPable design.

REFERENCES

1. Hyde JM. CIP spray devices. In: Society of Bioprocessing Institute Proceedings, Seattle, WA, 2005.
2. Greene D. Practical CIP system design. Pharm Eng 2003; 23(2):120–30.
3. Walker WH, Lewis WK, McAdams WH, Gilliland ER. Principles of Chemical Engineering. 3rd ed. New York: McGraw Hill, 1937:113.
4. NA-connect is a Registered Trademark of Millipore, 290 Concord Road, Billerica, MA 01821, U.S.A.

10 CIP Distribution Piping Systems

Samuel F. Lebowitz
Electrol Specialties Company (ESC), South Beloit, Illinois, U.S.A.

INTRODUCTION

The clean-in-place (CIP) distribution piping system provides the means of conveying cleaning cycle fluids from the CIP system to the process equipment and piping that it is intended to clean. CIP distribution piping can be as simple as two hoses to complete clean-in-place supply (CIPS) and clean-in-place return (CIPR) connections between a portable CIP system and a portable tank or as complex as miles of fixed piping in combination with transfer panels and automated valves to service 100 CIP circuits or more. A well-planned CIP distribution system is a key factor that affects the efficiency, economics, and ultimate success of CIP in a biopharmaceutical facility.

The CIP distribution system must provide for safe transport of CIP fluids. Personnel protection is an ever-present consideration when designing the means by which both hot and chemical-containing fluids are transported within a production facility. Piping design integrity is also crucial to eliminate risk of cross-contamination that might occur as the result of unintended paths from one piece of process equipment to another through CIP distribution piping.

CIP DISTRIBUTION PIPING FOR PORTABLE CIP SYSTEMS

A special simple case of CIP distribution piping is that of the portable CIP system. The portable CIP system is rolled in close proximity to the cleaning load, which may be a vessel or other process equipment such as a blender, centrifuge, or ultrafiltration (UF) system.

The distribution piping for the portable CIP skid is usually flexible hose, a length of hose for CIPS and another for CIPR. It is important that these hoses be selected properly and combine the requisite strength with adequate flexibility. While hoses with braided reinforcement are customary, it is not necessary (or desirable) to use stainless reinforced hoses. Stainless reinforced hoses are often used for safety during sterilize-in-place. However, for CIP, the inflexibility makes it difficult to route the hose and maintain proper pitch.

If given the choice of locating the CIP system between a longer more circuitous supply path or return path, then choose to minimize the return path length every time. The supply path hose is being fed under pressure by the CIPS pump. However, the return path is a critical zone where every inch of net positive suction head (NPSH) must be preserved and proper downhill slope of the CIPR line from the vessel outlet to the CIP skid or return pump is important to prevent air binding of supply or return pumps.

FIXED CIP SYSTEM LOCATION

The first step in designing CIP distribution piping for a facility utilizing a fixed-location central CIP concept is to determine the best location for the CIP system.

There are two guiding concepts to this decision: central location and lowest possible elevation.

Geometry dictates that locating the CIP as central as possible to the cleaning load will minimize the amount of CIPS and CIPR piping. With each foot of piping comes added installation cost and added water in every phase of every CIP circuit forever after. Also, a CIP system located near one end of a large process area will have greatly varied head loss conditions in both the CIPS and CIPR piping. This head loss variance may have significant negative consequences if multipath flow is required as part of an integrated CIP system.

Thus, while it may be convenient to locate the CIP system remotely in a utility area distant from the process, such a location is highly undesirable. Some facilities "affectionately" use the name "skid row" for the gray space utility area set aside immediately adjacent to process area where the CIP systems are lined up next to their respective process suites.

The best location for a CIP system is at the lowest possible level of the facility. The ideal CIP system location would be in the basement, below the lowest level of process equipment being cleaned. A basement location allows full benefit of gravity, eliminating or greatly reducing the need for CIPR pumps and enhancing their performance.

Second best is to place the CIP system on the same level as the process equipment being cleaned. In doing so, the length of CIPS and CIPR piping as well as the hydraulic discharge head encountered by the CIPS and CIPR pumps are minimized.

Lastly, the worst possible location for the CIP system is above the level of the process equipment being cleaned, sometimes called the "penthouse" CIP. In this case, the CIPR pump encounters the maximum static discharge head for every phase of the program. During pumped drain steps, after which there is no longer sufficient liquid in the process tank to fully sustain flow, the CIPR pump will maintain the vertical column of water in the return line, even when air locked. If the pump is stopped, then that column of water will backflow into the process tank being cleaned. For small vessels, such as 25 to 50 L bioreactors, this backflow volume can be significant and, possibly, even overflow the process tank. Thus, a CIP system located above the process requires added hardware and software fixes to overcome unfavorable hydraulic conditions.

CIPR PIPING DESIGN BASED ON RETURN FLOW MOTIVATION

Taking into consideration process design, facility layout, project budget, and other factors, the engineer decides on the optimum configuration for CIPR motivation. Substantial field experience reveals that it is most commonly easier to pump water into process tanks than to get it back to the CIP system through CIPR. Proper system hydraulic balance is greatly dependent on the choice of CIPR motivation in conjunction with the design of the required CIPR piping.

Solution return may be accomplished by gravity, return pump, eductor, eductor-assisted pumped return, or top pressure.

Gravity Return Flow

Gravity return flow is applicable only when the vessel being cleaned is at one or more levels above the CIP system (Fig. 1). Flush, wash, and rinse solutions must be

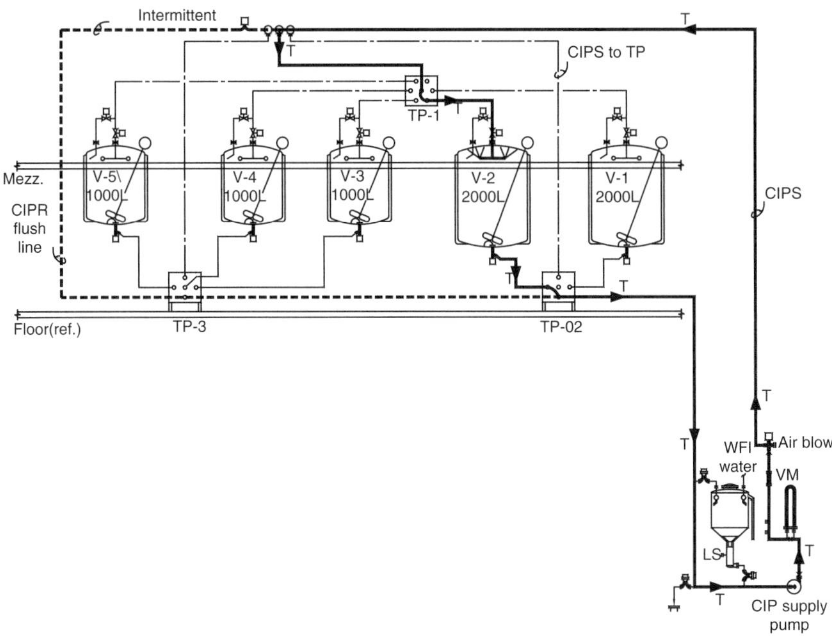

FIGURE 1 Illustration of gravity CIPR flow. *Abbreviations*: CIPR, clean-in-place return; CIPS, clean-in-place supply; LS, level sensor; VM, vortex-type flow meter; WFI, water-for-injection.

continuously removed from the vessel being spray cleaned at a rate equal to the solution supply. Tank outlets and return piping systems must be sized large enough to permit return by gravity alone based on the difference between available static head and friction losses. However, static head of 18 to 20 ft, often readily available, will overcome the friction loss at required design velocity in runs of more than 200 ft including 10 elbows/100 ft and two typical valves. Gravity is not only an excellent choice for return flow motivation, but, when properly engineered, is also more effective than any other method for removing the final traces of liquid from a circuit.

Pumped Return Flow

If the horizontal distance between the process tank and CIP system creates too much friction loss to utilize gravity alone, then a good alternative is to locate a return pump directly below the process tank, on the same level as the CIP system, and then pump return flow back to the CIP system. Figure 2 illustrates the use of a portable return pump, connected to the vessel outlet and discharging through a horizontal run to the CIP skid. The return pump benefits from the significant static head and provides the discharge head required to elevate the flush, wash, and rinse solutions to the return run height and overcome friction loss in the entire system.

Low-speed (1750 rpm) centrifugal return pumps of a standard volute type provide effective and reliable return flow *if* the return header pitches continuously from the tank being cleaned to the pump inlet. CIPR pumps are subjected to the probability of becoming air locked at the end of every rinse burst and/or program phase because of the need to fully evacuate the vessel. The pump will release the

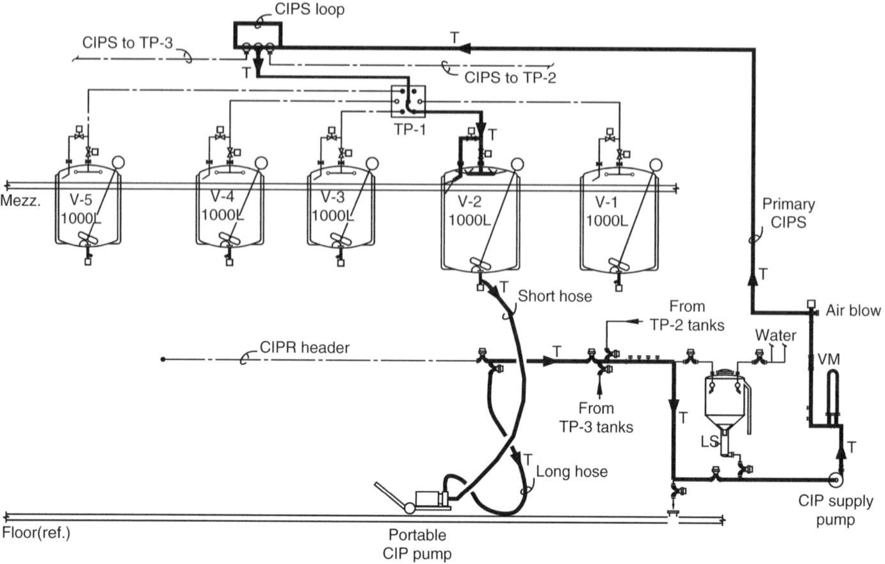

FIGURE 2 Illustration of pumped CIPR with a portable return pump. *Abbreviations*: CIPR, clean-in-place return; CIPS, clean-in-place supply.

entrapped air held at the eye of the impeller by centrifugal force if the line to the pump slopes upwards toward the source vessel in an uninterrupted manner. High-speed (3500 rpm) return pumps are subject to substantial cavitation when air locked and for this reason are generally not used, or needed, to overcome static head and friction loss in CIPR side piping.

Liquid-ring (self-priming) pumps can overcome some installation and operating deficiencies at increased cost and power requirements. However, if fitted with a casing drain and drained after each program, it will be self-priming only when the initial flush water sprayed into the vessel flows by gravity to and into the pump inlet. Also, as fluid temperature increases, liquid-ring self-priming pumps become less effective.

Figure 3 shows a CIPR pump mounted on a stainless steel dolly for portability. A flexible hose would connect the suction of the portable CIPR pump to the outlet discharge line from the process tank being cleaned. Another flexible hose will connect the portable CIPR pump discharge to the fixed-piping CIPR line. Figure 4 shows a fixed base-mounted CIPR pump, which is located in a utility area (gray space) adjacent to the process that it serves.

Eductor Return Flow

The high-speed motive pump and eductor incorporated in eductor return recirculating units with air separation tanks and used as the sole motive force will create 15 to 18 in. of vacuum. Note that eductor performance is limited by water temperature relative to the vacuum capacity being generated, decreasing to $\approx$12 in. of vacuum at 80°C. An eductor will pump both air and water. Whereas an eductor-based system appears less susceptible to air leaks in the return line, in practice this is not

FIGURE 3 Portable clean-in-place return pump mounted on a stainless steel dolly.

true and comparable attention must be given to proper gasketing of all openings except the one in use, to achieve the desired benefits of drawing air only through the vessel outlet, and hence accomplishing CIP with an essentially empty vessel. Whereas eductor return flow can be used as a sole motivating force for short CIPR runs involving minimal static head, an eductor-only return flow is most commonly used in conjunction with gravity return. As an adjunct to gravity return, the eductor will increase the flow rate achieved by gravity alone.

Eductor-Assisted Pumped Return Flow

Eductor-assisted return used in conjunction with either low- or high-speed return pumps is extremely reliable and effective. High-speed return pumps, without self-priming casings or air relief valves, can be used in combination with small-diameter

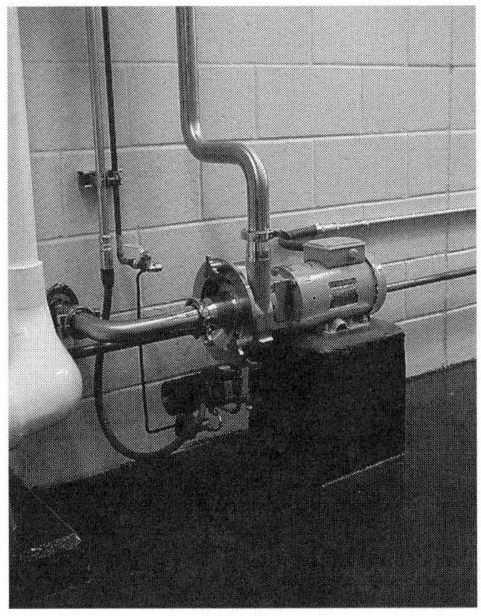

FIGURE 4 Fixed base-mounted clean-in-place return pump.

return line piping, substantially reducing the quantity of fluid required to fill the return system. The eductor continuously primes the CIPR pump, moving a roughly 50:50 air–fluid mixture back to the air separation/recirculation tank of the CIP system, where the air disengages from fluid. The choice between a low-speed and a high-speed return pump is generally made on need and fluid temperature. If the discharge head of a low-speed pump is adequate, then a low-speed pump should be used. If the added discharge head of a high-speed pump is advantageous *and* fluid temperature in the cycle will not exceed 60°C, then a high-speed pump is a good option. Beyond 60°C, experience has been that flashing can occur in the high-speed pump, and thus the low-speed pump is preferred in this circumstance.

Figure 5 illustrates a typical biotech facility application where media vessels (vessels MP-1 to MP-5) located on an upper level combine gravity and top pressure to feed to bioreactors (vessels BR-1 to BR-3) on ground level. In the lower right-hand corner of Figure 5 is a CIP system with an eductor, which can receive CIPR flow by gravity and RP-1 combined from the upper media vessels, and eductor-assisted pumped return from the bioreactors on the ground level. In a process area layout of this type, the eductor-assisted CIP system might be installed approximately 70 to 100 ft horizontally and 30 ft vertically from the far end of the CIP distribution piping, but require only 30 to 40 gal for recirculation through any circuit. The details of the installation and operation will be further described in a subsequent section.

Top Pressure Return Flow

The use of a slight top pressure added to the head space of the target process vessel being CIP cleaned can be used in combination with any of the above return flow methods when more conventional means need to be supplemented due to poor hydraulics. Top pressure is often a last-resort retrofit when other attempts to balance return flow fail. The top pressure should be kept to a minimum (not more than 10 psi) to avoid a pressure surging condition. A means to maintain a "seal volume" in the process tank must be incorporated. If the outlet of the process tank blows clear, the entire vessel will need to repressurize. For this reason, an analog control hold-back valve in the CIPR line can be added to always maintain a low-level volume in the process tank, except on program steps where complete drainage is required.

CIPS AND CIPR FLOW DISTRIBUTION CONTROL

As discussed previously for CIP distribution with portable CIP systems, there are some situations where flexible hose is a suitable means of branching flow, such as changeover of the outlet discharge configuration of a process tank from process to CIPR mode. This may be acceptable in a small pilot plant setting, but for large-scale good manufacturing practice facilities flexible hose allows no chance of automation to ensure that a correct flow path has been established. Also, hoses create a cleaning challenge due to difficulty in keeping track of their use and cleaning, as well as being a detriment to maintaining proper line pitch and drainability.

The next simplest approach to control connections is the make–break elbow. In this case, three adjacent lines come in proximity with a planned geometry so that a "key piece" or "swing elbow" can connect between different pairs. While make–break elbows do not allow automated tracking of process connections, they are a great improvement over hoses with regard to maintaining pitch and drainability.

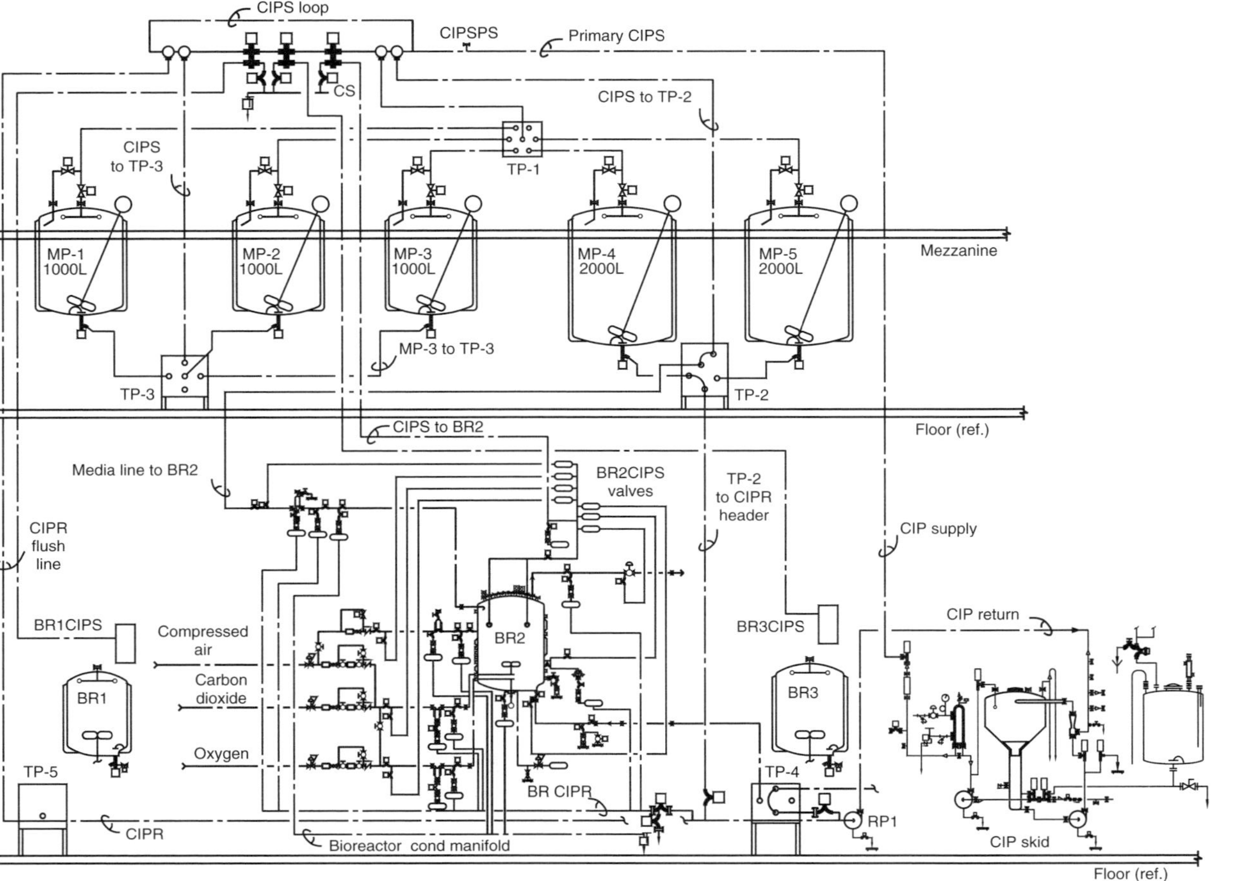

FIGURE 5 Illustration of integrated product and clean-in-place piping in a typical biotech facility. *Abbreviations*: CIP, clean-in-place; CIPR, clean-in-place return; CIPS, clean-in-place supply.

U-bend transfer panels bring many make–break connections to a central piping termination point, providing a maximum flexibility for the production function, yet assuring controlled sanitation through CIP cleaning and further guarantee the integrity of all individual product and cleaning and/or sterilizing flow path. Figures 1 and 5 illustrate the transfer panel concept. The use of manually positioned transfer panel U-bends for establishing processing, CIP, and steam-in-place connections in a highly automated system requires welding permanent magnets that are encased in stainless steel rods to the center of the U-bend connection. Proximity switches located behind the skin of the transfer panel detect the presence or absence of a U-bend magnet between any selected pair of ports.

Whereas manual "make–break" or transfer panel connections may be preferred or required between solution-containing piping and vessels and product-containing piping and vessels, valves are more typically used to control the flow within large-scale processing systems. Automatically controlled valves afford software-based control of both the process and cleaning program to assure proper cleaning of all vessels, equipment, and interconnecting piping. Common valves used in CIPS and CIPR distribution systems are diaphragm valves (preferred where positive barrier isolation is required), sanitary compression valves (as commonly found in the food and dairy industries), and double-seat mix-proof valves (used in CIPS and CIPR distribution to provide compact, conveniently installed double block and bleed assurance).

Elimination of CIP Distribution Dead Legs

A critical factor in any CIP distribution piping system is the elimination of dead-end flow branches. While this discussion focuses on CIP distribution, the same consideration is also due to many product piping applications, for both small-scale areas, such as in transfer panels, and facility-wide distribution piping.

There are three methods of preventing dead-end flow branches in piping systems. The first is by use of blocking valves at each flow branch. However, in piping systems of any substantial magnitude, it is impractical to include blocking valves at each point as needed along a complex header. It is especially difficult to design a "drainable" piping system using diaphragm valves, since in a horizontal run they must be inclined at an angle to meet this requirement and the actuators occupy considerable space and require support. Flow control generally requires the application of two valves, one in the run and other on the branch tee. Also, with each added valve in the CIPS or CIPR distribution system comes extra hydraulic static head loss, hardware/software controls, and maintenance requirements.

The second method, illustrated in Figure 6, permits the elimination of dead-end flow branches in CIPS and CIPR headers free of the additional blocking valves. When CIPS is brought through the header to a distribution valve and then to a two-port transfer panel (for isolation) before connection to the spray device, it fills the header downstream of the branch, and this is normally a dead leg. Now consider what would happen in the CIPR header. As flow discharges from the selected vessel through the transfer panel to the CIPR header toward CIPR pump RP-1, flow will back-up into the CIPR header dead-end toward MT-3. To eliminate this problem, a CIPR flush valve is added from the far end of the CIPS header beyond MT-3 to the far end of the CIPR header below MT-3. By pulsing the CIPR flush valve for brief intervals during the CIP cycle, the CIPS and CIPR header can be subjected to brief periods of flow, on a continuous but intermittent basis throughout the full program.

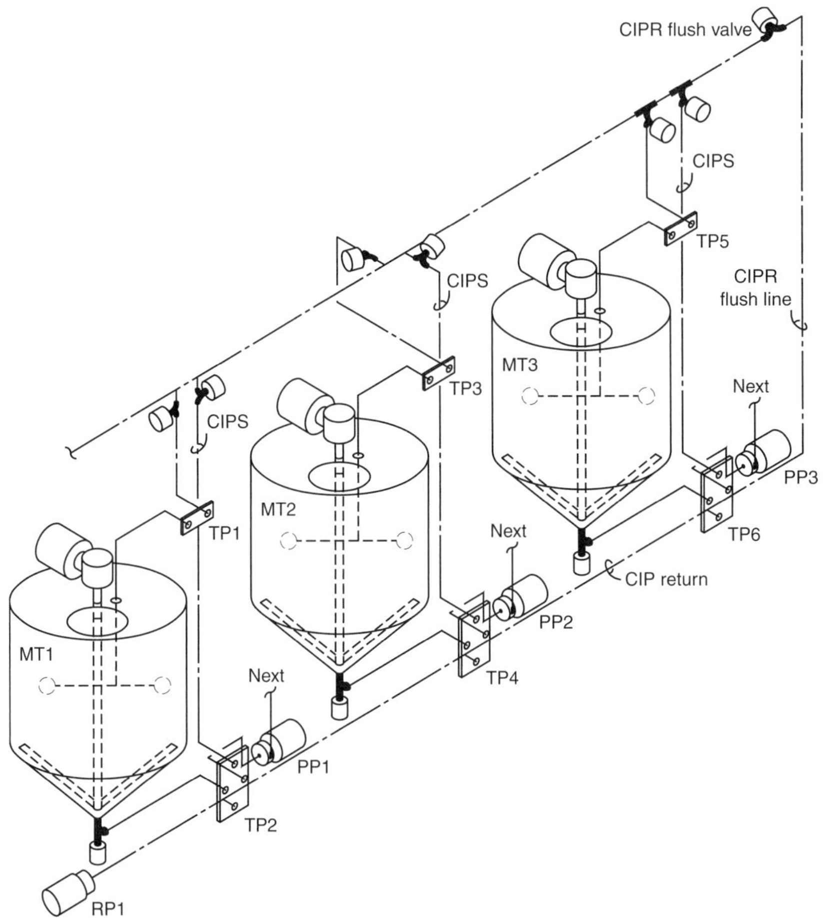

FIGURE 6 Illustration of the clean-in-place return (CIPR) flush valve concept as a means of eliminating clean-in-place supply (CIPS) and CIPR dead-end flow branches. *Abbreviations*: PP, product pump; RP, return pump.

This assures the movement of all flush, wash, and rinse solutions through the entire supply return piping system, eliminating all dead-end piping concerns. The CIPR flush valve and connection can be small, for instance 1 in. When the flush valve opens, the spray back pressure will cause a rapid flow of water through the piping, well above minimum required line velocity, and generally a flow of 2 to 5 ft/sec is sufficient.

The last method of preventing dead-end branches is a CIPS loop, such as shown at the top of Figure 5. The loop ensures live flow in both directions around the header, eliminating the potential for a dead-end.

Types of Distribution Valves and Positions

Figure 6 illustrates three different diaphragm valve orientations commonly found in product, CIPS, and CIPR piping configurations. The valves located above TP-1 are oriented vertically. This is the least preferred orientation since an uncleanable vertical dead-leg "trap" is created in the distance between flow traveling through the main header and the diaphragm of the valve.

The horizontally oriented valves located above TP-3 are an acceptable arrangement as long as the distance between the center of the main flow path and diaphragm valve weir is less than two pipe diameters.

The preferred orientation is the zero-static type diaphragm valves located above TP-5. In this case, the zero-static valves ensure that there is essentially no dead leg between the main flow path and diaphragm valve weir.

CIPS and CIPR Loops

The isometric drawing of Figure 7 illustrates a single CIP system that services various process areas in a biopharmaceutical facility. A composite of various methods of installing CIPS and CIPR is shown in this example. All concepts assure complete freedom of flow dead legs and provide essentially uniform operating volumes and times for all tank CIP programs. One of the two primary CIPS control valves located just beyond the CIP system branches toward the CIPS header supplying MT-1 through MT-3, the remainder of the area being in accordance with Figure 6 as described above. CIPR pump RP-1 discharges to a CIPR distribution valve before the final line to the CIP skid.

Loop-type piping is shown in Figure 7 for the CIPS and CIPR connections to tanks T-4 to T-9.

Permanent product pump PP/RP, also used as a CIPR pump (which might include variable frequency drive (VFD) variable speed capacity control), is located at the discharge of T-5. A portable CIPR pump (RP-2) is used for T-4 and T-6 (shown connected) and a permanent CIPR pump (RP-3) on low-level transfer panel TP-10 is used for T-7 to T-9.

An alternative "loop-type" CIPS/R concept is illustrated as the means of providing CIPS/R capability to tanks T-4 to T-9 inclusive in several different manners. The objective of this revised approach is twofold and includes totally eliminating the possibility of any "dead ends" in that portion of the system in use and providing a uniform supply and return path for each of the tanks in the several groups at various locations within the facility.

Four different methods of providing pumped return are illustrated in Figure 7. These include the following.

1. A permanently installed return pump RP-1 supplied by the header at vessels MT-1 through MT-3, as described above, engineered and specified to serve the CIP solution return need only.
2. A centrifugal pump PP/RP installed in the discharge line of vessel T-5 and equipped with a VFD for capacity control would be used for product transfers and CIPR.
3. A portable CIPR pump RP-2 would be used at tanks T-4 and T-6. This pump would connect via a hose directly to the tank outlet valve of T-4 or to the port on TP-9 for T-6, and via another hose to a CIPR port on TP-7 or TP-9. The suction-side hoses must be short and installed to maintain continuous pitch upwards from the return pump inlet to the tank outlet valve for efficient and reliable performance of the CIP system.
4. A permanently installed return pump RP-3 is shown as the means of returning solution from vessels T-7 to T-9, which connect to ports on TP-10. For CIPR, the

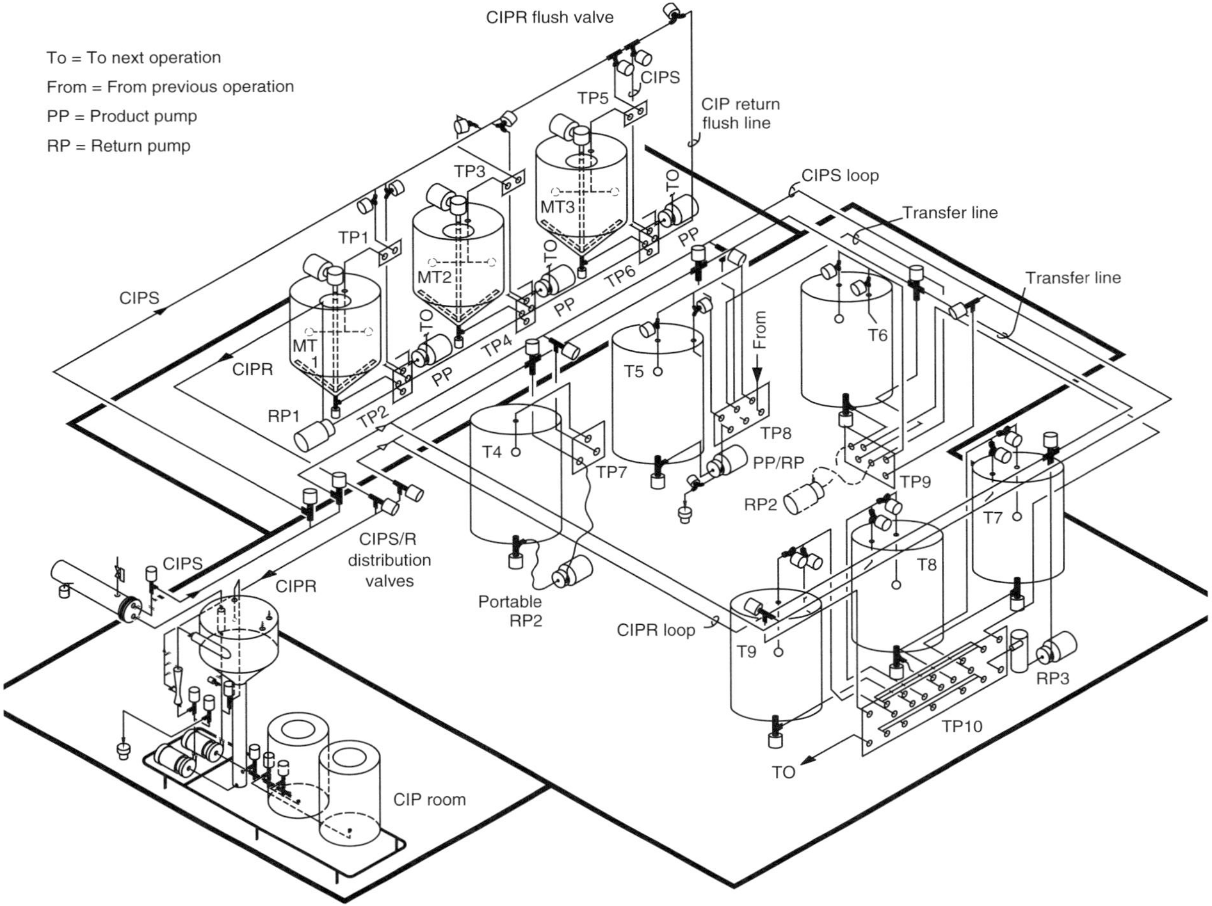

To = To next operation
From = From previous operation
PP = Product pump
RP = Return pump

FIGURE 7 Illustration of a composite of various methods of installing CIPS and CIPR. *Abbreviations*: CIPR, clean-in-place return; CIPS, clean-in-place supply; FROM, from previous operation; PP, product pump; RP, return pump; TO, to next operation.

suction-side port on TP-10 would connect to the lower header used as the CIPR header and the tank outlet line would also connect to this same header, thereby providing continuous pitch from the tank outlet to the pump inlet. The return collection tank on the inlet path to RP-3 would permit the cleaning of one, two, or all three vessels (T-7 to T-9) simultaneously, in a common circuit, by sequencing flow through the vessels by control of the spray supply valves.

The CIPS/R distribution valves control flow to CIPS transfer panels and from the various return pump locations via a CIPR loop.

Tanks T-5 to T-9 all use a common line to provide product or CIPS flow from the transfer panels to the tank, with diaphragm valves controlling flow to the product fill nozzle for processing and to the fill nozzle and spray for CIP operations.

Line Sizing for CIPS and CIPR Loops

The complete system may include most (but not all) of the following components, sized to handle a nominal flow of 80 gpm (≈300 Lpm) for this example.

1. Primary CIPS and CIPR lines will be installed from a point above the recirculating unit to supply/return distribution valves, which will preferably be located near the center of the various CIP loads. These lines will generally be 2 in. (50 mm) in size for 80 gpm (300 Lpm). To minimize total fall on individual runs, the distribution valves will be at the high point of the supply/return piping and the primary lines will drain from that point back to the CIP unit.
2. A secondary CIPS line, also 2 in. (50 mm) in size, may then be installed from the primary supply valve to the secondary supply valves via the CIPS header and should pitch to the CIPR flush valve. The secondary CIPS lines should all pitch continuously toward the transfer panels to drain at that point when the U-bend is removed.
3. Most spray devices have 1.5 in. (40 mm) connections and since the spray supply lines are relatively short, the head loss through a 1.5 in. (40 mm) line will be quite manageable. This requires only one 2 ×1.5 in. (50×40 mm) reducer at the CIP transfer panel connection. The spray supply line will preferably pitch from a point directly above the spray(s) to the port on the transfer panel.
4. The alternative means of supplying two or more different destinations (vessels T-4 to T-9) from a single primary supply valve is to install a 1.5 in. (40 mm) loop, which contains approximately the same volume and produces approximately the same head loss as a single 2 in. (50 mm) line from the common point to any other point on the loop. This approach totally eliminates "dead ends," for whereas the flow through the loop will not divide 50/50, it will seldom be less than 30/70 with good velocities in all portions of the supply loop. This 1.5 in. (40 mm) loop must be installed to drain backwards toward the primary supply valve and then back to the CIP system, or alternatively to drain to a selected secondary distribution valve. The secondary distribution valves can be 1.5 in. (40 mm) in size and in turn can supply the transfer panels and spray devices also through 1.5 in. (40 mm) tubing.
5. The CIPR header which connects a multiplicity of tanks to CIPR must be 2 in. (50 mm) in diameter and must pitch continuously from the most distant tank to the pump inlet. The header flush tie-line may be 1 in. (25 mm) in diameter for the design spray supply pressure will be available to drive solution through this tie-line for the required brief intervals to effectively flush the larger 2 in. (50 mm) return header.

6. The CIPR loop may also be 1.5 in. (40 mm) in diameter. This loop would receive the discharge from PP/RP, portable pump RP-2, or permanently installed pump RP-3 via lines which rise to a high point above the pumps, pitch to the loop, and then go through the loop to the primary return distribution valves and back to the unit. If necessary, the loop may drain in two directions both forward to the CIP system and backward to the return pumps, which in that instance would need to be fitted with casing drain valves.

Obviously, the various components of a complete CIPS/R piping system described above may be used in many different combinations to achieve the required operating capability. The design objectives should include (*i*) minimum line diameter to minimize line volume commensurate with (*ii*) an acceptable supply side head loss with the spray device representing 30% to 50% of the CIPS pump discharge head and (*iii*) a return side head loss which will permit application of 1750 rpm return pumps.

INTEGRATION OF CIPS AND CIPR PIPING AND PROCESS PIPING IN A CIP CIRCUIT

Every effort should be made during the design process to minimize the amount of CIPS and CIPR piping required, by maximizing the use of process piping and equipment skid piping for CIP purposes. Figure 8, derived from Figure 5 previously discussed, will be used to illustrate this concept by describing the piping used to CIP clean BR-2, by following the flow of CIP flush, wash, and rinse solutions through the circuit.

1. The circuit will begin at the flow control valve on the CIP skid and continue through the primary CIPS to the CIPS loop at the top of the process, delineated as a solid line.
2. The mix-proof CIPS valve will be open to provide continuous flow, delineated by a solid line, to the BR-2 sprays via the BR-2CIPS valve manifold on the BR-2 bioreactor skid.
3. A CIPS valve on the loop will also supply TP-2 and the U-bend will be positioned to provide flow to the media line to BR-2. Intermittent flow to the vessel will be controlled by the process valves in the media leg piping and is delineated as a dashed line.
4. Nine other bioreactor legs will be subject to intermittent flow controlled by sequencing of the BR-2CIPS valves and these are delineated by two short dashes in each line. These flows may vary from a one second "pulse" to a 5- to 10-second interval, depending on leg length and diameter. The flow through the legs will be either (*i*) to the vessel, (*ii*) the BRCIPR header which is valved to the main CIPR header, or (*iii*) the bioreactor condensate manifold, a direct discharge to drain provided only at the end of each phase to clean the seats of valves which are not exposed to product.
5. The continuous flow from the sprays and the intermittent flow through legs to the vessel will drain continuously via the outlet valve and harvest line to TP-4, where the U-bend will connect to the CIPR header port.
6. The CIPR flush valve on the CIPS loop also will open intermittently to flush the full length of the CIPR header to RP-1. A dashed line depicts this flow.

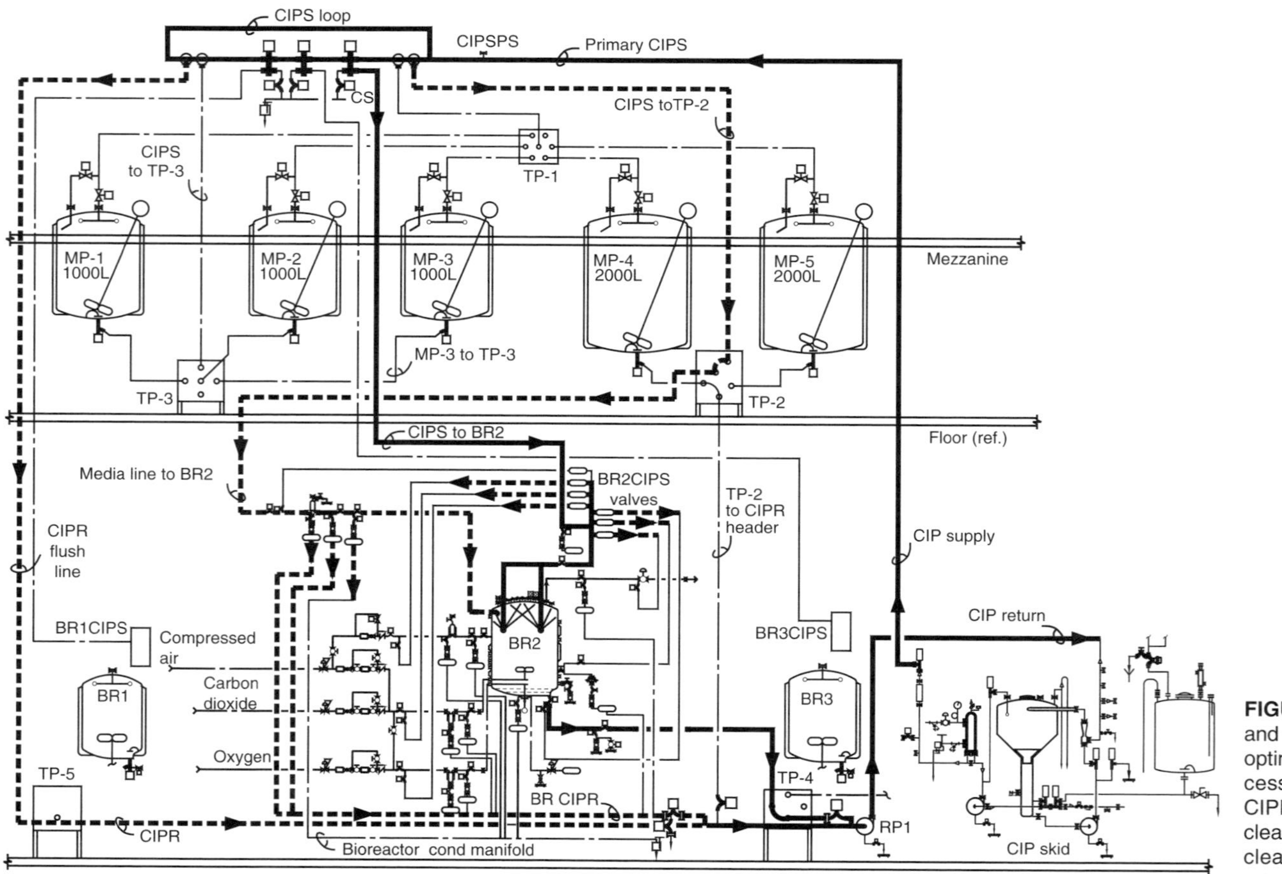

FIGURE 8 Heavy-line CIPS and CIPR illustration showing optimized integration of process piping with CIPS and CIPR. *Abbreviations*: CIPR, clean-in-place return; CIPS, clean-in-place supply; RP, return pump.

FIGURE 9 Photo of actual CIP supply loop and distribution valves for the process of Figures 5 and 8.

7. The combined return flows through or upstream of TP-4 will continue to RP-1 and then to the CIPR manifold on the CIP skid, and is delineated as a solid line.

Figure 9 shows the actual CIPS loop and distribution valves for the process design from which Figures 5 and 8 were extracted. Ten bioreactors, five media preparation vessels, and a series of small line circuits were cleaned by a single CIPS loop and two return headers with pumps. A CIPS pressure sensor installed on a tee at the far end of the loop could detect the change in pressure at the high point in this circuit that occurred when any of the many CIPS valves opened to provide a path parallel to the flow through the sprays. The insulated piping was to provide personnel protection as automated clean steam valves supplied steam through the bottom ports of the mix-proof CIPS valves to sterilize the bioreactors from this same point. Media preparation vessels could also be steamed via clean steam ports on the transfer panels (not shown).

Figure 10 shows the small transfer panel that supported each pair of bioreactors and Figure 11 is a view of the CIPR valve and ported diaphragm condensate drain valve close coupled to the CIPR header as shown to the right of TP-4 in Figure 8.

This example describes design and operating concepts applicable to the filling, emptying, cleaning, and sterilizing (if needed) of almost any kind of a vessel in any biopharmaceutical facility.

CIPS AND CIPR SYSTEM INSTALLATION CONSIDERATIONS

The CIP distribution piping system must be laid out with a design permitting pitch to a drain point. Frequently, the highest point of the piping system will be a primary

FIGURE 10 Photo of small transfer panel (such as TP-4) that supported pairs of bioreactors for the process of Figures 5 and 8.

FIGURE 11 Photo of CIP return valve and ported diaphragm condensate valve shown to the right of TP-4 in Figure 8.

CIPS valve group. On the inlet side of this valve group, lines slope back to the CIP system and on the discharge side, lines slope toward the process area destination. Likewise, for CIPR piping, there is a high point from where lines slope either back to the CIPR pump at the tank being cleaned or forward to CIPR at the central CIP system.

As with all sanitary piping systems, proper installation methods must be used to ensure that piping is not only "hung" but also supported, so as to retain the original intended slope. Permanently installed spray devices should be installed using a spray supply connecting manifold that allows for easy removal. The CIPS piping and the spray manifold should be supported so that stress is not present at clamps, which could cause poor alignment and leaking.

Areas requiring routine access by operators or maintenance personnel must be preplanned. There is need to access sample points in the CIPS and CIPR distribution piping. This includes both sample valves used to obtain solution during various points of the CIP cycle as well as piping access locations to be used for swabbing during cleaning validation. Also, permanently installed valves must be accessible for routine maintenance.

CIPS AND CIPR HYDRAULIC CONSIDERATIONS

Before starting to design a CIPS and CIPR piping system, an understanding of the hydraulic factors is important. Figure 12 illustrates the concepts discussed below.

CIPS hydraulic factors for tank circuits are as follows:

- CIPS static head loss is due to the difference in height between the CIP recirculation tank level and the spray device of the process tank being cleaned (Fig. 12, elevation REF A to REF B).
- CIPR static head loss is due to the difference in height between the process tank level and the CIP recirculation tank return point connection at a spray device, nozzle, etc. (Fig. 12, elevation REF C to REF D).
- CIPS and return friction head loss is due to the energy necessary to move flow through the piping system at a given flow rate. The manufacturers of sanitary components often publish data tables for friction head loss through valve, fittings, and sanitary tubing. Note that sanitary "OD tubing" has different internal diameter than industrial "schedule piping" and, therefore, industrial friction loss data tables are not correct for sanitary OD tubing design.
- The CIP spray device has a required operating pressure, both on the CIPS side in the case of the process tank being cleaned and on the CIPR side if a spray device is used for recycle of flow back to the CIP system.

Thus, the total CIPS or CIPR head requirement is the sum of static head, friction head, and spray device operating pressure.

For line circuits, (*i*) there is no spray device involved for a target vessel being cleaned, (*ii*) static head loss is minimal because flow both begins and ends at the CIP system, and (*iii*) a CIPR pump is not required. In this case, the hydraulic calculation is reduced to the sum of friction head loss for CIPS and CIPR piping, possibly plus the operating pressure of a spray device if one is used for recycle of flow back to the CIP system.

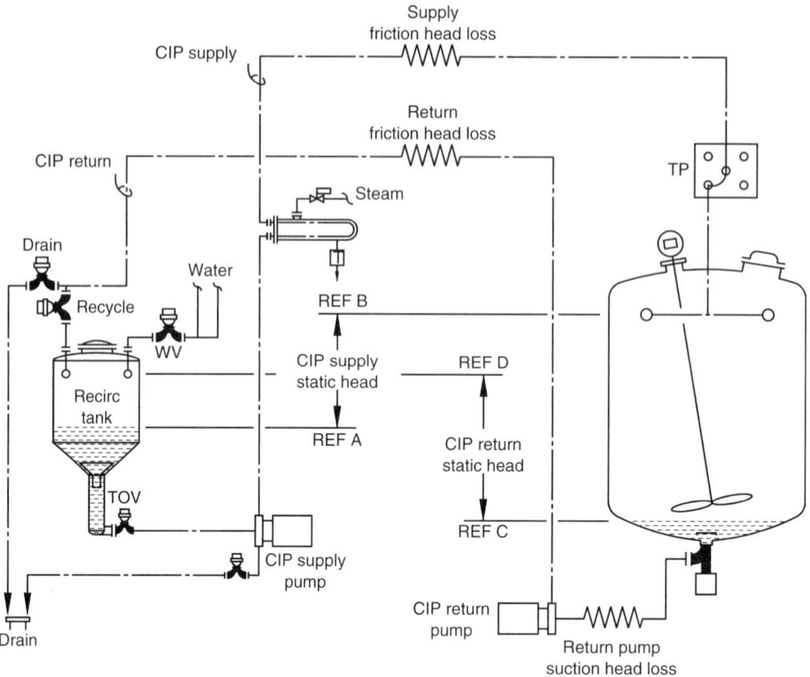

FIGURE 12 Reference elevation and head loss diagram for definition of CIPS and CIPR hydraulic factors. *Abbreviations*: CIPR, clean-in-place return; CIPS, clean-in-place supply; TOV, tank outlets valve.

One often overlooked critical hydraulic factor is the CIPR pump suction head. This value is defined as the elevation difference between the liquid level of the target process vessel being cleaned (Fig. 12, elevation REF C) and the suction inlet of the CIPR pump.

NPSH is defined as the difference between the pump suction head, including available atmospheric pressure, and the liquid vapor head. The required NPSH (NPSHR) is the NPSH that must be exceeded to avoid vaporization and cavitation at the pump impeller. By definition, NPSHR is always less than NPSH due to accounting for friction losses and pump hydraulics. NPSHR is usually incorporated in the pump performance curves provided by the pump manufacturer.

As liquid temperature approaches the boiling point, the available NPSH decreases in accord with the temperature-dependent vapor pressure curve of water. This is significant to the biopharmaceutical user, where high-temperature water (such as 80°C water-for-injection) is common. For this reason, to achieve the NPSHR it is important to have maximum possible CIPR pump suction head height—with minimal suction-side friction losses, such as from a long piping run, fittings, or hose.

The CIPR pump presents a unique pumping challenge. In most pumping situations, a vessel begins with a substantial fluid level and then is either pumped out or, perhaps, recirculated. For the CIPR pump, the goal is to pump out and continuously vacate a near-empty tank that is being sprayed by CIP spray devices. Thus, the contribution to NPSH by liquid head in the tank is negligible. At the same time, the opportunity for air binding of the pump is great. For this reason, a vortex breaker is customarily installed in the outlet of the process vessels to be cleaned. In addition, it is important to ensure that there is a continuous downhill pitch from the outlet of the process tank to the suction of the CIPR pump so that air in the return pump suction line can unload in reverse direction.

11 Materials of Construction and Surface Finishes

David M. Greene
Paulus, Sokolowski & Sartor (PS&S), LLC, Warren, New Jersey, U.S.A.

INTRODUCTION

Materials and finishes used in the biopharmaceutical industry must be easy to clean; corrosion resistant; nonreactive with the process, cleaning, or sanitizing elements; and they must not decompose or shed particles that could contaminate the process. Lastly they should have a pleasing appearance and be economic both in terms of materials and fabrication costs.

Although many materials comply with most of these requirements, the one material that consistently satisfies all of these requirements is stainless steel.

STAINLESS STEEL

Adding 12% chromium (along with some nickel and molybdenum) improves the corrosion and stain resistance of steel. This is due to the formation of a thin chromium oxide layer on the surface that prevents oxidation of iron in the base metal. In the absence of the tightly packed chromium oxide passive layer, the iron would oxidize to form loose iron oxide flakes (rust) on the surface.

Stainless steels can be characterized as ferritic, marstenitic, and austenitic.

Ferritic

Plain chromium (15–30% chromium, low carbon) stainless steels are magnetic; they have good resistance to high temperature environments; and they are moderately resistant to corrosive environments. They cannot be hardened by heat treatment although they can be moderately hardened by cold working. Their ductility is good but their poor weldability limits their use to thinner applications.

Marstenitic

Marstenitic stainless (12–20% chromium, controlled carbon, and other additives) steels are magnetic; they can be hardened by heat treatment; and they have adequate corrosion resistance to mildly corrosive environments. They can be hardened by heat treatment that also results in a high tensile strength but ductility and toughness coincidentally diminish.

Austenitic

Austenitic stainless steels contain chromium, nickel, and manganese. Replacing some of the manganese with nickel enhances corrosion resistance; molybdenum increases resistance to localized (crevice and pitting) corrosion while titanium prevents sensitization and intergrannular corrosion.

Type 300 austenitic stainless steels are preferred for process contact in the biopharmaceutical industry because of their corrosion resistance and ease of welding. They exhibit good ductility and are easily welded but they are not hardened by heat treatment. Sometimes called 18-8, they are available in a low carbon (<0.03%) grade where the reduced carbon minimizes chromium carbide precipitation that can lead to stress corrosion.

Today, austenitic stainless steels are often made from recycled sources of lower grade alloys such as ferritic or marstenitic to which nickel, additional chromium, and, in the case of 316, molybdenum is added. Aluminum may also be added to remove oxygen and prevent porosity. However, the aluminum or other impurities may also form inclusions that then become sources of corrosion.

316L SS
Non-corroding
Non-contaminating—very low extractables
Can be polished to very smooth finish
Strong & rigid
Low coefficient of thermal expansion
Economic and readily available
Can withstand heat and chemical sanitation
Long service life

Chemical Composition of 316L Stainless Steel

The bulk phase of the 316L stainless steel has the following composition:

316L SS composition

Element	Concentration	Notes
Iron	67–69	Base metal
Chromium	16–18	Increases oxidation resistance. May be depleted in the heat-affected zone because of chromium carbide formation
Nickel	10–14	Increases resistance to mineral acids
Molybdenum	2–3	Increases resistance to chlorides
Manganese	2.0 (max)	Manganese combines with carbon to form carbides and sulfides that help eliminate chromium carbides
Silicon	1.0 (max)	Silicon is a graphite stabilizer but is neutralized by manganese
Phosphorus	0.04 (max)	Phosphorus and sulfur stimulate carbide formation but can form inclusions that may lead to pitting
Sulfur	0.03 (max)	
Aluminum	Trace	Aluminum will reduce porosity and increase strength but may also form inclusions that can initiate pitting
Carbon	0.03 (max)	Carbon stabilizes the liquid melt and is important in the formation of austenite but can form chromium carbide depleting the chromium in the bulk phase of the alloy

The chemical composition of stainless steel varies from the bulk phase to the metal phase (50–200 Å below the surface) with the chromium, molybdenum, and nickel concentration increasing toward the surface. The final surface, or passive layer (50 Å to the surface), of oxidized compounds is still higher in chromium, carbon, and oxygen, and lower in iron and nickel.

CORROSION

Galvanic

Galvanic corrosion occurs when a less noble metal, such as zinc, is electrically connected to a more noble metal such as iron in a corroding electrolyte. The dissimilar materials form a battery in the presence of an electrolyte causing a current flow that removes the material of higher potential. Because stainless steel is nobler than most engineering materials, galvanic corrosion is uncommon, but it can be eliminated by electrically insulating dissimilar materials.

Pitting

Pitting occurs when the protective oxide breaks down in small areas. One cause is sulfide or aluminum inclusions that prevent the complete formation of a passive layer. Sulfur can be removed with nitric acid during passivation but aluminum added either intentionally to improve the melt or unintentionally during grinding is not removed by passivation and can lead to pitting.

Crevice

Crevice corrosion is a localized form of corrosion that may occur when stagnant residues accumulate producing localized differences in oxygen concentration that can lead to aggressive corrosion especially in chloride environments. It can also result from dissolved sulfur concentration gradients. Crevices can occur under deposits of biomass, biofilm, or inorganic material as well as under components such as washers and bolt heads or in welded joints. Crevice corrosion begins in hidden focus points with no signs on the surrounding exposed surface. It can be prevented by proper design and fabrication and by ensuring good cleaning velocity to prevent or remove deposits. Resistance to crevice corrosion can be improved by higher nickel, molybdenum, tungsten, and nitrogen concentrations.

Intergranular

Localized corrosion may occur at the microscopic grain boundaries particularly in the heat-affected zone (HAZ) where carbon can react with chromium to precipitate complex chromium carbides. If the metal matrix around the carbide precipitate is depleted in Cr content it becomes anodic to the affected stainless steel surface and sensitive to intergrannular attack.

Stress Corrosion Cracking

Stress corrosion cracking (SCC) increases in a corrosive environment when the stainless steel is subject to high stress. It is a particular problem in the presence of chlorides. Points where the stress is amplified become the origin or destination of cracks. To prevent cracking, it's important to have a high degree of surface integrity. For example, corner welds should be ground flush and polished to remove stress points.

As resistance to SCC improves with increased nickel content, 316/316L is more resistant to SCC than 304/304L.

Fatigue

As stainless steels are not as fatigue resistant as other steels, they are vulnerable to cyclic and repetitive stress, and it is important to minimize stress to reduce the possibility of fatigue failure.

Microbial Influenced Corrosion

Microorganisms may adhere to surfaces because of gravity, surface charge, hydrophobicity of the surface, or the presence of nutrients. Although bacteria may be present as individual cells, they generally adhere to surfaces within biofilms. Their presence can lead to contamination, fouling, and cleaning difficulty. Although there is no consensus on the relationship between surface finish and bacterial adhesion, there are regulatory requirements such as US 3-A and ASME BPE that recommend minimum requirements. It is important to understand the nature of the soil to be removed in order to implement an effective cleaning program that includes selection of construction material and finish.

Rouge

As the main constituent of stainless steel is iron, it should not be a surprise that rust can appear on a stainless surface. Because the rust is generally reddish, it is referred to as rouge. Although rouge can be caused by the presence of foreign iron or elevated temperature, it usually results from a breach in the passive film that allows oxygen to contact iron in the base metal. The resulting discoloration is called rouge. If untreated, the rouge will self-catalyze and attack the base metal phase and contaminate the process fluids with corrosion products (iron oxide, aluminum oxide, and silicates). The surface will become roughened and cleaning will be more difficult.

Class I Rouge

Class I rouge originates from an external carbon steel source that produces iron oxides and hydroxides ranging in color from orange/red-orange to magenta. One source may be pump impellers that have had silicon added to improve fluidity during the casting process but may also increase the delta ferrite content. The delta ferrite may not be dissolved by heat treatment, and because it erodes more easily than austenitic stainless, cavitation may remove portions of the passivated surface and expose iron that will soon rust.

Class II Rouge

Class II rouge is caused by a reaction with chlorides or other halides. It can be removed with grinding/polishing or citric acid or mitigated by using alloys with increased molybdenum or nickel. It's less likely to occur at a $pH>7$ or with electropolished surfaces that are more resistant to corrosion because of the higher Cr/Fe ratio and the absence of surface scratches.

If the process contains chlorides, particular care should be taken when cleaning to ensure the removal of chlorides during cleaning, possibly by adding a strong surfactant to the rinse water.

Class III Rouge

Class III rouge is caused by an external iron source such as the erosion of pump impellers and exacerbated by low Cr/Fe ratio, lack of passivation or coarse surface finish.

Class III rouge occurs in the presence of high-pressure steam, purified water, or pure steam at elevated temperature. Deposits are initially blue but turn black as the thickness of the layer increases and oxygen diffusion decreases. Ultimately, it becomes glossy black on non-passivated mechanically polished surfaces and powdery black on electropolished surfaces. It can be removed by oxalic acid.

SURFACE FINISH

Introduction

A surface is a boundary between a workplace and its environment, and it should be selected to perform the desired function in an economic manner. For biopharmaceuticals this application selection should consider:

- Cleanability
- Sanitization
- Sterilization
- Impermeability
- Product Contact
- Aseptic Requirement

Although surface finish may be improved by the addition of material (plating), this method is not generally accepted because of concern that the additional material may become disassociated during use and end up in the product. Instead, surface finish is improved by removing superfluous metals using grinding and polishing.

Historically, finishes have been described by an inexact numbering system representing a processing method rather than a result. In recent years, ASME has developed a standard (ASME BPE) that includes a definition of surface finishes in terms of roughness for tubing, fittings, valves, and vessels as shown in the following summary table

ASME BPE finish designations

	Ra (ave), μin. (μm)			
Component	10 (0.25)	15 (0.375)	20 (0.5)	25 (0.625)
As drawn or mech polished				
Tubing	—	SFT1	SFT2	SFT3
Fittings	—	SFF1	SFF2	SFF3
Valves	—	SFV1	SFV2	SFV3
Vessels	—	SFVV1	SFVV2	SFVV3
Mechanical or electropolished				
Tubing	SFT4	SFT5	SFT6	—
Fittings	SFF4	SFF5	SFF6	—
Valves	SFV4	SFV5	SFV6	—
Vessels	SFVV4	SFVV5	SFVV6	—

Polishing

Why Polish?

A pleasant appearing cleanable stainless steel surface can be obtained from the mill by specifying a 2B mill finish. The standard sheet 2B mill finish is a uniform, clean, durable finish that would meet most pharmaceutical industry requirements for cleanliness if this finish could be maintained during shipping, handling, fabrication, construction, and service. This finish is produced by rolling sheet through highly polished rolls but, unfortunately, this finish is not applicable for plate, pipe, castings, and machined components. As the production processes for plate, bar stock, and castings do not produce a similar finish, these components need to be finished to produce an equivalent surface.

Polishing is specified to facilitate cleaning, inhibit corrosion, and improve appearance. It may or may not improve "cleanability" as mechanical polishing produces an infinite number of microscopic "scratches" that may provide refuge for microscopic particles including bacteria.

On the other hand, electropolishing will remove peaks, may remove corrosion-causing impurities, and will reduce both surface roughness and area making it more difficult for dirt, bacteria, or proteins to "cling" to the surface or hide in the scratches. The reduction in surface area also reduces the number of sites where corrosion can originate.

Polishing improves the appearance of a surface. If there is a need to sanitize exterior surfaces, electropolishing of the exterior may be justified but often exterior polishing is done solely to improve the appearance and is an unnecessary cost particularly when the exterior surface is covered by insulation.

Cost of Polishing

The following table provides a comparison of polishing costs for a range of typical vessel sizes:

	Vessel budgetary pricing (Oct 03) [150 psig, 350°F (10 barg, 180°C)]		
Vessel volume (L)	SFT 3 finish	SFT 5 (add)	Passivation (add)
100	$11,000	$14,000	$2,000
1,000	$17,000	$23,000	$3,000
10,000	$60,000	$79,000	$4,000
100,000	$300,000	$375,000	$23,000

Types of Polished Finishes

Mechanical cleaning using blasting, grinding, and/or wire brushing is the primary cleaning method used by fabricators to clean stainless steel.

Sandblasting may be used to create a uniform finish on stainless steel and remove fabrication soils, smearing, and heat tint caused by welding. The surface will appear dull with a silver gray appearance and have a roughened texture that will soil easily. The force of the blast can cause buckling and warping particularly when finishing lighter gage materials. As shot, grit, or sand might damage the surface, they should not be used for sanitary applications.

Wire brushing with stainless brushes is the next most common method of cleaning. Carbon steel brushes should never be used, as iron particles will become embedded in the surface and cause rouging. Even stainless steel brushes may leave a layer of dissimilar material on the surface that can lead to rouging. Heat tint and other surface contamination can be removed using clean abrasive discs or flapper wheels to lightly grind the surface but these methods may also leave a smeared layer of lower corrosion resistance. This mechanically smeared surface layer is a potential source of both rouging and batch-to-batch contamination caused by components adhering to the smeared surface.

Mechanical Polishing

Aluminum oxide of various grit sizes is used as the polishing medium. Unfortunately, this material and associated impurities permeate the stainless microstructure possibly providing sources for future corrosion initiation.

Mechanically polished surfaces have scratch patterns caused by the action of the grit. For a pleasant appearance, scratches should run in parallel straight lines. This can be done with machine polishing but is hard to do with hand polishing because of the difficulty of maintaining straight lines.

An alternate approach uses an orbital sanding motion to produce random scratches. The resulting surface is dull yet uniform. It is less costly to produce and particularly useful for removing heat tint on a rolled mill finish. The random scratches tend to blend and usually result in an acceptable finish when followed by glass beading.

Grinding and chipping should be limited to removing defects prior to rewelding and to removing weld reinforcement as grinding can cause surface damage that cannot be removed by pickling, blasting, or electropolishing.

The surface layer left by common mechanical cleaning operations is heavily cold-worked and highly stressed. It is full of microfissures and often contains marstenitic that has lower corrosion resistance than austenite.

Vibratory Finishing

Another finishing process places articles to be cleaned or polished into a mechanically vibrated tub with abrasive media. By starting with a more aggressive media and then successively using finer and finer media, it is possible to refine the finish to a high polish.

The process is not labor intensive since the batches can be run unattended. It is ideal for small parts particularly those with many intersecting surfaces and difficult-to-reach corners. The resulting finish is normally quite uniform and cosmetically pleasing. However, deep scratches, major imperfections, and unsightly welds must first be hand ground to improve the surface condition and this method is not suitable for interiors of hollow components.

Stainless steel can be easily cleaned with soap and water and there are proprietary cleaners for removing stubborn soils. Operations, maintenance, or cleaning personnel must be careful to avoid damaging stainless steel surfaces particularly on the highly reflective polished surfaces where small defects are very noticeable. Restoration, while possible, is expensive.

Although stainless steel is considered to be relatively tough and strong, 300 series stainless steels do not have a very hard surface and can be scratched, nicked, or gouged when contacted by harder materials.

Glass Beading

Glass beading, generally limited to noncontact surface finishing, is similar to sand blasting but uses glass beads to produce a smooth uniform surface without the roughness of sandblasting. If a blasting chamber has been previously used for carbon steel, the chamber must be thoroughly cleaned of used media before attempting to blast stainless steel in the same chamber. Only clean unused blast media can be used as foreign material may embed itself in the stainless surface and cause future rusting and corrosion problems. It may also be inappropriate for thin surfaces that can be damaged by the force of beading.

Electropolishing

Although mechanical or chemical polishing methods can provide a pleasing appearance, these methods alone are insufficient to achieve the desired cleanability and corrosion resistance. Mechanical polishing may leave a smeared surface with reduced corrosion resistance and microscopic scratches large enough to conceal bacteria and other impurities. Chemical cleaning, traditionally pickling and/or passivation, produces a rough surface with similar deficiencies. To overcome these problems, electropolishing has become the preferred finish for product-contact surfaces in the biopharmaceutical industry.

Electropolishing removes metal ions from the surface by passing an electric current through the metal while the work is submerged in an electrolytic solution. Metal from the anode is dissolved in the bath forming a soluble metallic salt with the quantity of metal removed being proportional to current and time. Microscopic shards of metal are dissolved because they experience a high current density while flatter sections experience a lower current density and show minimal change. The result is a reduction in surface profile and a smooth, bright surface appearance. Because iron atoms are more easily removed than chromium and nickel atoms, the surface becomes deficient in iron and the Cr/Fe ratio increases. Electropolishing will also remove hydrogen that, if left on the surface, might lead to hydrogen embrittlement in the future.

Electropolishing produces an extremely smooth surface that is easy to inspect and clean. Corrosion resistance is improved by the increased Cr/Fe ratio on the surface and the surface is cleanable because of the reduction in roughness.

Electropolishing may be followed by passivation to remove sulfur and aluminum inclusions or surface contaminants.

In order for electropolishing to be effective, the surfaces must first be mechanically polished to a fine finish, as the final mirror-like finish will amplify any surface imperfections.

Electropolishing preferentially removes iron improving the Cr/Fe ratio to 1.3 to 1.6 with an oxide layer depth of 15 to 25 Å. There are limits as to how much "smoothing" can be achieved by electropolishing. A "rule of thumb" is that electropolishing can improve a root mean square finish by about a factor of 2 (a SFVV3 finish can be improved to a SFVV5).

Electropolishing is not inexpensive and is also not suitable for items with high dimensional tolerances. It is labor intensive and may require custom-made molds for complex shapes. The final appearance will depend on the initial condition and pretreatment. If surface defects are present or contaminants not removed, the final finish will likely magnify the blemishes and present a poor appearance.

PASSIVATION

The passive film is an insoluble, nonreactive, thin chromium oxide film on the metal surface. Although the layer is a very thin (1–10 molecules thick), corrosion will not occur as long as the surface remains intact. If the protective film is breached, passivity is lost and the metal becomes active and may corrode.

What Is Passivation?

In the biopharmaceutical industry, passivation is the process used to remove iron and iron compounds from stainless steel surfaces and expedite the formation of a passive surface film. Free iron can occur from improper cleaning at the mill; contamination during storage or it may have been introduced during a grinding or machining processes. The iron particles, although invisible to the naked eye, will oxidize when exposed to humidity. Passivation also helps repair heat-affected weld areas and may remove or reduce sulfide and aluminum inclusions or surface oxides and contaminants.

Traditionally, passivation has used a combination of nitric acid and oxidizing salts to treat parts that have previously been cleaned of organic and metallic residues. By adjusting the concentration, residence time, and temperature of the solution, the iron and sulfides are dissolved without affecting the substrate alloy. As the ratio of chromium to iron (Cr/Fe) at the surface increases from 0.25 to 0.75–2, the potential for pitting attack and rouge formation is reduced.

Today, citric acid/chelant techniques have replaced nitric acid processes as the chemicals are easier to handle and the corrosion resistance is improved (as measured by the Cr/Fe ratio).

Surface Cleanliness

A clean surface is free of oil, grease, embedded iron, and inclusions and covered by an intact oxide layer. Welds must be free of significant heat tint and the smearing of layers caused by wire brushing.

The cleaning process typically involves general degreasing and cleaning using vapor degreasing, solvent cleaning and/or alkaline soaking to remove oils, greases, forming compounds, fingerprints, films, lubricants, coolants, cutting fluids, and other undesirable organic and metallic residue left behind during fabrication and machining.

Alkaline cleaning proceeds using a combination of chemical and physical reactions involving displacement, flotation, penetration, wetting, emulsification, and saponification of the contaminants. As all these reactions are sensitive to both time and temperature, care should be taken to ensure the cleaning conditions fit the properties of the soil to be removed. Some soils may require little or no alkaline soaking while others may require substantially higher attention to achieve the desired level of cleanliness. Cleaning effectiveness should be verified by borescope and/or visual inspection.

Pickling

Mechanical cleaning methods leave a smeared surface layer of lower corrosion resistance that provides a potential source of future rouging unless mechanical cleaning is followed by pickling or electropolishing.

Stainless steel that has been chemically cleaned, or pickled, at the mill will naturally form a passive layer. Pickling is controlled corrosion of the thin surface layer to remove embedded iron, heat tint, the smeared layer left by mechanical cleaning and manganese sulfide inclusions in the surface that act as preferential sites for pit initiation on stainless steels.

A common pickling bath contains 10% nitric and 2% hydrofluoric acid. Pickle paste can be used for equipment that can not fit in the pickling vat. Pickling fluid should be rinsed with water within 30 minutes of application to avoid etching the surface. Although, nitric acid will remove sulfide inclusions, it will not remove aluminum or calcium particles that may also provide sites for preferential pitting.

Pickling is hazardous and technicians must be well trained and wear protective equipment when performing pickling operations. Disposal of spent pickle liquor is a problem that tends to limit pickling by immersion to those fabricators that have pickle tanks and to chemical cleaning contractors who have approved arrangements for disposal.

Both pickling and electropolishing clean a "dirty" surface and reform a uniform, defect free, protective oxide film. However, pickling leaves a rough surface that may be more difficult to clean than the electropolished surface.

How Is Passivation Performed?

A typical passivation process cleans and degreases with 5% NaOH at 160°F to 180°F (70–80°C) for 30 minutes followed by a rinse and an acid passivation bath. The passivated metal is then rinsed, neutralized with an additional 5% NaOH bath at 160°F to 180°F (70–80°C) for 30 minutes, rinsed and dried. After the pickle liquor or electropolishing fluids are washed away, the passive oxide film reforms uniformly and instantaneously over the cleaned surface.

During the passivation process, the conductivity of the rinse fluids is monitored to verify the effectiveness of flushing and ensure that all residual chemicals are removed.

Passivation Variables

Time, temperature, and concentration of the bath are selected based on the type of alloy processed. Improper bath and process selection and/or process control will produce unacceptable results that may include catastrophic failure, including severe pitting, etching, and/or total dissolution of the entire component. Typical immersion times are 20 to 120 minutes; bath temperatures range from room temperature to 160°F (70°C); and nitric acid concentration is in the 20% to 50% (vol) range.

Mineral Acids

Nitric acid and other mineral acids are effective for removal of iron; however, they may also remove nickel and chromium components producing a low Cr/Fe ratio and a thin passive layer. Typical conditions might be 120°F to 140°F (50–60°C) for 20 minutes using 50% HNO_3. Alternatively, a system using sodium dichromate ($Na_2Cr_2O_7$) and 20% HNO_3 might operate at 120°F to 140°F (50–60°C) for 20 minutes.

Citric Acid

The use of citric acid for passivation originated in the beverage industry as surface iron causes a bad taste and nitric acid passivation could not achieve the desired iron surface content. It has gained wide acceptance in the semiconductor industry where the degree of passivity is measured by the Cr/Fe ratio and the thickness of the chromium-rich layer. More recently, the biopharmaceutical industry has begun using citric acid.

Citric acid is organic, safe, and easy to use. When formulated correctly, citric acid provides better results at less cost than mineral acids as it preferentially removes iron without removing significant amounts of nickel and chromium thus producing a higher Cr/Fe ratio than mineral acids.

Because of citric acid's high reactivity with free iron and low reactivity with other metals, passivation is done at lower temperatures and shorter residence time. Typical passivation conditions might be 150°F (65°C) for 30 minutes using 10% citric acid. Citric acid is nonhazardous, does not produce nitrogen oxides, does not require special handling equipment or safety devices, and does not corrode other equipment, finishes, and structures.

Chelants

Iron can be removed from stainless surfaces by citric acid alone but inclusions require additional treatment. Chelants are added to enhance free iron removal and help remove other impurities such as sulfides, calcium, aluminum, manganese, silica, carbon, and silt. Five or more ingredients (surfactants, acid chelants, buffer agents, and stronger reducing acids) may be added to aid in dissolving and removing contaminants.

Welding and the HAZ

General

Welding produces heat tint discoloration as alpha ferrite is converted to delta ferrite. The resulting magnetism can lead to galvanic action and then corrosion. The welding process changes the surface chemistry from the weld bead through the heat affected zone (HAZ) and may dramatically reduce the corrosion resistance throughout the weld area because the Cr/Fe ratio is lowered and manganese concentrates in both the weld area and the HAZ.

Conventional passivation will not adequately protect the HAZ because the depth of the HAZ tinting and disruption is greater than can be addressed by passivation alone. Although the surface layer will be passive, the surface just below the chromium oxide is ripe for attack if the passive layer is breached.

Resolution of heat tint problems requires avoiding the formation of heat tint. This is done using orbital welding procedures with appropriate high-purity gas purge and avoiding the welding of dissimilar materials. Poor quality welds should be rejected, cut out, and replaced by acceptable welds to prevent future problems.

Dissimilar Materials

Gaskets, valve packing, diaphragms, and hoses must be carefully specified, stored, and installed to avoid introducing iron or other foreign material into a process system.

Dissimilar metals particularly at welds may cause an attack on the structure and integrity of stainless systems by forming corrosive galvanic cells.

Machining and Heat-Treating Techniques

Contamination introduced during manufacturing or thermal processes may lead to corrosion. Manufacturing processes should be reviewed to minimize the possibility of cross-contamination during manufacturing and increase the chances of successful passivation and tests results.

Grinding wheels, sanding materials, or wire brushes made of iron, iron oxide, steel, zinc, or other undesirable materials, which may cause contamination of the stainless steel surface, must not be used. Stainless tools used on other metals must not be used on stainless steel.

Use only clean, unused abrasives such as glass beads or iron-free silica or alumina sand for abrasive blasting. Steel shot, grit, or abrasives that have been used to blast other materials should never be used.

Thorough cleaning prior to thermal processing is critical. Stress relieving, annealing, drawing, or other hot-forming processes should be avoided as they can draw surface contaminants deeper into the substrate, making them impossible to remove during passivation.

The passivation process is both an art and a science. It will enhance the corrosion resistance of stainless steels but it is important to understand that machining, fabricating, and heat-treating practices can also impact the corrosion resistance of the metal.

Repassivation

Purified water, clean steam, and oxidizers such as bleach are extremely corrosive and will attack the passive layer. In addition, there is a continuous natural migration of iron from the base metal where iron is present in high concentration to the surface layer where it is present only in minute concentration. Periodic repassivation may be necessary to regenerate the chromium-rich passive layer.

TESTING

Compliance requirements and procedural verification are critical in the performance of chemical treatments in validated facilities. The certification and validation package for passivation processes should contain specific procedural documentation, certificates showing passivation and acceptance, quality control testing logs, chemical batch records, certificates of chemical analysis, and a detailed scope of included systems and designations of treated equipment.

A spectrophotometer is generally used to measure the passivation effectiveness by determining the surface composition. These measurements may be made by electron spectroscopy for chemical analysis (ESCA), argon electron spectroscopy (AES), or scanning electron microscopy (SEM).

Other, less sophisticated methods may be used to test the iron content of the surface. The usual methods are the ferroxyl test (ASTM A-380) and the copper sulfate test. If iron is present, it will show up as a deep blue color when the ferroxyl test is performed. With the copper sulfate test, passivated parts are immersed in a solution for six minutes, rinsed, and visually examined. A pink color indicates the presence of free iron and the test is considered unacceptable.

Still other validation tests include a 2-hour salt spray or a 24-hour high humidity test. These tests are performed by placing passivated parts in a highly controlled chamber that creates an accelerated corrosive environment. After subjecting the test pieces to the corrosive atmosphere for the prescribed exposure periods, the parts are removed and evaluated. ASTM B-117 provides a reference for determining acceptability.

HOW MATERIALS AND FINISHES AFFECT RESIDUE REMOVAL

The first step in establishing a cleaning regimen is to understand the nature of soil including chemical composition, moisture level, and temperature. Microorganisms can either be in the soil or attached to it. They will grow and multiply in the presence of nutrients, moisture, and temperature and the absence of antimicrobial agents. In the biopharmaceutical industry, higher temperature and low moisture cause soils to become "baked" and makes cleaning more difficult.

Soil adherence is related to a combination of intermolecular forces and mechanical adhesion that contributes to the strength of the bond through the physical interlocking of the soil into the irregularities of the surface. The chemical and physical bonds are dependent on a number of factors including species, pH, ionic strength, surface finish, and construction material. If soils are not adequately removed, microbial survival is enhanced and contamination and corrosion may occur.

Although surface finish contributes to soil adherence, other factors such as the cleaning process and chemical treatment may be more important as they affect the strength of the intermolecular bonds.

DESIGN FOR CLEANING

Functional Requirements

The objective of cleaning is to restore the surface to its original clean, non-rusting surface, and optimize corrosion resistance by removing surface contaminants including soil, bacteria, process fluids cleaning chemicals, and metallic impurities that result from exposure to the processing, ambient, or cleaning environment.

Designing for cleaning involves understanding and addressing both the process and mechanical requirements of the system. The system design must address the specific chemical and physical characteristics of the soil as well as the interaction between the soil, cleaning compounds, and surface materials.

The construction materials must be selected for the normal and abnormal operating conditions including the variations that will occur during startup, shutdown, cleaning, and maintenance. Consideration must also be given to protecting the equipment components against the external environment such as providing chloride-free insulation and following supplier's instructions when using cleaning chemicals.

The mechanical design must be in accordance with the correct codes and standards. A quality plan must be developed and followed to ensure that surfaces are not damaged during maintenance and cleaning operations. Regular inspections must be made to ensure that the cleaning objectives are being achieved, the surfaces are undamaged, and the passive layer is intact.

Surface

A cleanable surface need not be made of stainless and need not be polished. It must be selected to minimize adhesion of soils, be easy to clean and dry, and it should be smooth.

Ra is the arithmetic average height of roughness component irregularities generally measured in microinches (μin.) or micrometers (μm). For successful cleaning in typical biopharmaceutical applications, the surface roughness should be 0.8 μm or 30 μin. Ra or less to eliminate imperfections that would provide hiding places for dirt and bacteria.

Welds

Welds should be continuous, smooth, crevice free, and ground flush. There should be no cracks, inclusions, weld splatter, or irregularities.

ASME BPE provides a standard acceptance criteria for interior surface finishes and classifies the particular treatment of welds ranging from no treatment (Class III), smoothing (Class II) to making the welds smooth, flush, and defect free (Class I).

Ribbon polishing can be used on weld areas to remove heat tint. This technique involves grinding and polishing the welds and a narrow zone on either side with the scratch pattern running parallel to the weld. This zone will terminate with a clean line of demarcation and leave the rest of the surface untouched. This treatment can be applied to either the 2B mill finish or the SFVV3 mechanically polished plate. The appearance can be pleasing but the grain or scratch pattern will not be running parallel on all portions of the surfaces.

Metal Thickness

As thin materials may distort during welding, a minimum thickness of 16 gauge (0.06 in. or 1.5 mm) should be specified for tubing and it is common to use a minimum of 3/16 in. (5 mm) for tank walls.

Joints

Design for cleaning; avoid butt, concave, and square joints.

Ferrite Number

Although composition influences the formation and stabilization of ferrite, the cooling rate of the casting or weld appears to be the major factor in establishing the ferrite content.

The amount of ferrite, particularly in austenitic stainless is indicative of the strength, corrosion resistance, and cracking tendency of castings and welds. For a particular application or form of stainless, there is a desirable level of ferrite, generally in the range of 3 to 8. The main variables are the composition (chromium and nickel equivalents) and the cooling rate. Materials with lower ferrite numbers are more susceptible to SCC particularly if welded to materials with significantly higher ferrite numbers. Higher ferrite values provide increased strength but may result in cracking during machining operations.

CONCLUSION

Electropolished 316L stainless steel continues to be the preferred material of construction for process-contact applications in the biopharmaceutical industry. Although the use of polymers is gaining acceptance for disposable components, there is little interest in changing materials for permanently installed components.

BIBLIOGRAPHY

ASME-BPE-2002. Bioprocessing Equipment. New York, NY: ASME, 2002:88.

ASTM 380. Standard Practices for Cleaning, Descaling and Passivation of Stainless Steel Parts, Equipment, and Systems—Copper Sulfate Test, 2006.

Avery RE, Tuthill AH. Guidelines for the Welded Fabrication of Nickel-Containing Stainless Steels for Corrosion-Resistant Services, 1992.

Barbosa-Canovas, Gustavo V, Rodriquez JJ. Microbial Attachment and Sanitizer Effectiveness on Different Stainless Steel Surfaces (Presentation).

Barnes LM, Lo MF, Adams MR, Chamberlain AHK. Effect of milk proteins in adhesion of bacteria to stainless steel surfaces. Appl Environ Microbiol 1999; 65(10):4543–8.

Chmielewski RAN, Frank JF. Biofilm formation and control in food processing facilities. Compr Rev Food Sci Food Saf 2003; 2:22–32.

Cluett JD. Cleanability of certain stainless steel surface finishes in the brewing process. Dissertation, October 2001.

Dillon CP, Rahoi DW, Tuthill AH. Stainless steels for bioprocessing. Biopharm 1992; 5:32.

Flair-Flow Europe Technical Manual F-FE 377A/00, May 2000.

Fleming JR, Kemkes D, DeVoe DW, Crenshaw L, Imbalzano JF. Material of construction for pharmaceutical and biotechnology processing: moving into the 21st century. Pharm Eng 2001; 21(6):34–44.

Flint SH, Brooks JD, Bremer PJ. Properties of the stainless steel substrate, influencing the adhesion of thermo-resistant streptococci. J Food Eng 2000; 43:235–42.

Frank JE, Chmielewski RAN. Effectiveness of sanitation with quaternary ammonium compound or chlorine on stainless steel and other domestic food-preparation surfaces. J Food Prot 1997; 60(1):43–7.

Gibson H, Taylor JH, Hall KE, Holah JT. Effectiveness of cleaning techniques used in the food industry in terms of the removal of bacterial biofilms. J Appl Microbiol 1999; 87:41–8.

Gonzalez M. Stainless Steel Tubing in the Biotechnology Industry. Biotechnology/Pharmaceutical Facilities Design 2001; 21(5):48–63.

Hilbert LR, Bagge-Ravn D, Kold J, Gram L. Influence of surface roughness of stainless steel on microbial adhesion and corrosion resistance. Int Biodeterior Biodegradation 2003; 52:175–85.

Holah JT, Thorpe RH. Cleanability in relation to bacterial retention on unused and abraded domestic sink materials. J Appl Bacteriol 1990; 69(4):599–608.

Jack TR. Biological corrosion failures. ASM Handbook. Vol. 11. Materials Park, Ohio: ASM International, 2002.

Jenkins L, Libert S. Lusvardi V. Consider using fluoropolymers in biological applications. CEP 2004; 100:39–41.

Katsikogiannim M, Missirlis YF. Concise review of mechanisms of bacterial adhesion to biomaterials and of techniques used in estimating bacteria-material interactions. Eur Cell Mater 2004; 8:37–57.

Kobrin G, Lamb S, Tuthill AH, Avery RE, Selby KA. Microbiologically influenced corrosion of stainless steels by water used for cooling and hydrostatic testing. In: 58th Annual International Water Conference, Pittsburgh, PA, November 3–5, 1997.

Lelièvre C, Legentilhomme P, Gaucher C, Legrand J, Faille C, Bénézech T. Cleaning in place: effect of local wall shear stress variation on bacterial removal from stainless steel equipment. Chem Eng Sci 2002; 57:1287–97.

Lelièvre C, Antonnini G, Faille C, Bénézech T. Modeling of cleaning kinetics of pipes soiled by *Bacillus* spores assuming a process combining removal and deposition. Trans IChemE 2002; 80(4):305–11 (Part C).

McWhirter MJ, Bremer PJ, McQuillan J. Direct infrared spectroscopic evidence of pH- and ionic strength-induced changes in distance of attached pseudomonas aeruginosa from ZnSe surfaces. Langmuir 2002; 18(5):1904–7 (see also 2002(26):365–72).

McWhirter MJ, McQuillan J, Bremer PJ. Influence of ionic strength and pH on the first 60 min of *Pseudomonas aeruginosa* attachment to ZnSe and to TiO_2 monitored by ATR-IR spectroscopy. Colloid Surf B 2002; 26(4):365–72.

McWhirter MJ, Bremer PJ, Lamont IL, McQuillan AJ. Siderophore-mediated covalent bonding to metal (oxide) surfaces using biofilm initiation by pseudomonas aeruginosa bacteria. Langmuir 2003; 19(9):3575–7

Milledge JJ. The hygienic design of food plant. In: Institute of Food Science and Technology Proceedings, 1981:74–86.

Milledge JJ. Liquid hold-up on stainless steel surfaces: I—effect of surface finish. J Food Eng 1982; 1:43–53.

Milledge JJ, Jowitt R. The cleanability of stainless steel used as a food contact surface. In: Institute of Food Science and Technology Proceedings, Vol. 13, 1980:57–62.

Milledge JJ, Jowitt R. Cleaning and descaling stainless steel. In: Institute of Food Science and Technology Proceedings, Vol. 13, 1980, 57–62.

Moller GE, Avery RE. Fabrication and metallurgical experience in stainless steel process vessels exposed to corrosive aqueous environments. In: Air Pollution Seminar, Buffalo, NY, U.S.A., October 1987.

Morison KR, Thorpe RJ. Liquid distribution from Cleaning-In-Place Sprayballs. Trans IChemE 2002; 80:270–5 (Part C).

Stevens RA, Holath JT. The effect of wiping and spray wash temperature on bacterial retention of abraded domestic sink surface. J Appl Bacteriol 1993;75.

Taylor JH, Rogers SJ, Holah JT. A comparison of the bacterial efficacy of 18 disinfectants used in the food industry against *Escherichia coli* 157:H7 and pseudomonas aeruginosa at 10 and 20°C. J Appl Microbiol 1999; 87:718–25.

Tuthill AH, Avery RE. Specifying stainless steel surface treatments. Adv Mater Processes 1992; 142(6).

Tuthill AH, Avery RE, Covert RA. Cleaning stainless steel surfaces prior to sanitary service. Dairy Food Environ Sanit 1997; 17(11).

Tuthill AH, Covert RA. Stainless steels: an introduction to their metallurgy and corrosion resistance. Dairy Food Environ Sanit 2002; 20:506–17.

US 3-A Sanitary Standard 01–07, International Association of Milk, Food and environmental Sanitarians, 1990.

Verran J, Boyd RD. The relationship between substratum surface roughness and microbiological and organic soiling: a review. Biofouling 2001; 17:59–71.

Verran J. Biofouling in food processing biofilm or biotransfer potential. Trans IChemE 2002; 80:292–8 (Part C).

12 Cleanable In-Line Components

Lyle W. Clem
Electrol Specialties Company (ESC), South Beloit, Illinois, U.S.A.

INTRODUCTION

Hygienic processes must be developed to allow both repeated and repeatable cleaning procedures. These cleaning procedures must be performed in a manner that assures the removal of materials that could adversely impact the process or in some manner contribute to contamination of the products manufactured. While some alternatives, such as disposable materials for containers and transfer hoses, etc., may offer flexibility and elimination of the need for cleaning, both before and after processing, there is little question that these approaches cannot address many process operations. These would include thermal transfer, handling and storage of large fluid volumes, and sensor applications.

Thus, the compilation of equipment and components in a hygienic process must comply with a variety of specifications developed with regard to the needs of cleaning and sterilizing. This chapter focuses on the practice of cleaning this equipment in place, clean-in-place (CIP) within the process, and those specific features of the equipment that allow this practice to achieve the desired results.

APPLICABLE STANDARDS

Many practitioners of hygienic process design rely on the availability of equipment and components that can be assembled to create a system or unit operation(s) that can be cleaned in place. In turn, fabricators of these items rely on industry standards and guidelines to assure inter-operability, cleanability, and often sterilizability. In this manner, the manufacture of equipment and components by a large number of companies can be assembled from a variety of sources to perform as a system. Fundamental to this capability is the existence of several industry standards and practices, developed and promulgated by multiple standards-developing organizations (SDOs).

Each SDO has developed a structure and procedures to create standards, guidelines, or practice-type documents, with specific focus on a particular process topic, component fabrication, or fabrication technique. The following SDOs are known to have developed one or more documents that have impact on hygienic process equipment.

- *3-A Sanitary Standards, Inc.* (1). This SDO is comprised of multiple stakeholders with traditional focus on the food, dairy, beverage, and quality sectors. This group has over 65 standards focused on equipment and related hygienic fabrication methods, and, additionally, has a large compilation of practices ranging from CIP methods, air-in-contact-with-product to culinary steam preparation and pasteurization operations. The standards development activity is accredited under the guidelines of the American National Standards Institute (ANSI) as a consensus process.

- *P3-A*. This entity is a subset of 3-A Sanitary Standards, Inc. with a focus on the development of standards for equipment for the manufacture of active pharmaceutical ingredients (APIs). This SDO is comprised of multiple API manufacturers, equipment fabricators, and technical interest groups. The procedures are also ANSI accredited and contact information is through 3-A Sanitary Standards, Inc.
- *American Welding Society* (2). The American Welding Society (AWS) has developed multiple standards relative to welding and joining of stainless steels and high-alloy materials. The D18 committee has developed D18.1 and D18.2 standards for stainless steel welding and discoloration levels respectively. This SDO is ANSI accredited with many of the current standards recognized by ANSI/AWS, meaning these standards are recognized as American National Standards.
- *American Society of Mechanical Engineers (ASME) International* (3). Formerly organized as the ASME, this SDO receives primary recognition for its work and standards relating to pressure-vessel design and fabrication. "The Boiler and Pressure Vessel Code" contains sections regarding materials, design, testing, and certifications, and Section VIII contains rules for pressure vessels. Often adopted by public entities as code, these rules are of interest to hygienic processors when pressure/vacuum ratings are required by specific equipment. These rules also apply to repair and recertification of equipment following modifications.
- *American Society for Testing and Materials (ASTM) International* (4). Formerly organized as the ASTM, this SDO has developed and promulgates hundreds of individual standards relative to specific testing methodologies and criteria for materials. Two such standards of interest to hygienic process system designers include ASTM A269 and ASTM A270 relative to criteria for bright annealed and polished stainless steel tubing, respectively.

APPLICABLE GUIDELINES

In addition to specific equipment or component fabrication standards, multiple entities worldwide are actively developing and promulgating hygienic process guidelines. These documents are utilized by hygienic systems designers and specifiers to provide guidance to fabricators of equipment and to establish common practices for facility planners and operations personnel.

- *European Hygienic Engineering Design Group* (5). The EHEDG is primarily focused on food processing facilities and processes. While EHEDG develops and publishes many guideline documents relating to equipment and building design and cleanability functions, a major effort of this organization is also related to certification of specific equipment compliance to a repeatable cleaning regimen. This also includes certification of third party entities to perform this testing of equipment cleanability. The EHEDG is comprised of processors, equipment manufacturers, and individual members.
- *ASME International–Bioprocessing Equipment Main Committee*. This committee's function is focused on development of guidelines for bioprocessing equipment, the including sections on materials and joining, construction, inspection, and testing of vessels and piping, and references to valves, pumps, and fittings for use in the biopharmaceutical industry.

- *The International Conference on Harmonization of Technical Requirements for Registration of Pharmaceuticals for Human Use (ICH)* (6). This international conference includes regulatory agencies from Europe, Japan, and the United States along with industry experts. The primary effort is to harmonize these regions' pharmaceutical processing guidelines and product registration with a focus to minimize duplicate testing of R&D for new medicines. This conference is also actively developing guideline documents to provide guidance to industry. These guidelines also offer guidance to equipment fabricators and third party verifiers as to needs for equipment compliances in specific manufacturing sectors.

FITTINGS, TUBING, AND PROCESS-LINE CONNECTIONS

A significant aspect of hygienic process systems and equipment is related to application of special finishes to metallic product contact surfaces. The goal of these specialized finishes is to eliminate or significantly reduce areas that might harbor soils or process residues in ways that would preclude their removal by CIP operations.

Interior Surface Characteristics

Mechanical Polishes

The primary method of applying a special finish involves sequential abrasive polishing, starting with abrasives of coarse grits and progressing to grits that ultimately yield the desired surface finish. Generally accepted minimum surface finish for a product contact surface to be cleaned in place would be produced by a 150-grit abrasive. This surface would typically be evaluated by use of a stylus-tipped profilometer to provide results of 32 μin. average roughness (R_a) (0.8 μm R_a). While this surface finish has found acceptance in processes involved in the production of comestibles, base levels of abrasive polishes applied to pharmaceutical and biotech equipment are generally required to yield a finish ranging from 25 μin. R_a to 15 μin. R_a (0.4 μm R_a).

Electropolish

Often these mechanical surface treatments are followed by an additional electrochemical process known as electropolish. Here, the surface being treated is generally submerged in an acid bath and connected to an electric current while an electrode is passed in close proximity to the surface. The electric current causes an accelerated dissolution of the metal yielding a significantly higher quality surface finish and, as an associated benefit, a significantly passive surface to improve corrosion resistance.

Plating

An alternate electrochemical process is sometimes used to apply a metal overlay and also yields an extremely smooth surface. This process of electroplating is frequently used to apply layers of nickel to surfaces used in thermal transfer duties. Although electroplating is able to yield smooth surfaces, these surfaces are also subject to impact chipping and are also typically less resistant to acid exposure.

Nonmetallic materials are generally not able to have post-manufacturing surface modifications, with the possible exception of machining certain types.

Thus, high-quality, extremely smooth surfaces in nonmetallic materials are most often created in their manufacturing processes including extrusion, casting, and rotational molding.

FABRICATION OF PRODUCT AND CIP SUPPLY/RETURN PIPING SYSTEMS

Stainless steels, typically austenitic and duplex alloys, have been applied with wide acceptance in the pharmaceutical and biotech sectors. Generally, these alloys are formed into tubing shapes in compliance with standards developed by the ASTM as ASTM A269 and ASTM A270. The fabrication of tubing is generally in lengths of 20 to 21 ft. Fittings including tees, and elbows are then in turn fabricated from the tubing material in special forming and cutting machines. The final surface finishes are then applied to the tubing and fittings as a combination of mechanical polishing and, if required, electropolish is performed as a final step.

Tubing and fittings manufactured in the above manner are the preferred material of construction for permanently installed piping for all needs in the pharmaceutical and biopharmaceutical industries. Tuthill and Brunkow (7) published an in-depth review of the use of stainless steel tubing in the biotechnology industry in 2004. This chapter will briefly review the most commonly used methods of fabrication.

Permanent Joints

Welding is the preferred method of assembly piping used for both product handling and conveying flush, wash, and rinse solutions for CIP. The proper control of welding procedures is essential to create a resulting surface finish that can be repeatedly cleaned-in-place. A properly performed weld will not have excessive surface deformities (concavity or convexity) and the weld zone will not have inclusions or presence of impurities that will later promote or contribute to corrosion.

There are certain situations, primarily related to fabrication of equipment, where a permanent joint is created through the use of press fits or shrink fits. These methods are generally not acceptable for processes that utilize high-temperature sterilizations. The wide temperature excursions can cause the nonpermanent joint to gradually move, thus creating a junction that is not easily cleaned.

Nonpermanent Joints

Flanged, Cam-Type, and Acme Threaded Connections

Nonpermanent joints are often required to accommodate periodic maintenance, replacement of certain components, and to allow flexibility of process through make-break swing connections. These nonpermanent joints can occur in process utilities that are not subject to CIP requirements. In these instances, industrial flanges (circular-shaped components with bolt circles) or threaded male-female joints are often utilized. The resulting sharp edge grooves in threaded fittings or the lack of close alignment of a flange gasket preclude repeatable flushing and thus cannot be used for CIP applications. Cam-type fittings are often utilized for non-hygienic ingredient handling or for utility connections on portable equipment. These fittings lack close alignment of sealing surfaces and can collect soils or process residues in the gasket areas. Thus, these fittings are also not desirable where CIP is the sole cleaning method.

Alternatively, a variety of threaded fittings have been utilized in food processing systems. These typically involve the use of American Standard stub Acme-type threads (bevel seat and John Perry) or DIN405 knuckle-type threads. In each of these applications, an elastomer is utilized to perform the final seal duty. Although the use of these types of joints can meet the needs of a system using CIP procedures, they are most often challenged by the need to manually clean the threaded areas and by the lack of close alignment of the sealing gaskets to the inside diameter of the tubing or component. Further, these threaded joints often lack a means to control the compression of the junction and thus overtightening of the hexagonal nut will cause extrusion of the gasket into the product zone.

Clamp-Type Connections

The method of joining tubing or process equipment that has gained almost universal acceptance utilizes a clamp-type connection of two, identically fabricated end pieces and a self-centering elastomer gasket. This joining method relies on a tapered, circular flange on each end piece to be drawn together when a clamp with similar tapers is closed concentrically around the connection. The primary advantage of this connection is the identical nature of mating connections and the ease of inspection and cleaning of the gasket groove. This clamp connection system has been marketed as Tri-Clamp™ and S-Clamp™ as well as simply referred as a sanitary clamp (Fig. 1). Recent gasket developments have begun to address the concern of gasket extrusion into the product zone by utilizing a hard plastic to control the face-to-face distance of the end pieces coupled with a softer elastomer that functions to create the fluid seal at the inside diameter of the connection. Analysis of flow conditions using computational flow dynamics software indicate that optimal fluid turbulence at the gasket interface with the inside diameter of the tubing is obtained when the gasket is recessed approximately 0.2 mm, i.e., the gasket inside diameter is 0.4 mm greater than the inside diameter of the tubing.

Other variations of the clamp-type connection have been developed and marketed including heavy wall versions and uniquely shaped ends that ultimately create proprietary solutions. Further, male–female clamp-type fittings have been developed to address high-pressure applications that could cause the gasket to be blown between the clamp hinge points.

Installation of hygienic equipment in process systems also requires consideration for the interconnection via process piping (primarily tubing materials) and the associated needs for support, expansion, drainage, and cleaning/sanitizing.

Sealing Materials

Generally, the hygienic aspects of a process are most impacted by sealing and joining methods other than welding. The use of mating surfaces of similar materials is not an acceptable method of joining two segments of tubing or components when a make–break capability must be maintained. The presence of a noncontrollable groove at the junction, without a sealing method, will create a zone or area where soils and process residues cannot be flushed or cleaned by a mechanical means or recirculation of cleaning solutions. Thus, it becomes important to design joints with a seal or specific means of preventing entry of soils into the junction area. Further, it

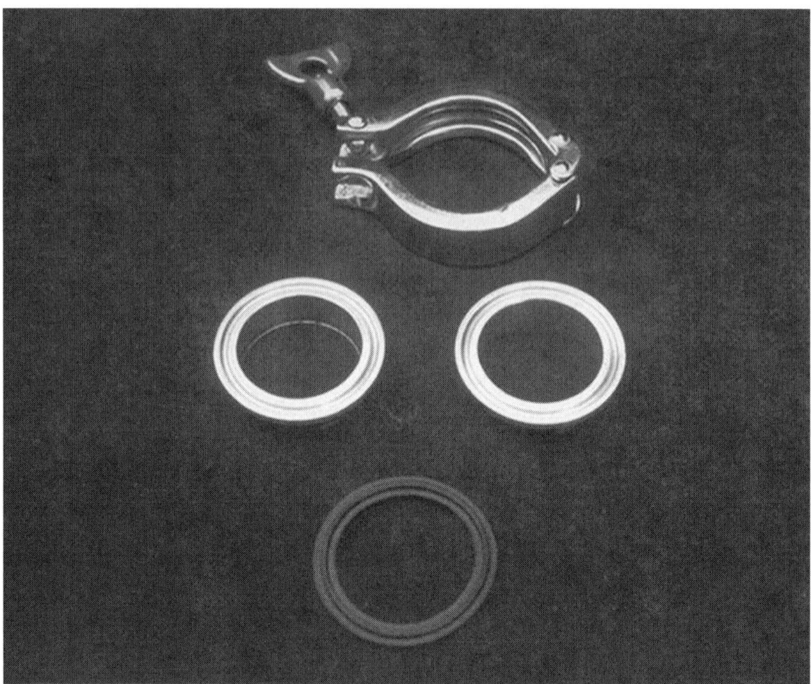

FIGURE 1 Hygienic clamp, ferrules, and gasket.

is also important to design the seal method to prevent uncontrolled extrusion into the product zone.

Gasket Extrusion

This resulting extrusion will not only create sharp edged areas that are not easily flushed, but also create possible breakage of the seal materials and the resulting blockage of spray devices, etc.

The most effective sealing mechanisms include an elastomeric component, with characteristics to allow conformity to the junction area, and a mechanical means to control the compression or approach of one seal surface to another. The design of sealing surfaces using elastomers must also consider the possible contamination of a process through the presence of leachables or extractable compounds from the elastomer. For this reason, it is often a design criteria that elastomers comply with criteria developed by the United States Pharmacopeia.

Pressure Limitations

Pressure limitations of fittings and joining methods are mostly controlled by the materials of construction of the joints and seals, although temperature will often contribute to further limit the operational characteristics. If welding is the joining method, the pressure/temperature limitation of the resulting joined segments will often approach that of the parent materials.

Expansion Considerations

The use of interconnecting piping materials must consider the temperature range of operation of the process. Typically, this range occurs between 32°F (0°C) and 260°F (127°C). The coefficient of expansion for austenitic stainless steel and related alloys is approximately 9.2 in./in. per $°F \times 10^{-6}$. Thus, a length of 100 ft of tubing could expand 2.5 in. in length over this temperature range. Similarly, a nonmetallic piping material such as perfluoro alkoxy (PFA) has a coefficient of expansion of approximately 7 in./in. per $°F \times 10^{-5}$. This would result in an expansion of over 19 in. in 100 ft for this same temperature range. This demonstrates a need to consider the needs of a piping system to expand and contract and yet maintain the critical need to be fully drainable. Often this need is accommodated by changes in direction or elevation, thereby shortening the overall length of individual segments.

Support Systems and Maintenance of Slope

The support system must allow for the expansion/contraction of the piping while maintaining the slope of the piping for drainage. Similarly, the support system needs to allow for inspection access, and insulation when required.

The slope of the piping system is perhaps the most critical aspect of the installation with respect to impact on hygiene and associated cleaning and sterilization procedures. The requirement to be freely draining, either entirely to a single low point or to a series of low points equipped with drain valves, is essential to assure that no residual cleaning or rinsing solutions remain to impact the process. Further, in the case of steam-sterilized systems, the drainage is critical to assure that condensate can be constantly removed from the piping and thus not act to insulate a surface and prevent it from achieving the sterilizing temperature.

FLEXIBLE HOSES AND CONNECTIONS

The use of flexible piping sections is often a preferred method to connect portable equipment that may not have exact, repeatable positioning. The flexible piping section or hose allows for non-exact connections. Further, these hose sections are also utilized to isolate a source of vibration from the remainder of the process. Typical construction of these flexible sections include a flexible center, usually an elastomeric segment, with ends of hard plastic or metallic fittings that can be used to mate with clamp or threaded connection points. The materials and fabrication methods of hose sections must yield an interface between the elastomer and the metal ends that is cleanable, sterilizable, and, when positioned correctly, totally drainable.

Often an overbraid of stainless steel or a spiral overwire is used to provide backing to the elastomer and thus allow the flexible section to withstand higher internal pressures. Use of overbraiding can create a standalone problem in that the fine wires are most difficult to adequately clean. Thus, an additional outer coating or sheathing is used to prevent exposure of the overbraid to soils from the environment of use.

Another fabrication-related need is to assure a crevice-free interface between the lining and the end connections. If the elastomer is a rubber or rubber-like material, heat vulcanization is possible to adhere the lining to the fitting metallic surfaces. Alternatively, a compression method may be utilized to crimp or maintain the lining in close contact with the end fitting. Regardless of the method, this junction must be crevice free and completely drainable.

Flexible piping sections must be adequately supported to assure drainage and relieve bending or fatigue of the lining at the connection end points. Also, if the function of the piping system includes some exposure to hydraulic shock or water hammer, the flexible section will require use of restraints to prevent uncontrolled movement during the hydraulic event.

Flexible piping sections are sometimes applied to select differing flow paths. While it is desirable to use an easily moved component, the position detection and verification of flow path integrity are difficult to accomplish with flexible components. The section on "Transfer Panels" that follows offers a solution to these issues.

TRANSFER PANELS

U-bend transfer panels have been included in many locations within the fluid handling of process and cleaning solutions in many pharmaceutical and biotech operations. This approach provides a maximum flexibility for the production function, yet makes it possible to assure controlled sanitation through mechanical/chemical cleaning and further guarantees the integrity of all individual product and cleaning and/or sterilizing flow paths. Transfer panels are the result of continued modification and development of the component commonly referred to as a "flow-verter" or "cleaning hook-up station" used in the past primarily to control CIP solution distribution.

Typical Port Arrangements

Figure 2 shows the most basic of design concepts used as the basis for development of both small transfer panels and very large transfer panels. These may be constructed with 1.5 in. (40 mm) ports and 5.5 in. (140 mm) long U-bends or 2 in. (50 mm) ports and 6 in. (152 mm) U-bends, with proximity sensors to monitor U-bend positions. The *two-port* panel depicts a means of manually connecting an inlet or outlet line to a product transfer, clean-in-place supply (CIPS), or clean-in-place return (CIPR) line. In the latter instance, the isolated port would be above the header.

Figure 3 describes a *three-port* arrangement that provides for a break in the transfer line following which either end may be connected to the supplemental port for flow from or into the line, or a header, at that point.

The triangular pattern of the *three-port* arrangement can be extended to the diamond pattern in the *four-port* panel in Figure 4. In this application, the arrangement permits a tank outlet to be connected to the process, or to CIPR, and provides also for a CIPS line to the process.

The *six-port* hexagonal pattern (Fig. 5) with a common center port is based on the original six-port cleaning hook-up station, expanded and modified for many process transfer purposes. The addition of the two solid ports is the basis of the diamond pattern shown in the *four-port* panel, and the further expansion of this concept provides substantial flexibility for making product transfer connections and cleaning supply or return connections, always isolating the process from the cleaning circuits.

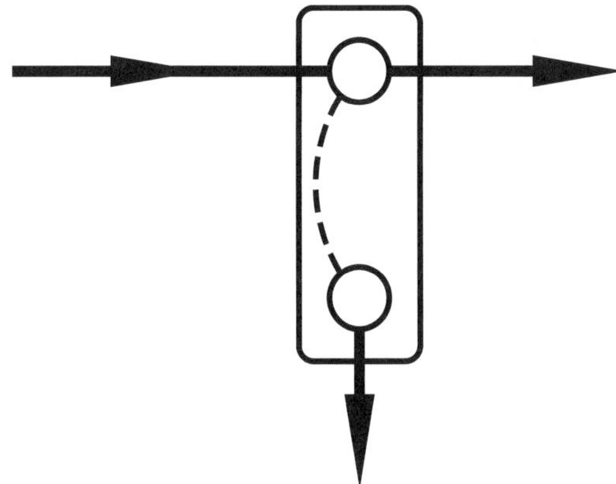

FIGURE 2 Two-port transfer panel.

Shown isometrically in Figure 6 is the left-hand portion of a transfer panel illustrating the use of *primary* U-bends of variable but long length to connect a tank to fill and discharge headers, and 5.5 in. (140 mm) or 6 in. (152 mm). *Secondary* U-bends to connect these headers to internal tie lines or manifolds provided for steaming the process piping and tanks.

Elimination of Dead Ends

Single-piped headers on a large transfer panel would create "dead ends" of considerable magnitude. Such dead ends can be eliminated via use of either the looped-type header or double-tube header shown in Figure 6. Both have been applied successfully in transfer panels up to 28 ft (8.5 m) in length. A 1.5 in. (40 mm) loop-type header has the same flow rate capability, at equivalent line losses, as a single 2 in. (50 mm) tube. To conserve space and/or improve appearance, 2 in. (50 mm) tubes with 1.5 in. (40 mm) pullouts for the ports may be fitted with a 1.5 in. (40 mm) internal tube to provide the flow split to eliminate dead ends.

Headers of double-tube construction or loop-type design as shown in Figure 6 will assure movement of solution in all portions of the piping at all times. The length of the branch from a tee or pullout on such a header must be limited to

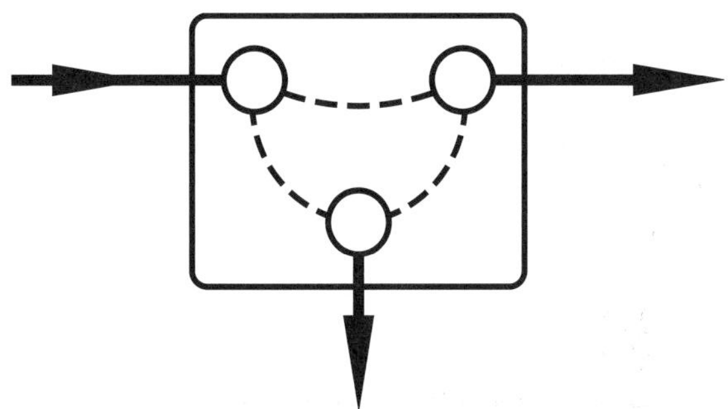

FIGURE 3 Three-port transfer panel.

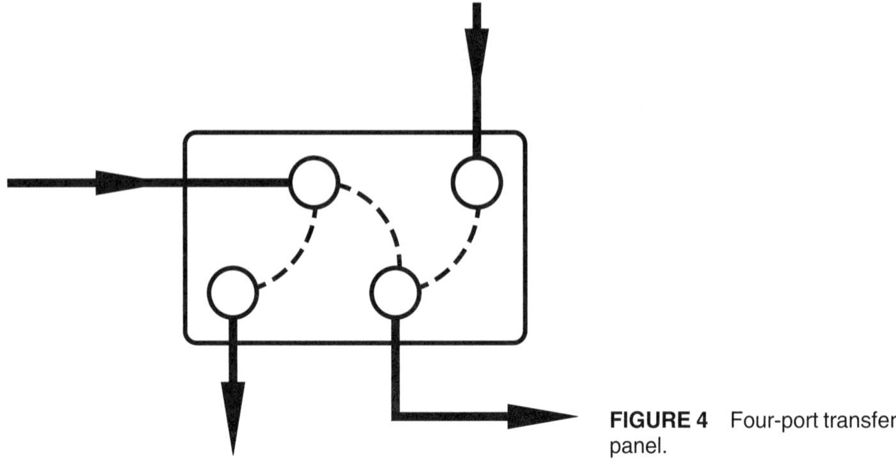

FIGURE 4 Four-port transfer panel.

approximately 1.5 times the pipe diameter to allow recirculation cleaning at normal velocities.

Secondary transfer panels may be used to mount special equipment such as filter housings in addition to providing the CIPS, CIPR, and steam-in-place (SIP) ports.

Proximity Sensors

The use of manually positioned U-bends for establishing processing, CIP, and SIP connections in a highly automated system requires some means of verifying the integrity of the required flow path. This has been accomplished in practice by installing permanent magnets in stainless steel enclosures welded to the center of the U-bend connection as shown in Figure 7. Proximity switches located behind the skin of the transfer panel may then be used to detect the presence or absence of a U-bend between any selected pair of ports. Figure 8 shows the reverse side of the transfer panel including multiple proximity switches. Each proximity switch will monitor a possible U-bend position. Thus, when a proximity switch is activated, the associated automation system

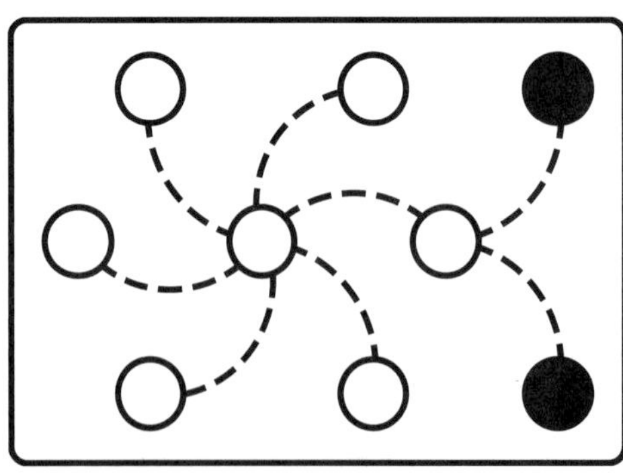

FIGURE 5 Six-port transfer panel.

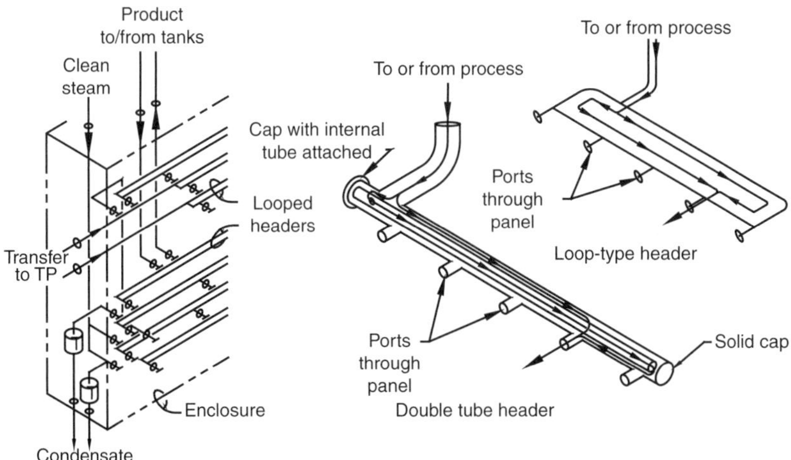

FIGURE 6 Transfer panel with double and loop headers.

will associate this event with the position of a U-bend on the front of the panel. The computer or programmable controller database is developed to include the "allowed" or "required" connection for every established flow path necessary for processing, cleaning, or steaming procedures.

FIGURE 7 U-bends with magnet assemblies.

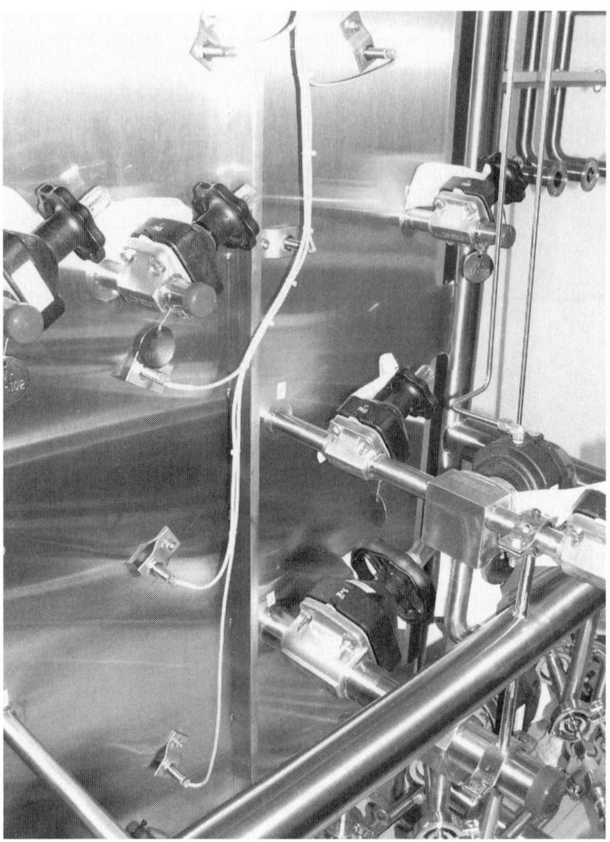

FIGURE 8 Transfer panel with U-Bend monitoring proximity switches.

VALVES

Flow control for process, cleaning, and sterilization operations depends on selection and proper application of valves within the process piping system. These flow control valves are available for a wide range of duties from non-hygienic utilities to ultraclean aseptic processes. Similarly, the world of valves includes a diverse selection of materials, operating methodologies, actuations, end connections, position sensing, and control technologies that bring automation to the valve in a variety of ways. Our focus will be for those valves that can be utilized for flow control of hygienic processes, both for product and for cleaning and sterilization functions.

Like the piping systems that interconnect with process equipment, the installation of hygienic valves must consider supports, drainage requirements, and service access. The support method, like that of the piping, must assure that the orientation of the valve can be maintained to assure free drainage. Further, the support structure near the valve must also consider service access to assure that no obstructions exist that would preclude the removal of the valve or its actuator. Appropriate service or maintenance access is also needed for periodic inspection and calibrations of valve-control components and to allow a safe working area that does not require personnel to climb onto and over an installed component.

Valve Actuation and Position Sensing

Flow-control is implemented by opening and closing valves or by allowing a valve to be positioned in a partial open condition. These functions are controlled through the operation of an actuation means ranging from a handle on a manual valve to an automated positioning motor on an electric valve or a networked pneumatic positioner. In all cases, the actuation means allows an external force to be applied to a control surface exposed to the fluid being controlled. Hygienic process systems can and do use a variety of actuation methods. As this implies, some control-related need is converted to motion that in turn applies a mechanical force to the fluid. Hygienic processes also require an isolation of the process fluid from the mechanical aspects of valve actuation. This is often accomplished by an elastomer seal or elastomer diaphragm located between the process fluid and atmosphere.

The vast majority of hygienic valves utilize either manual or pneumatic actuation. Although electric actuation is available, pneumatic operation of valve actuators has proven to be an extremely reliable control method. Valve actuators can also be equipped with position indicating means. In the case of a manually actuated valve, an extension of the valve stem is often utilized to indicate the valve status. Automated actuators can be equipped with position switches to sense full open or full closed conditions. Modulating or throttling valves often utilize a positioner to maintain a precise position of the valve, and these positioners can be equipped to provide a control signal that varies with the valve position. In all cases, the presence of electrical controls in the process environment requires attention to safety as well as a need for isolation from the often wet, wash-down conditions of a fluid process facility.

Rising Stem Valves

Flow-control valves for hygienic applications generally fall into two major categories. The first category of rising stem valves refers to valves with a central valve stem, traversing through one or more valve bodies, with plunger seats and mating seat rings. Rising stem types are available in both single- and double-seat varieties.

Single-Seat Rising Stem Valves

Rising stem single-seat valves have a long history of application in the food, dairy, and beverage sectors. The typical valve includes a body with two or more ports and a valve stem with plunger that can open or close a port to effect flow control. Typically, the plunger includes an elastomer seal to conform to a mating seat ring and thus causes a stoppage of flow. There are versions of this type of valve that use only metallic plungers to create a metal-to-metal seal. These are less prevalent as the precision of the mating surfaces must be maintained in order to assure that a shutoff condition can be achieved. The valve stem usually extends from the valve body adjacent to the actuator via a seal zone that can utilize either a static or a dynamic elastomer seal. In many valves, this is an "O" ring although many different cross sections of elastomers have been used in this duty. The valve stem also passes through an open area prior to attaching to the actuator. This open area or "yoke" serves to isolate the stem seal from the actuator, preventing any possible transfer of actuator lubricants to the seal zone. Additionally, this open spacing serves as a leak detection zone in that process fluids leaking past the stem seal are easily noted.

CIP procedures for rising stem single-seat valves have been well defined by the comestible product sectors. These procedures call for frequent cycling of the valve actuator to raise and lower the valve stem in the presence of rinse and wash solutions. This causes a constant exchange of cleaning solutions in the valve stem seal area and thus the valve components in this area are very adequately cleaned. The remainder of the valve body and product contact surfaces include appropriate radii to minimize collection and retention of process residues. Thus, these areas also promote cleanability by their fabrication criteria.

Since the valve stem of a rising stem valve must, by design, raise and lower through the stem seal area, it is possible for the atmosphere on the exterior of the valve to be drawn into the product zone. While this condition is of lesser concern in the processing of comestible products, it is of great concern where an aseptic product condition must be maintained. In these applications, a diaphragm seal is fitted to the stem and effectively isolates the product zone from the valve exterior. One side of this diaphragm is a product contact surface while the opposite side remains exposed to possible atmospheric conditions. A similar isolation mechanism involves the use of a bellows (either metallic or semi-rigid plastic) to function as the separation between the product zone and atmosphere. These types of valves are most often applied in ultraclean processes where aseptic conditions must be maintained and where repeated steam sterilizations are performed.

Double-Seat Rising Stem Valves

Another variation of rising stem valves is the rising stem double-seat valve (Fig. 9). This valve configuration is functionally that of a double block and bleed assembly. A double block and bleed arrangement is often created by configuring three valves where two valves are used to block an in-line flow path and a third valve is used to open or "bleed" the interstitial space between the two primary valves. A similar condition is created in the double-seat valves in that two product zones (Fig. 10) are separated by two separate valve seats and an opening to atmosphere exists between the seats. Thus, if either seat experiences a leakage condition, the fluid leaking past the seat is free to escape before it can pressurize the opposite seat and possibly contaminate the fluid in the opposite valve body. The configuration of a double-seat valves with the leakage or vent chamber open to atmosphere means that the surfaces of the leak chamber are not able to maintain an aseptic condition. Thus, rising stem double-seat valves are often utilized for routing of ingredients or cleaning solutions throughout hygienic process systems and are not utilized for product handling where sterility must be maintained.

CIP for rising stem double-seat valves is similar to procedures for single-seat valves with the major exceptions that it is possible to utilize seat lifting features to separately actuate the upper and lower seats and valve stems. Thus, it is possible to clean the upper valve body and its associated upper seat separately from the lower valve body and lower seat. This offers significant flexibility in scheduling cleaning operations for valve groups or matrices wherein it is not possible to fully open and close the passage between upper and lower valve bodies due to ongoing process paths using one or the other valve bodies. Utilizing seat lifting also offers cleaning of

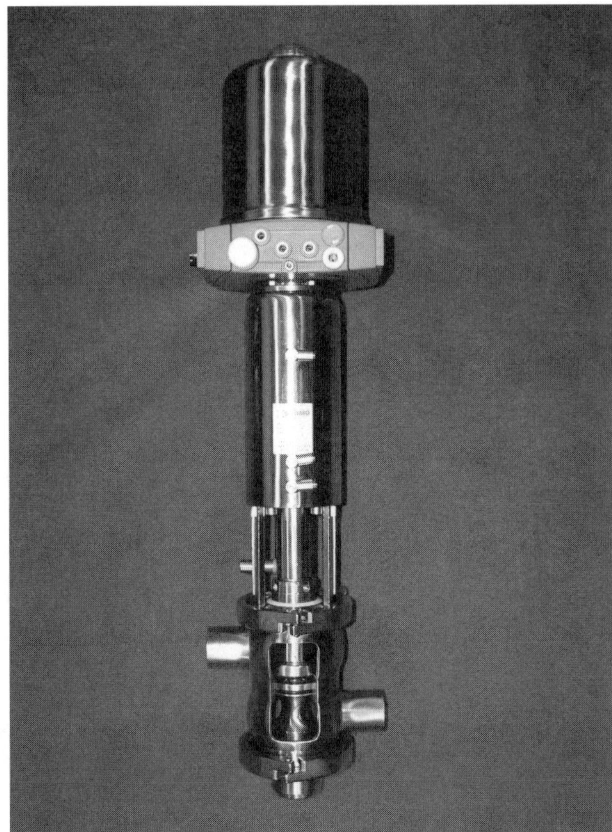

FIGURE 9 Rising stem double-seat valves.

the vent chamber when either seat is lifted, thus assuring that this area that is exposed to atmosphere during vent or leak detect function is always cleaned to the same level as the remainder of product contact surfaces.

Diaphragm Valves

The second major group of valves for hygienic duty are classified as diaphragm type.

Single-Port

These valves are generally configured with two ports separated by a raised weir (Fig. 11). A flexible diaphragm is retained between the actuator and the valve body and is attached to a backing plate and valve stem. As the stem is lowered, the diaphragm is pressed against the weir, effectively closing the passage between the valve ports and thus stopping flow. (Note: This should not be confused with a stem diaphragm commonly applied in rising stem valves to

FIGURE 10 Double-seat valves internals showing two seats and leak detect zone.

provide isolation of the product zone from the atmosphere around the valve stem.) The diaphragm movement can be either via a manual, multi-turn, handwheel, or via a pneumatic piston-type actuator. In either case, the diaphragm compression on the weir can be limited by a mechanically adjustable stop on the valve stem to prevent premature failure of the elastomer material. Another feature of the pneumatic actuation is such that valves may be configured for either normally open or normally closed operation.

Multi-Port

Using the basic building block of a diaphragm sealing against a mating surface, many specialty configurations have been developed to address specific process needs. Among these configurations, multi-port valves, ported or tapped boss valves, and zero dead-leg valves are most common.

Multi-port valves generally utilize a central valve block assembly, often with a common or central chamber, surrounded by two or more diaphragm/actuator assemblies, and capable of connecting the ports of the individual diaphragms to the

FIGURE 11 Diaphragm-type valve—disassembled.

common chamber or port. This configuration has the benefit of minimizing the space required for multiple, stand alone, valves and also minimizing the fluid volume when compared with a multiple valve arrangement.

Ported or tapped boss valves generally combine two or three individual diaphragm valves in a close coupled arrangement. The point of attachment usually involves boring a port into the thickened casting boss on either side of the weir and attaching a second valve perpendicular to the primary valve flow path. It is possible to install two valves, one on either side of the weir, on a primary valve. The process benefit to this configuration is that it is possible to fully drain the primary valve, up to and including the cavity on one side of the weir. This would also improve the cleaning operations in that typical distances from the end of a valve to the weir often exceed the desired maximum L/D of 1.5. Ported valves are often used for cleaning circuit isolation from vessels that may contain product, while cleaning operations are in use to clean adjacent piping or equipment.

Zero Dead-Leg

Zero dead-leg valves use the concept of a diaphragm valve with the exception that the diaphragm is seated on a port that is flush with the inside of the primary process line (Fig. 12). Generally configured as a tee, these valves include a main body that functions as the run of a piping tee with the diaphragm sealing the branch. The primary benefit of this approach is to allow piping to be configured without any branch dead-legs that could contribute to loss of product, cleaning challenges, or traps for condensate during steaming operations.

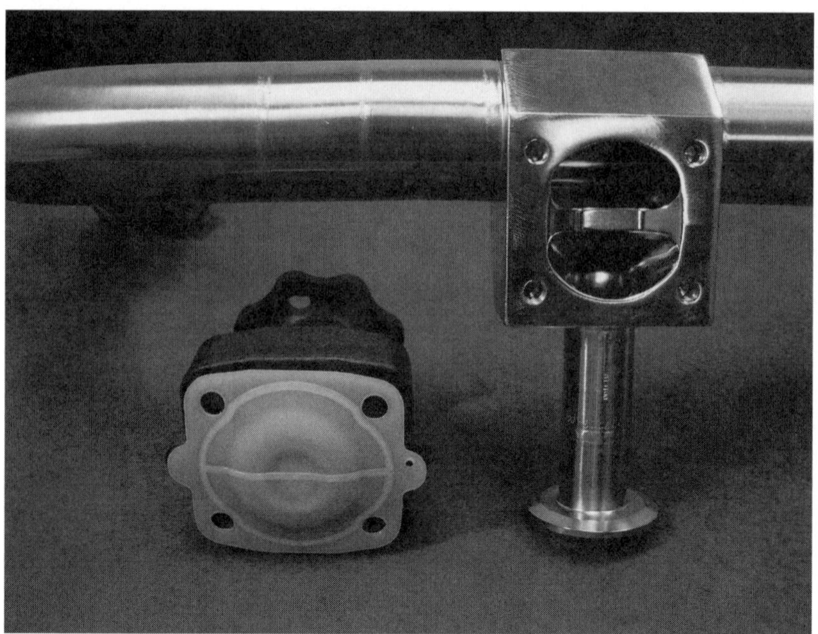

FIGURE 12 Zero dead-leg diaphragm valve—disassembled.

Other special duty valves with capability of installation in a CIP process would include sample valves, radial seal tank outlet valves, pressure relief valves, and modulating flow control valves. In every case, it is desirable to utilize automation and actuation of these valves and those described above to assure that cleaning solutions can contact all internal surfaces. However, in some sample valves and radial seal tank outlet valves, it is desired to clean the outward surfaces of the valve stem and seals while product is in the vessel or system. In these situations, it is important to prevent excessive pressure in the cleaning solution that could cause leakage past the valve seat. Similarly, these valves would also allow sterilization to be performed in the valve body while isolating the sterilant from the product.

Other Valve Types

Other valves often found in process facilities include ball-type and butterfly-type. These are most often applied in flow control of utility fluids and gases. These are not suggested for use in product contact applications. It is important to note that ball and butterfly-type valves are not considered to meet the requirements of CIP procedures due to the rotational aspect of their valve stems and the inability to adequately flush the stem seal zone.

PUMPS

Pumps are often used as the mechanical motivation for fluid product transfer operations. They are also used to perform CIP for the process piping and equipment. While it is possible to use gas overpressure to transfer fluids from a vessel, the ability to use speed controls to alter flow rates, pressures, and levels make the use of pumps very common in process systems.

Installation of pumps needs to consider accessibility for maintenance such as seal replacements, lubrication, and general housekeeping. It is also important to consider physical location as the prime movers for pumps include electric motors or hydraulic motors, both of which require attention to cooling as well as the fact that they often contain many crevices and openings that are not desirable from a clean process environment perspective. Electric motors with fan cooling are often considered to be sources of particulates not desired in clean process spaces. These issues are sometimes addressed by placing pumps in service or "gray" space and allowing the piping to penetrate walls to/from the process. It is also important to consider the needs for system drainage, line pitch, and vibration potential. In many process systems, pumps are equipped with low-point casing drains to enable complete evacuation of fluids from the system.

The hygienic aspects of pumps need to include appropriate materials of construction for the main pump chambers and, perhaps more important, all sealing points. With the exceptions of air-operated double diaphragm (AODD) and peristaltic type pumps, all other pump types include some sealing method on a rotating shaft. The complex nature of rotating seals is inherently difficult to flush and clean. Often the best design option is to place the seal mechanism outside of the product contact zone, relying on a minimal elastomer exposure to product and cleaning solutions. Another design method utilizes double mechanical seals, with two rotating seal surfaces, and the option to use a flushing liquid in the area between these sealing faces. In all cases, the rotating seal faces rely on either a flush liquid or the process liquid to lubricate the rotating mating surfaces. Although this lubrication uses extremely small quantities, it nevertheless presents a need for cleaning solutions to also be utilized to clean these surfaces.

Centrifugal Pumps

Centrifugal pumps designed for hygienic applications often include impeller retention methods that completely isolate the driving shaft from the liquid being pumped (Fig. 13). Static seals are used to prevent fluids from touching threaded fasteners such as impeller retaining bolts or the impeller includes an integral shaft that can extend beyond the fluid chamber (Fig. 14). The result is that process residues can be easily rinsed and cleaning solutions effectively cover all product contact surfaces. Similar construction and impeller retention methods are used for rotary lobe pumps. In fact, where it was once necessary to remove the lobes or rotors from rotary lobe pumps to assure access of cleaning solutions, there is now available, positive rotary lobe pumps that do not require disassembly prior to using CIP procedures.

Progressive Cavity Pumps

Progressive cavity pumps are also available for CIP applications. It should be noted that extreme care must be taken to prevent dry running of the rotor inside of the stator as this can cause rapid destruction of the mating surfaces of the stator.

FIGURE 13 Hygienic impeller retaining nut.

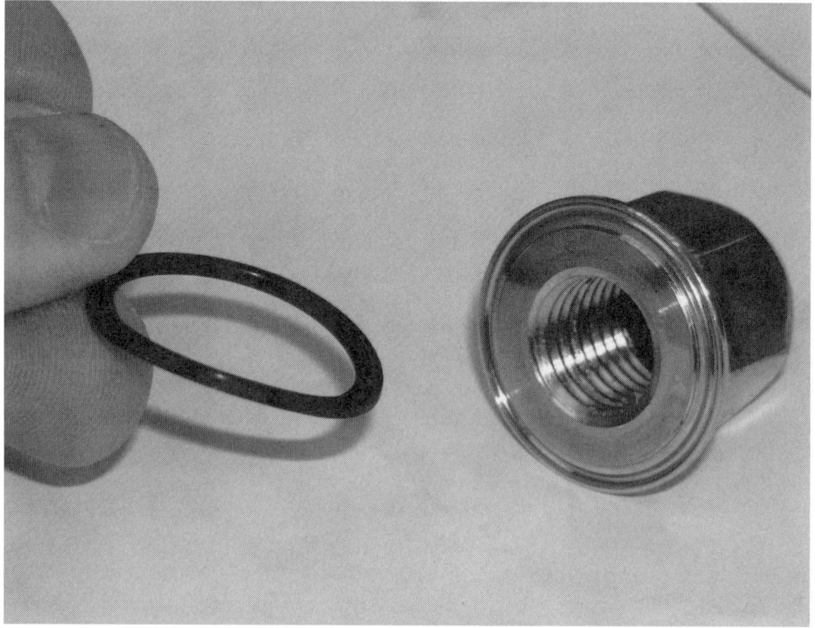

FIGURE 14 Hygienic retaining nut and seal.

Thus, for CIP duty, it is often preferred to provide instrumentation to assure presence of liquid or cleaning solutions before allowing the motor to start thereby providing the required lubrication of the rotor and stator.

Double Diaphragm Pumps

AODD pumps are often applied in locations where explosive atmospheres exist or where electrical apparatus is not desired. Alternatively, these pumps are excellent for self-priming and thus find favor where the suction line is frequently taken out of the source fluid, and thus air that enters the suction line must be easily evacuated from the pump. Their construction is such that there are no dead ends or extremities to make flushing and cleaning difficult. Also, the nature of the inlet and outlet check valves are such that these AODD pumps can be placed in lines where cleaning solutions are pumped through without a need to cycle the pump. Design and use of AODD pumps must consider the cycle rate of the alternating diaphragm positions as this can cause pressure and flow fluctuations in the fluid being pumped.

Peristaltic Pumps

Peristaltic pumps are available in a variety of sizes and are also excellent where self-priming is needed. The flexible hose in these pumps is used in two primary fashions. One is to consider the hose disposable and non-cleanable. This is often the case with small diameter hoses where cleaning would be very difficult or where the fluid being pumped presents a significant cleaning challenge such as coloring fluids or high potency flavorings. Alternatively, the hose can be considered for multiple duty. Either the hose is allowed to remain in place during cleaning operations and the pump is operated to move cleaning solutions through the hose or the hose is removed from the pump housing and is made part of an in-line cleaning path.

There are also applications for pumps involved in the manufacture of APIs that require transfer of volatile solvents, potent compounds, and fluids containing hazardous intermediates. These duties may require special seals to prevent emissions of the materials and/or features that are not common to traditional hygienic pumps. However, these applications do still require cleaning procedures to eliminate hazardous materials and to prevent carry over to subsequent processes and thus, when possible, the construction of these pumps applies the same features that allow CIP to be effective.

INSTRUMENTS

Process systems include instrumentation for monitoring and controlling of a wide variety of variables including flow, pressure, temperature, conductivity, level, and mass to name a few. These instruments typically include a sensing element that requires contact with the process fluid or material in order to measure a given variable. This invasive nature of the sensor will accordingly require cleaning and sterilization in the same manner as the rest of the process system. Thus, it is important to consider the connection methods, service accessibility, ongoing metrology needs, drainage of insertion points, and the materials of construction such that they are compatible with the process materials and cleaning solutions.

Connections for process sensors most frequently utilize the hygienic clamp-type ferrule system to retain a sensor in a pipeline (Fig. 15). This method also frequently utilizes the branch of a tee or a tee at a direction change to allow insertion of long sensor devices. Often, the accessibility for maintenance and calibration requires use of the run of a tee, and thus the position must also consider direction of fluid flow to assure that appropriate pitch can be maintained for system drainage. There are some sensors that are fabricated with a threaded system that utilizes an adapter to accommodate various line sizes and hygienic connections. In these cases, an elastomer seal is utilized to isolate the threads from the process and cleaning solutions.

Ultimately, the sensor must be fabricated from materials suited to the environment of use and in a manner that assures ease of cleaning without manual intervention. 3-A Standard 74 has been developed to offer guidance to instrument manufacturers for radii, materials, and connection methods.

Technologies exist to accommodate most sensor types in hygienic process systems, including pH. pH has traditionally been the most challenging parameter to measure, given the frequent need for calibration, inclusion of glass in sensors, and perhaps most significant is the exchange of electrolyte with the fluid being measured. Use of ion-sensitive field-effect transistor technology has allowed elimination of both glass and electrolyte from pH sensors, creating a significant improvement in hygiene and removal of contaminants from a process system. It is also possible to use multivariable sensors to reduce the number of devices installed in a process.

The use of instrumentation also infers the need to communicate between the instrument/transmitter and a control or automation system. Frequently, this

FIGURE 15 Temperature sensor and well clamped into hygienic process tee.

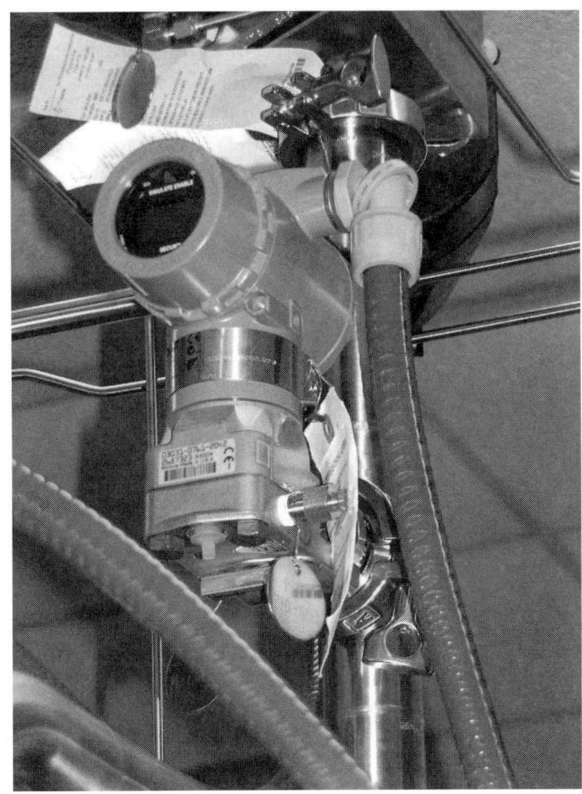

FIGURE 16 Pressure sensor with nonmetallic conduit.

communication is via a hardwire connection. Although wireless technologies are frequently used in industrial processes, and pneumatic signals are still utilized in some classified environments, the hardwire connection is most prevalent in hygienic process control. Whether this is analog, Ethernet, or a fieldbus signal, it is important to consider the wiring methods and their impact on housekeeping in a good manufacturing practice environment. Use of nonmetallic conduit and fittings is encouraged to minimize opportunities for corrosion and particulate generation (Fig. 16). Use of all welded, sanitary tubing as low-voltage conduit is frequently an accepted method to maintain hygienic conditions in and around instrumentation. This system may require use of plenum-rated cables, but is ultimately a zero maintenance installation.

REFERENCES

1. 3-A Sanitary Standards, Inc., 6888 Elm Street, Suite 20, McLean, Virginia 22101, U.S.A. Tel.: 703 790 0295; Fax: 703 761 6284; www.3-a.org
2. American Welding Society, 550 N.W. LeJeune Road, Miami, Florida 33126, U.S.A. Tel.: 800 443 9353; www.aws.org
3. ASME International, Three Park Avenue, New York, New York 10016-5990, U.S.A. Tel.: 800 843 2763; Fax: 212 591 8676; www.asme.org

4. ASTM International, 100 Barr Harbor Drive, P.O. Box C700, West Conshohocken, Pennsylvania 19428-2959, U.S.A. Tel.: 610 832 9500; Fax: 610 832 9555; www.astm.org
5. EHEDG Secretariat, Avenue Grandchamp 148, B-1150 Brussels, Belgium. Tel.: 32 02 761 7408; Fax: 32 02 763 0013; www.ehedg.org
6. ICH Secretariat, c/o IFPMA, 15 ch. Louis-Dunant, P.O. Box 195, 1211 Geneva 20, Switzerland. Tel.: 41 22 338 32 06; Fax: 41 22 338 32 30; www.ich.org
7. Tuthill AH, Brunkow R. Stainless steels for bioprocessing. BioProcess Int 2004; 3(11):46–53.

13 Cleanable Liquids Processing Equipment and Systems

Dale A. Seiberling
Electrol Specialties Company (ESC), South Beloit, Illinois, U.S.A.

INTRODUCTION

Chapter 1 of this book identified the various parts of a cleanable liquid process and described the operation of clean-in-place (CIP) and steam-in-place (SIP) for a simple two tank train. Chapters 2–8 of this book have provided detailed information about the many issues to be considered when designing a new pharmaceutical or biotech process, i.e., project planning, water for the CIP system, composition of cleaning agents, cleaning cycle sequences, CIP system components and configurations, and CIP system instrumentation and control and cleaning agent injection systems. Chapters 9–12 have elaborated on design of CIP supply (CIPS) and return (CIPR) piping, spray devices, materials of construction and cleanable in-line components. This chapter will focus on methods of combining the in-line piping components, process vessels, and affiliated process systems, i.e., chromatography systems, ultrafiltration (UF) systems, centrifuges, and filters, with sprays and the required CIP piping to develop a CIPable process for liquid products or components, varying in the degree of complexity. Subsequent chapters will consider application to a dry solid process and the active pharmaceutical ingredients process. The book will continue with consideration of CIP system troubleshooting, waste handling, commissioning, and validation. The final chapter will address regulatory compliance requirements.

The figures and photographs used to illustrate this chapter have been extracted from successful projects, the figures often in a generic format. The objective is to illustrate how basic concepts can be utilized in different combinations to solve a variety of different problems.

Design Criteria

Flexibility vs. Full Automation

The final design of a liquid handling process will be affected by decisions about the degree of flexibility required versus the degree of automation desired. Any process intended to be fully cleaned by CIP procedures must be developed giving significant consideration to the number of individual CIP circuits. Product containing vessels and piping must be separated, meaning isolated, from vessels and piping containing CIP fluids. This isolation may be accomplished by manual "make–break" connections or valves. The valves may be manually operated or automatically controlled for the process with provisions being made in the cleaning program software to properly "pulse" or "sequence" the valves to assure proper cleaning of all internal parts and the interconnecting piping. In either case, the design must provide for easily organized control of all flow paths for processing and cleaning, and in processes which involve computer-based control, some means of monitoring the flow path is essential.

Flexible Tubing, Transfer Panels, or Automated Valves

Flexible tubing provides the maximum flexibility for making process and CIP connections but lacks a means of organizing and verifying connections. Flexible tubing is also less capable of meeting the CIP requirement for drainability. Flex-lines should be limited to use for temporary connections for liquid ingredient additions, or final transfer to a portable tank.

U-Bend transfer panels and/or automatically controlled valves are the preferred components to be used in a CIPable process design for either product development work or full-scale production.

HISTORICAL DEVELOPMENT

Chapter 1 provided historical references to CIP as applied in dairy, brewing, beverages, and food processing by the mid-1970s. These processes shared a unique operating requirement in that many large tanks contained product for extended periods of time, and required connections to piping and process systems that generally ran less than 14 to 16 hours. A traditional practice was to clean all piping after each period of use, and clean vessels whenever they were emptied (with restriction on maximum period of use). This operating requirement led to the development of extensive CIP supply/return (CIPS/R) piping systems to facilitate connections to establish a large number of CIP circuits.

Chapter 1 also defined the typical pharmaceutical/biopharmaceutical process as being comprised of a train of vessels, and as the product moved from vessel to vessel the soiled equipment generally consisted of the empty vessel and a portion of transfer piping and associated components. Seiberling therefore suggested in 1986 (1) and 1987 (2) the consideration of "Integrated piping system design for product and CIP solution use." This concept was further reviewed by Seiberling in 1992 (3) and Seiberling and Ratz in 1995 (4). The concepts first described in those publications are the basis of the generic concepts to be described in the remainder of this chapter. Marks further explained other aspects of the "integrated approach" in 2001 (5). Chrzanowski, Crissman, and Odum published a treatise on valve assembly use in 2001 (6) that further explains the means of achieving the functional capability suggested in some figures of this chapter. The use of CIP transfer panels as described by Shnayder and Khanina (7) is more typical of the dairy, brewery, and beverage facility applications from 1960 onwards, but applicable to some product development facilities that may favor flexibility over automation.

TYPICAL PROCESS APPLICATIONS

A Generic Process Train with U-Bend Transfer Panels

Figure 1 illustrates a basic concept applicable to almost all multi-tank process trains. The vessels could be variable in volumes, and as many as are required for a complete process can be organized in this manner. The indicated valves must be automated to achieve validatable CIP.

Mixproof valves are shown for CIPS only. U-Bend transfer panels are the suggested method of configuring flow paths for all other process and CIP/SIP requirements. Whereas the generic process train discussed in Chapter 1 used a single panel at the bottom of the tank, this concept shows high-level and low-level panels. A preferred approach for multiple vessels serving a common function, i.e., media prep, fermentation, or buffer prep, would combine all upper level

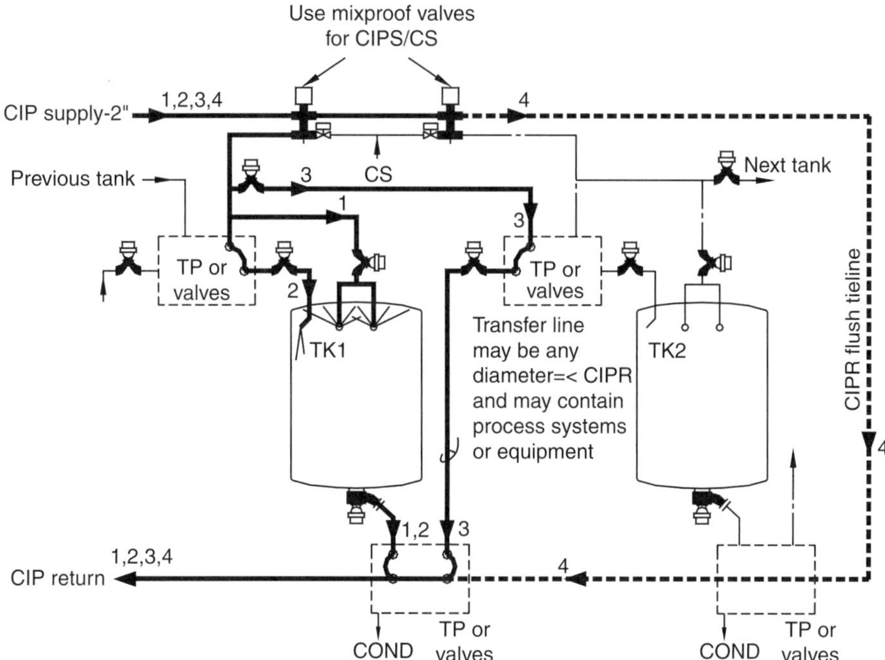

FIGURE 1 This generic two-tank process train illustrates a CIPable and SIPable design, with sub-circuits for clean-in-place of the vessel and transfer line identified by number for reference in the narrative.

connections in a single panel, and all lower level connections in a single panel, thereby reducing the required number of CIPS and CIPR drops. Each vessel and it's outlet transfer line would be supplied with CIP fluids from a single mixproof valve in the CIPS header. All other valves in the vessel piping can be diaphragm type, close coupled, or zero-deadleg type. Device sequencing software would control flow to the vessel sprays full time (1), pulsed flow through the inlet nozzle (2), supply to the discharge end of the transfer line (3), and the CIPR bypass (4) to eliminate CIPS/R deadlegs with minimal valve requirements. The flows would all combine at the vessel outlet for return to the CIP skid. The transfer line can be any size required, may incorporate other process equipment, and would be cleaned in reverse flow in parallel with spraying the tank, with flow through the line perhaps 10% to 15% of the total time for each phase (see also Chapter 1, "Tank and Line CIP in Combination").

Valve Pulsing and Split Flow Considerations

The CIP circuit described in Figure 1 includes the need to clean through an inlet line to the vessel (Path 2) in the direction of the process flow and a transfer line to the next vessel in reverse flow (Path 3). Both of these paths can be leaned in parallel with the operation of the sprays to the vessel. For purposes of discussion, assume that the split paths originate within a few feet of each other in the 2-in. CIPS line to the transfer panel and that:

1. The sprays (Path 1) require 60 gpm (230 Lpm) at 25 psi (170 kPa) and are supplied by a single 1.5-in. (38 mm) line consisting of 15 ft of tube, 5 elbows, and 1 diaphragm valve. Friction loss before the sprays at a flow of 60 gpm would be 25.8 ft and for Path 1 including the sprays would be 83.8 ft.
2. The inlet line (Path 2) includes 10 ft of 2-in. (50 mm) line, one 2-in. (50 mm) U-Bend, five 2-in. (50 mm) elbows, and one 2-in. (50 mm) diaphragm valve. Friction loss at a flow of 60 gpm (230 Lpm) would be 6.9 ft.
3. The transfer line (Path 3) includes 30 ft of 1.5-in. (38 mm) tube, 3 elbows, and a diaphragm valve to the TP and an additional 30 ft of tube, two 1.5-in. (38 mm) U-Bends, five 1.5-in. (38 mm) elbows, and a diaphragm valve. Friction loss at a flow of 60 gpm (230 Lpm) would approximate 47.8 ft.

When the valve controlling flow through Path 2 is opened in parallel with Path 1, the flow will split with approximately 46.5 gpm (184 Lpm) through the inlet line and 5 gpm (50 Lpm) through the spray, at a head loss through both paths equal to about 4.2 psi (28 kPa), as determined by successive approximations. The inlet line contains a volume of about 2.5 gal, and at 0.8 gpm, will flush completely in three seconds. A pulse of five seconds duration once each minute, throughout the program, will generally be sufficient to assure good results.

When the valve controlling flow through Path 3 is opened in parallel with Path 1, the flow will split with approximately 34.2 gpm through the transfer line and 25.8 gpm through the spray, at a head loss through both paths equal to about 15.5 psi. The transfer line has a volume of about 4.6 gal and at 0.58 gpm, will flush completely in 7.9 seconds, and a pulse of 10 seconds duration once each minute will generally be sufficient to assure cleaning the full length of line. A device sequencing program would be required to supply Paths 1 and 2 for 5 seconds, Paths 1 and 3 for 10 seconds, and then spray the vessel alone (Path 1) for 45 seconds. The transfer line will remain full and though the inlet line will probably drain, the chemical cleaning activity will continue through the period of no flow. Considerable experience has demonstrated that valves should be controlled to assure flow through each path three to four times during the pre-rinse, four to six times during the solution wash, and four to six times during the combination of the post-rinse and acidified final rinse. If a pure water final rinse is used, the valves should be moved two to three times for this step also.

The above calculations suggest that the results to be expected when placing a long small diameter line or a short larger diameter line in parallel with the spray supply path, and the numbers for this example were chosen to be representative of worst-case conditions in real-world applications. The friction loss estimates through piping were calculated via a program by Domanico (8). The split flow approximation was done by a program created and used by the author for many years.

Mixproof Valves and U-Bend Transfer Panel Combinations

The design mandate for a large biotech R&D facility project was to provide for use of six to eight vessels in each of several suites in combination with equipment in four support rooms, for short- or long-term periods of use for product development.

The large U-Bend transfer panel shown in Figure 2 enabled the configuration of the entire process, following which all process and CIP operations were fully automated. A second mandate was to permit any two larger tanks in any of five suites to be used in similar combination with the larger tanks in other suites

FIGURE 2 This larger process/clean-in-place transfer panel utilized loop-type headers and two different U-Bend lengths to configure a product development suite for a multiplicity of operations, all fully automated. *Source*: Courtesy of Electrol Specialties Company.

to configure an alternative automated process for clinical trial production. All U-Bends were monitored by proximity sensors for validated process and CIP requirements. Figure 3 illustrates a typical rising stem mixproof valve group or manifold used in combination with the transfer panel to control CIPS and CIPR flow, product transfers, addition of solvents, pure water, etc., to any process vessel in the suite. Figure 4 is a plan view of the valve group that illustrates the "looped headers" in plan and elevation view and establishes the size at approximately 3.5 ft wide by 7 ft long, for seven valves each for eight vessels. Figure 5 shows how the valve group, vessels, and transfer panel were installed in a multi-floor process to maximize the use of gravity to drain vessels and piping to the support suites and to the CIP skid.

This panel is mounted on the floor beneath a group of eight tanks on a pad with some piping continuing from the panel through the floor below to support suites. The eight pair of lines leaving the top of the panel connected to the fill valve of a loop for each tank and to the outlet valve of each tank. The large primary U-Bends on this panel would connect any of the eight tanks to any of seven fill headers, any of seven discharge headers, or to a common transfer header.

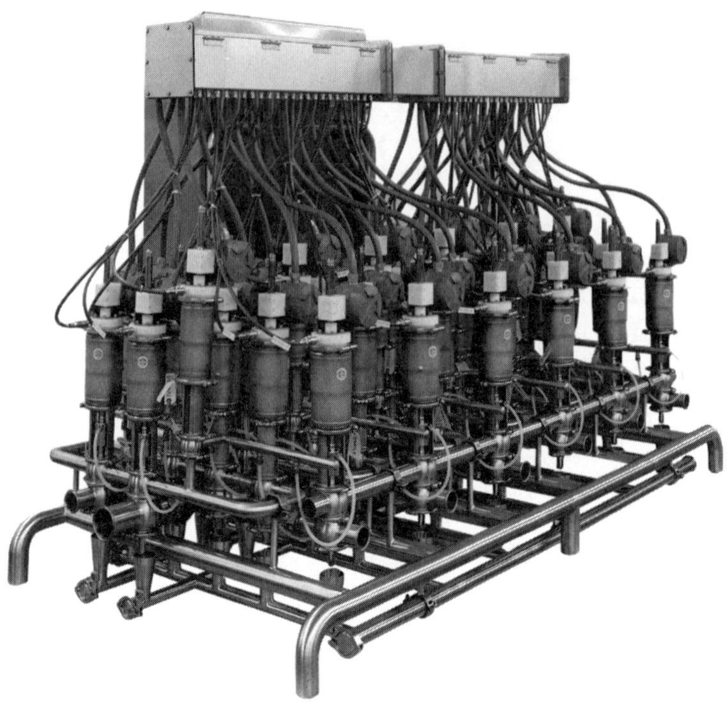

FIGURE 3 This shop-fabricated manifold, consisting of 42 mixproof, rising-stem valves, interfaced six vessels to a transfer panel similar to Figure 2 and provided CIP supply/return pure water and a solvent, all under automated control. *Source*: Courtesy of Electrol Specialties Company.

These headers in turn are connected via secondary U-Bends at either end of the panel to other processes such as fixed or portable chromatographic columns, fixed or portable evaporators, or portable equipment in other processing suites below. All U-Bends were proximity sensor equipped to permit verification of position, unchanged during normal use of a process configured for a particular product development function.

Chapter 12 provides pertinent information about valve and line sizing and design of the "looped headers" in both the transfer panels and valve groups to minimize space, cost and product holdup, and most importantly, avoid any deadlegs in the process piping.

Mixproof Valves in a Dedicated Production Process

Figures 2–5 described a product development process that required a maximum degree of flexibility in the use of the equipment. The next example is a large-scale production process dedicated only to buffer makeup and transfer, and supply to two dedicated process trains that included chromatography columns. Figure 6 shows a group of mixproof valves applied to control product flow from three buffer/chemical preparation vessels and CIPS flow to 10 downstream hold tanks, which in turn supported two chromatography trains.

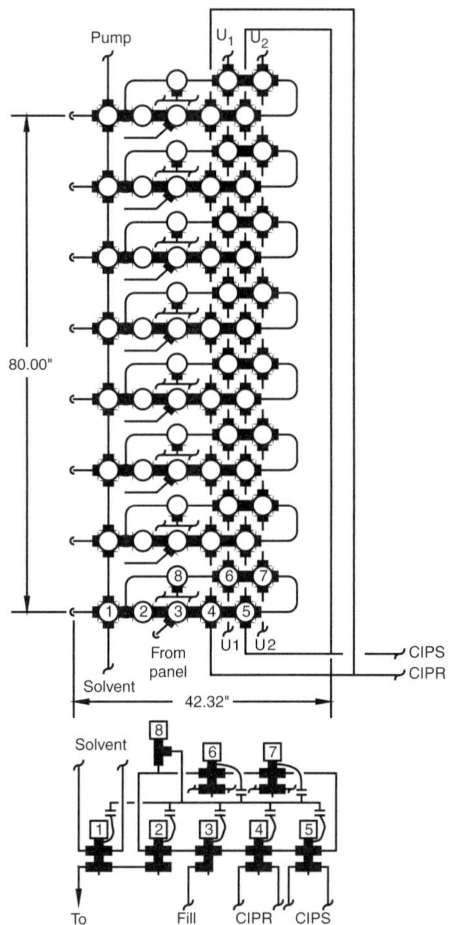

FIGURE 4 The eight-vessel valve manifold used with the transfer panel in Figure 2 is shown here in plan and elevation view. Note the looped headers on lines to and from the manifold.

Downstream of the 10 hold tanks, and one floor beneath, the buffer distribution valves shown in Figure 7 controlled flow of buffers to two chromatography trains of two columns each.

A partial schematic of 4 of the 10 buffer hold vessels, one pair of feed and collection tanks and two columns in series is provided in Figure 8. The valves for the second train of identical equipment are in the same valve group shown in Figure 7.

Mixproof valves were also utilized to bypass the column(s) during CIP of a buffer supply vessel and it's piping to the valves in the mixproof valve group. A buffer vessel and it's associated transfer line and mixproof valves would be cleaned whenever the vessel was empty. The flow of CIP fluids for any vessel would be continuous from the supply valve to the vessel sprays (1), would be pulsed through the inlet valve (2), and would be delivered via a common line from the supply valve group to the distribution valve group and return through the selected vessel outlet line in reverse flow to the vessel CIPR connection (3). This illustration is an extraction of a validated commercial installation.

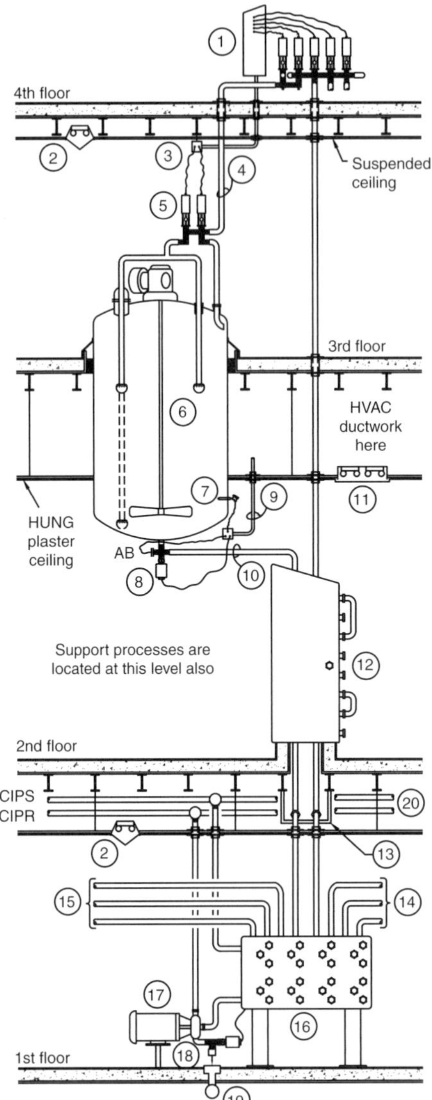

FIGURE 5 This elevation drawing illustrates the preferred arrangement of valve manifolds, vessels, and large transfer panel, and support rooms for a highly automated, CIPable process development facility.

Rising Stem Valves for Mixproof Isolation of Two Tanks in a Train

Figure 1 included rising stem mixproof valves for CIPS isolation and control and used transfer panels to configure the process for production or CIP. This approach required the repositioning of U-Bends between production and CIP, a common practice. Figure 9 uses a rising stem mixproof valve in combination with two close coupled diaphragm valves to isolate the destination vessel from the source vessel, eliminating all requirements for manual changes between hot water sanitizing, production and CIP of the vessel and its transfer line ti the next vessel

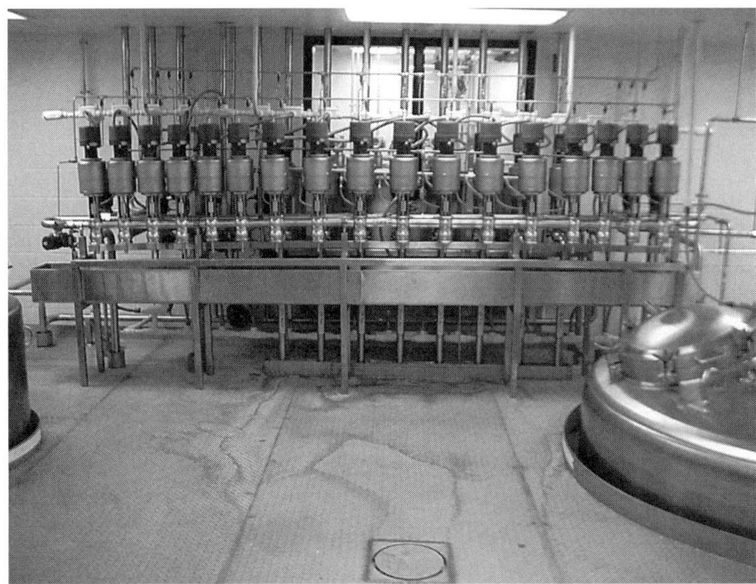

FIGURE 6 This mixproof valve manifold interfaced three Media Prep vessels to 10 hold tanks to support four chromatography columns in a highly automated CIPable process.

in the train. The sub-circuits for CIP, identified by the numbered arrow heads, include vessel sprays full time (1), pulsed flow through the inlet nozzle (2), supply to the discharge end of the transfer line (3), and the CIPR bypass (4). This schematic provides the same function operation and requires the same control scheme as Figure 1, but no manual changes are required between process and the CIP/SIP procedures.

Some process operations require a brief period of recycle from the source tank through a downstream process and back to the source tank. This requires the addition of a second mixproof isolation valve adjacent to the valve shown in Figure 9, a recycle line back to the source tank, and the required diaphragm valves for process flow and CIPS needs.

Diaphragm-Type Valves for Mixproof Isolation of Two Tanks in a Train

The traditional rising stem mixproof valve is not suitable for sterile processes, as actuation causes stem movement into an unsterile environment and the mixproof cavity is briefly exposed to the same environment in the moments required for the double disc seats to close together. As the valve transitions from closed to open. Fortunately, new developments in fabrication of diaphragm valve assemblies as described by Chrzanowski et al. (6) make it possible to achieve the same functional capability by properly combining diaphragm valves in a shop fabricated assembly.

Figure 10 illustrates the use of an assembly of six diaphragm valves for product in combination with two small "bleed" valves to control all product, CIP and SIP flows, and to isolate TK1 from TK2. The diaphragm valves require considerably more space and careful design to assure proper drainage, but only

FIGURE 7 A second lower-level mixproof valve manifold controlled flow from the ten hold tanks to the four columns. The design provided for CIP of all piping to and from a selected hold tank in combination with the vessel under fully automated control.

one additional I/O point is needed to replace the non-sterile rising stem mixproof isolation capability of Figure 9 with valves which can maintain sterility during all process operations. The sub-circuits for CIP, identified by the numbered arrow heads, include vessel sprays full time (1), pulsed flow through the inlet nozzle (2), supply to the discharge end of the transfer line (3), and the CIPR bypass (4). During flow through Paths 2 and 3, the associated bleed valves would be pulsed to clean through the valve seat.

The reader should now understand that the CIP concept for cleaning a discharge transfer line with a vessel can be identical for two tanks interfaced via U-Bend transfer panels Figure 1, rising stem mixproof valves Figure 9, or diaphragm type mixproof valve manifolds Figure 10.

Bioreactor with Multiple Legs

The vessels in the above examples included only a spray inlet, a product inlet, and an outlet line. A bioreactor of any size will include many additional legs that must receive both CIP and SIP flows to clean and sterilize the vessel and the legs in two separate operations.

A preferred method of meeting this need is to provide a manifold or group of CIP valves on the bioreactor skid (or field installed piping on large vessels) and supply these valves with the CIP fluids and clean steam (CS). CIP fluids from the same source would be supplied to the media line at it's origin, to these valves, and in some cases, to the destination of the harvest line. Alternatively the outlet (harvest) line would connect to CIPR, as would a skid mounted manifold to collect condensate from all traps and filter housings. This line would be valved to CIPR for

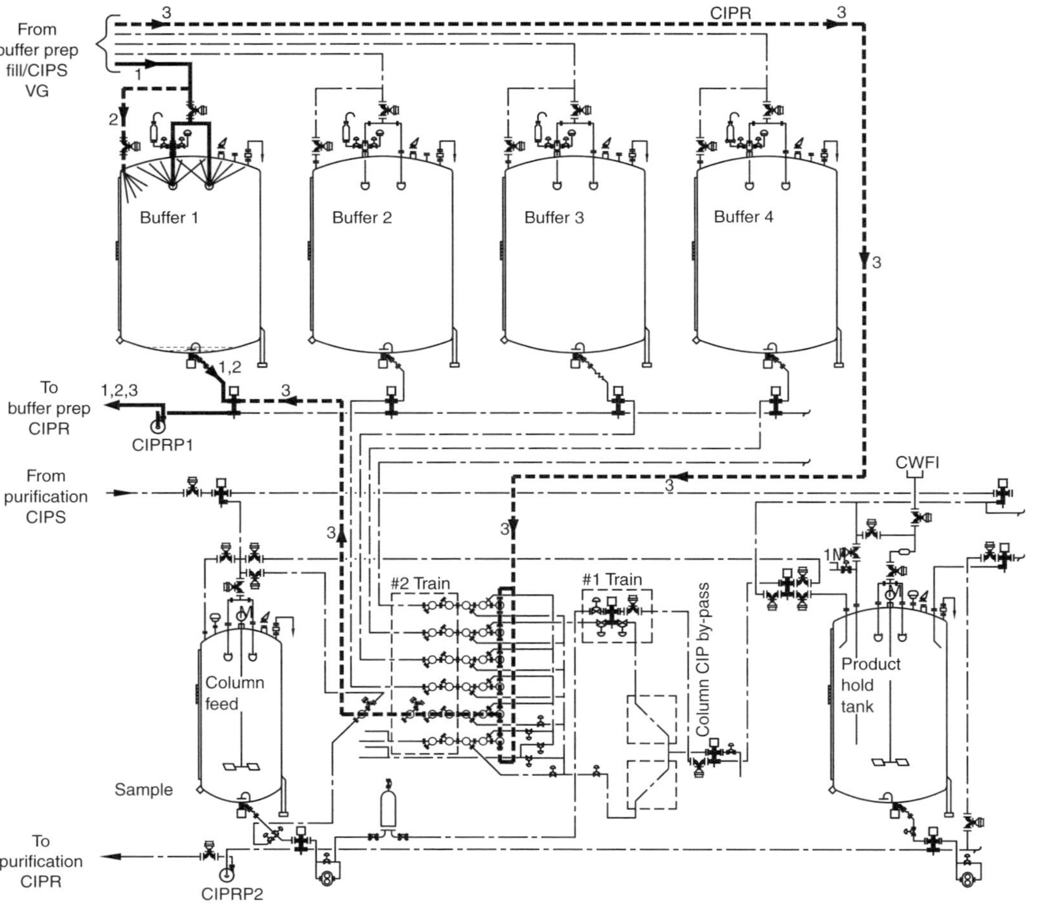

FIGURE 8 A partial extraction of the process schematic illustrates how Circuit 3 flow of flush, wash, and rinse solutions was sequenced through the piping and the vessel.

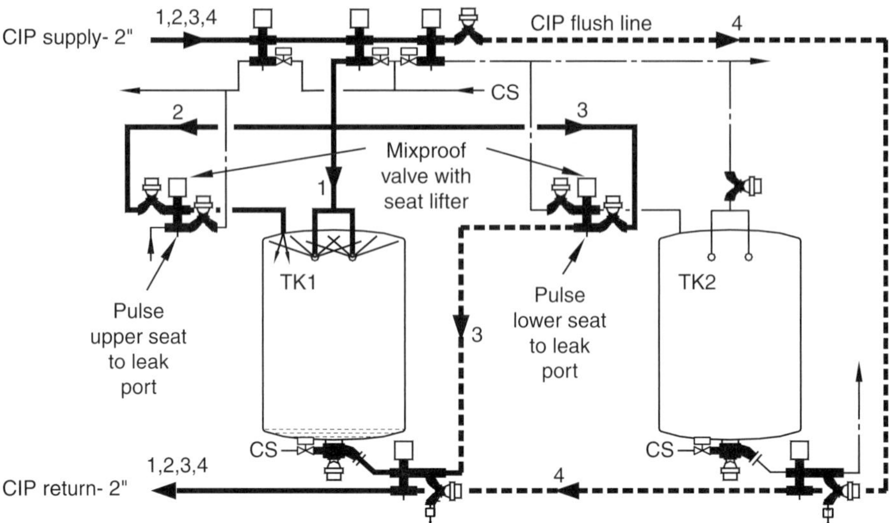

FIGURE 9 Rising stem mixproof valves, applied three-per-vessel, provide automated control of all product and CIP fluids and isolation of all fluid streams. A process requiring recycle to the source vessel would require a fourth mixproof valve. The solid and dashed lines and numbered arrow heads define the CIP sub-circuits described in the narrative.

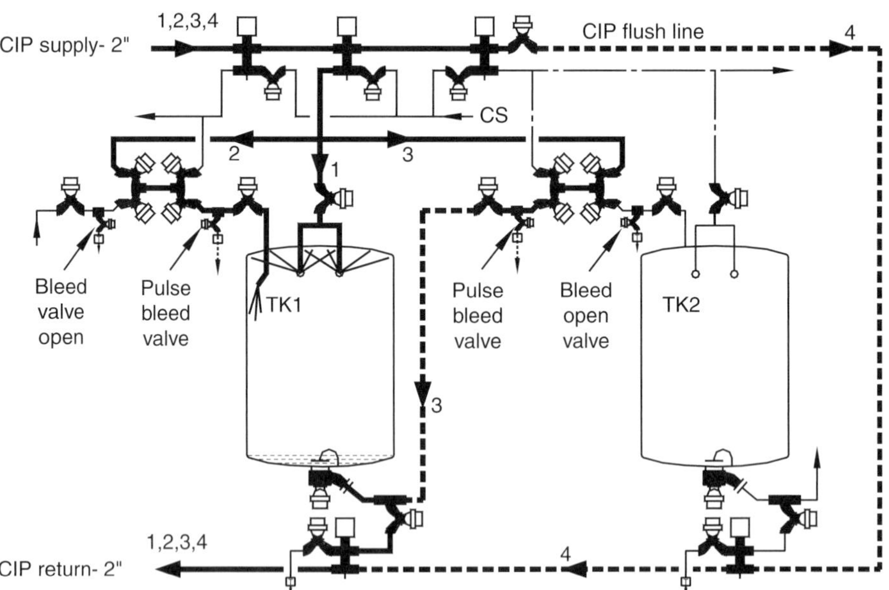

FIGURE 10 Diaphragm valves in valve assemblies can be applied to isolate product and CIP fluids and provide the identical operating capability of Figure 1 or 9. The solid and dashed lines and numbered arrow heads define the CIP sub-circuits described in the narrative.

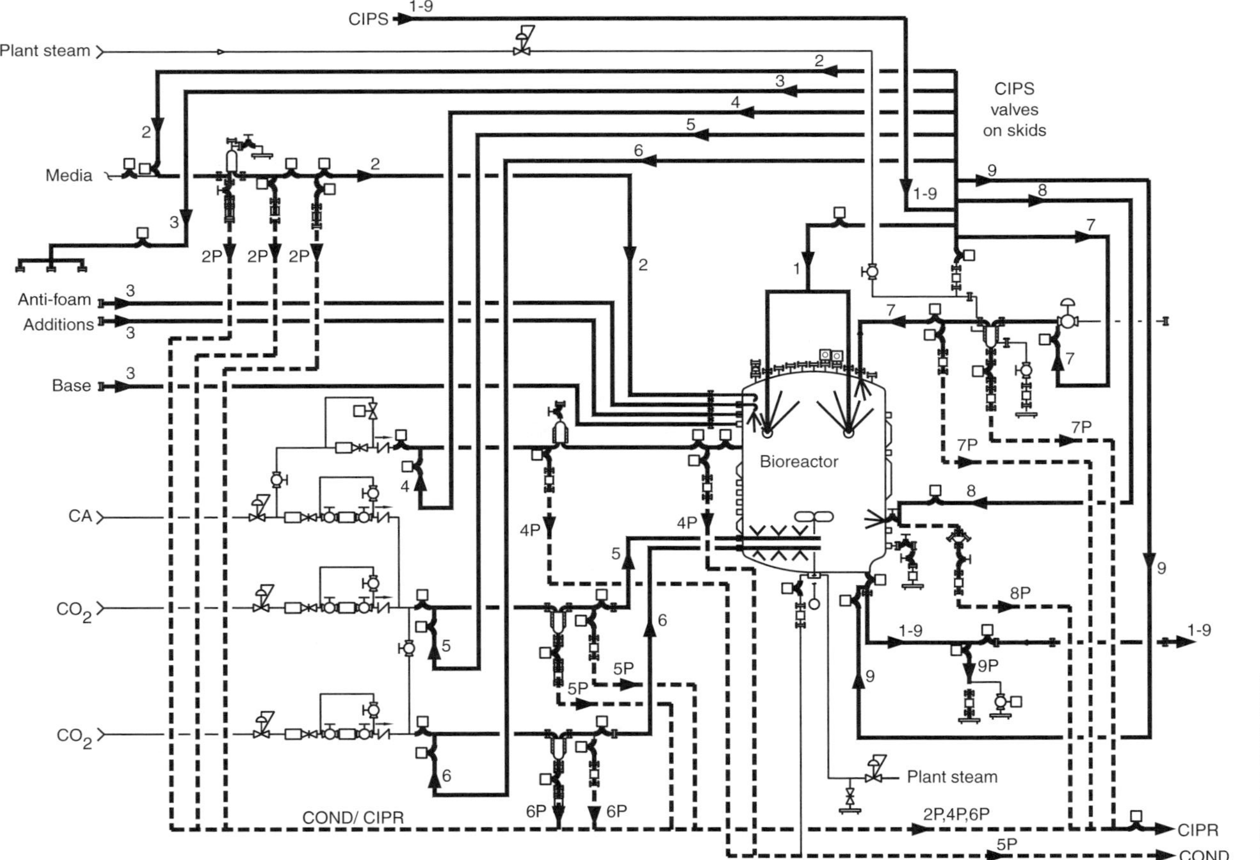

FIGURE 11 The CIP of a bioreactor and its multiple piping legs is a complex CIP engineering challenge, which, however, is best resolved via the same principles discussed previously in this chapter.

CIP and to drain for SIP, permitting the pulsing of CIP fluids through all valves and traps from the beginning to the end of the CIP program. In most applications, the flow required to supply the sprays will require a pressure sufficient to permit sequencing through the media, additive, gas, and exhaust lines, and sample valve in parallel with the sprays, and still maintain a viable flow through the sprays, at, of course, reduced pressure. Figure 11 suggests a total of nine paths including vessel sprays (1), media line (2), additive lines (3) by installation of a CIP jumper as three in parallel, gas lines (4–6), exhaust line (7), sample valve (8), and the steam line to harvest valve (9). Paths 2 to 8 all include automatic valves through traps to a condensate (COND)/CIPR header, and these valves would be "pulsed," i.e., opened for only two to three seconds when flow was through the associated leg.

The design objectives should be to achieve high velocity flow through each leg individually for sufficient time to replace all fluid in the leg, and "pulse" CIP fluids through the traps which would otherwise be deadlegs, recovering this solution during the chemical wash phases via the COND/CIPR header. A repetitive sequence of 60 to 90 seconds that *first* provides flow through each individual leg in parallel with the sprays, and then full flow through the sprays for 30 to 45 seconds has provided excellent results. The number of times through the sequence during each program phase will vary with vessel and line size and line length. The author's experience on multiple projects involving perhaps 60 vessels varying in size from seed reactors of 100 L cleaned at 40 gpm (150 Lpm) to production reactors of 15 kL cleaned at 100 gpm (380 Lpm) has confirmed the efficacy of this method. Leg diameters have ranged from 0.5 (12 mm) to 2 in. (50 mm) and more.

Though not fully illustrated in Figure 11, the media line to this bioreactor in the actual application was cleaned separately from the vessel and legs via CIPS at

FIGURE 12 This photograph shows two small bioreactors depicted schematically in Figure 11 connected to a common transfer panel for harvest/CIP flow selection.

the source transfer panel and double block and bleed valving to the COND/CIPR manifold and thence to CIPR.

Figure 12 shows two small bioreactors, each designed in accordance with Figure 11 connected to a common transfer panel via the upper (harvest) outlet line. The upper port on the panel connected to a transfer panel for distribution to six larger bioreactors. The lower port was to a CIPR header. The lower line from the bioreactors is the COND/CIPR header from the skid, and terminated with valves to drain, or to the CIPR header as described above. The manual preparation for CIP of either vessel and it's legs consisted of: (*i*) installing the CIP jumper for the additive lines, (*ii*) removing all filter housings and cartridges and reinstalling the empty housing, and (*iii*) positioning the U-Bend on the outlet transfer panel to connect the harvest line to CIPR.

Integration of Blenders and Mixing Tanks

Large-scale biopharmaceutical processes require many vessels for media and buffer preparation. Powder blenders used to add dry solids to water recycled from the vessel are often cleaned individually, and even by manual means, a labor intensive procedure. Figure 13 shows a powder blender based on the use of a centrifugal pump, capable of receiving large volumes of dry ingredients from portable intermediate bulk containers (IBCs) placed on the support frame by lift truck and then connected to the transition funnel on the mixer inlet. Following powder addition and rinsing of the IBC by a spray in the funnel, the IBC could be removed, and the funnel capped.

The powder mixer pump was then used to transfer the mixed liquid through filters to the next vessel, either bioreactors of buffer hold tanks. Then, for vessel CIP,

FIGURE 13 This photograph shows a high-capacity dry powder blending system in the outlet leg of a buffer prep vessel. The blender pump serves also as the filter supply pump and CIP return pump.

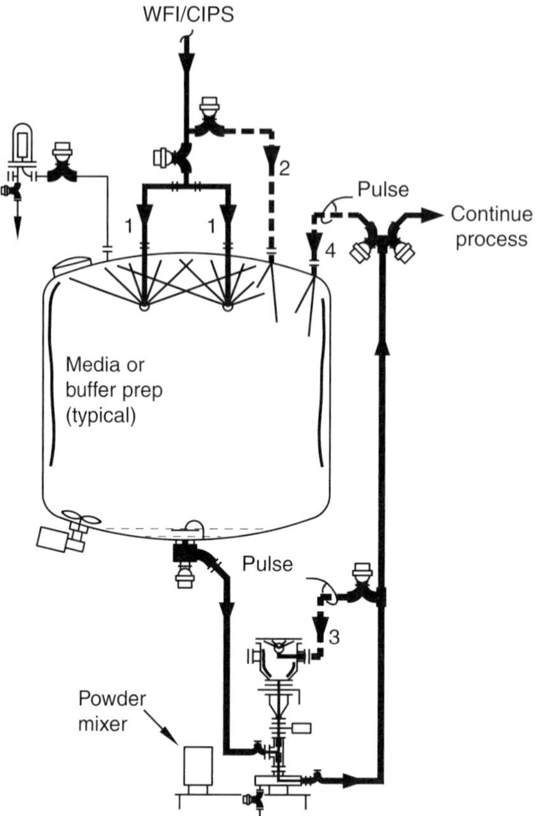

FIGURE 14 This schematic illustrates clean-in-place sub-circuit flow for CIP. The narrative describes the flows depicted by the numbered arrowheads.

the powder blender pump operated at the required revolutions per minute (RPM) by a variable frequency drive (VFD), served as a CIPR pump, providing the motive force to spray clean the large filter housing in the transfer path.

A schematic of the above is presented as Figure 14. All components are permanently installed except the CIP cap, installed following removal of the IBC. In accordance with the numbered arrow heads, CIP fluids are delivered full time to the sprays (1). The inlet line received only "pulsed" flow (2), as did the very short recycle connection to the spray in the funnel (3), and the recycle line (4) required for mixing.

The presence of the powder mixer (a centrifugal pump, actually) in the outlet line and the need for a high filtration flow and hence 2-in. (50 mm) line diameter made it possible to use the transfer line forward as a CIPR line for the vessel, and incorporate the filter housing CIP downstream.

UF Filter CIP in Transfer Line

Many UF skids are of portable design and are CIP'd cleaned manually by use of portable tanks and flex line hoses. The level of automation is low, and the manual

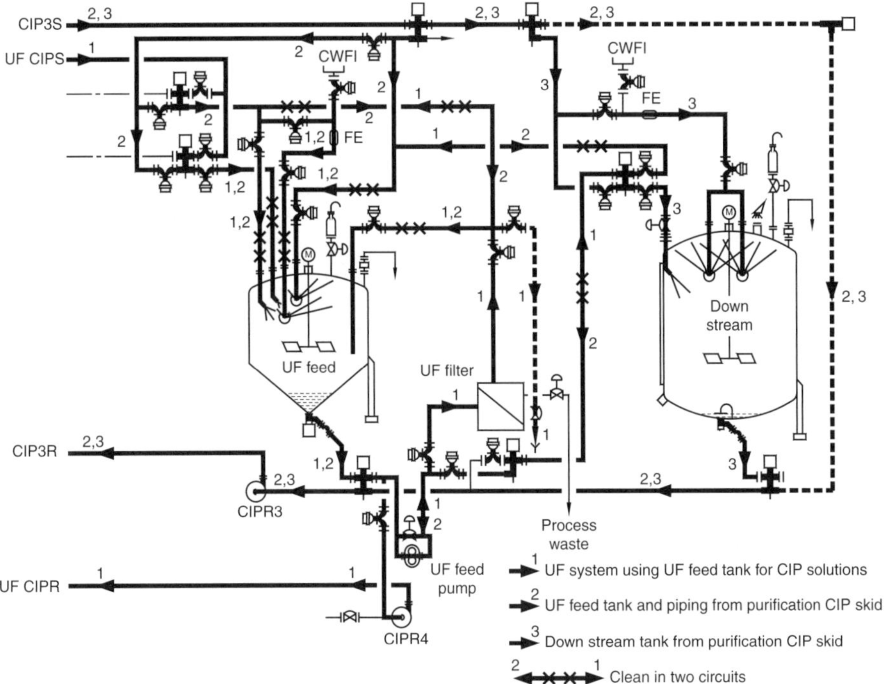

FIGURE 15 The incorporation of a UF filtration system in the transfer line and the need for a vastly different CIP program is described by the narrative with reference to the numbered arrowheads on this drawing.

labor is high. Figure 15 is a schematic of a larger and thus permanently mounted UF system designed to be cleaned using the UF feed vessel for supply to the UF feed pump in combination with added components on a "dual" CIP skid to supply ambient water for injection (AWFI), hot water for injection (HWFI), and heat for the CIP program. A "dual" CIP skid is shown in Figure 14 of chapter 6 and described in the accompanying narrative.

This process is an expansion of the concept of use of *rising stem valves for mixproof isolation of two tanks in a train* illustrated in Figure 9. TK1 of the generic schematic is the UF feed tank and TK2 is the Downstream hold tank. Some basic design and operating features of this system include

- The CIP skid supply pump delivers flush, wash, and rinse solution through a spray in the UF feed tank at a flow rate higher than required by the UF feed pump when operating at the CIP speed. The membrane system is thus subjected to fluids supplied by the feed pump only for Circuit 1, using all protective features incorporated in the UF system control hardware and software during CIP.
- The UF feed tank is cleaned twice, first with the program required to clean the membrane system, via the "dual" CIP skid provision, and then again by the purification area CIP skid which cleans all other vessels and piping as Circuit 2. Following CIP of the membrane system it is filled with storage solution and then isolated by automatic valves. Some of the interconnecting

FIGURE 16 This tank top photograph shows the application of a rising stem mixproof valve in combination with diaphragm valves as shown schematically in Figures 9 and 15.

piping is, of necessity, cleaned in both circuits. Note the recycle line from the membrane to the UF feed vessel, the extension of the retentate line to the second rising stem mixproof valve to permit recycle also to the previous hold tank, and the final transfer line between vessels.

- On completion of circuits 1 and 2, and after the downstream tank is emptied, it is also cleaned from the purification area CIP skid as circuit 3.

Figure 16 is a photograph of a typical downstream tank included at this point to illustrate the arrangement of the inlet isolation mixproof valve and associated diaphragm valves.

Filter Housing CIP in Transfer Line

The use of filters as single or multiple units, and often on fixed or moveable carts is a part of most recent biopharmaceutical process designs. Some of these filters are very large, requiring hoists to remove the housing. New projects often include the desire to clean these housing in place, as part of the product piping circuit. A section of Chapter 9 discussed the means of accomplishing this.

Figure 17 is a photograph of a permanently mounted cart supporting two filters used for media filtration enroute to a bioreactor selected by use of the U-Bend transfer panel upper right. A connection from the common port at bottom center to one of five ports in the circle above established the desired path. On completion of the transfer and clearing of the line, a special U-Bend from the upper right port to the common port introduces CIP flush, wash, and rinse solutions to the filter

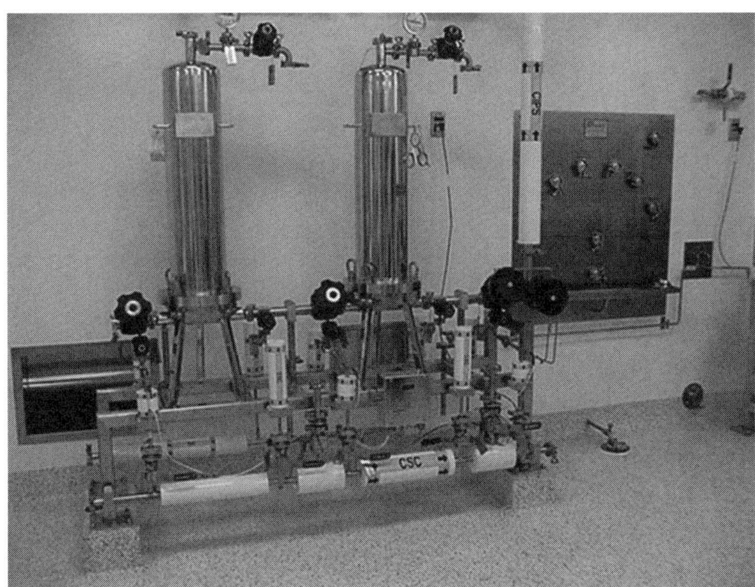

FIGURE 17 This tandem sterile filter for media is in the line to the transfer panel upper right. The U-Bend is installed to the selected bioreactor and the filter housings and transfer line are steamed-in-place prior to the transfer and filtration.

housings and the piping from the media vessels, via connection the end of this line to CIPR.

Consideration of filter housing CIP came late in the subject project, and the budget provided for only hand valves on the several filter carts. By replacing only two manual valves with air-operated valves in a strategic manner, and opening all other manual valves for CIP, fully automated CIP was possible at a modest added cost (Fig. 18).

The vent and condensate piping was all sized 0.5 (12 mm) to 1.5 in. (38 mm) header with an elbow to drain for production operations. A portable trap was added for SIP. A 1.5-in. CIPS header was connected to the fixed sprays by flex-lines to the two filter inlets for CIP. Following removal of the cartridges, the reassembled housings were spray cleaned in reverse flow at 45 gpm (22–23 gpm each) the velocity required to achieve 5 ft/sec (1.5 m/sec) in the 2-in. (50 mm) media supply line. Brief opening of V1 and V2 provided full flow through the filter outlet header and through seven 0.5 ft (12 mm) paths of nearly equal length in parallel. When both V1 and V2 were closed full flow was divided between the two housings.

STEAM-IN-PLACE

The design criteria to develop a CIPable process meet most of the requirements for SIP. Additional components for SIP might be limited to a source of CS, condensate traps at the low drain points, and temperature sensors to monitor the SIP procedure.

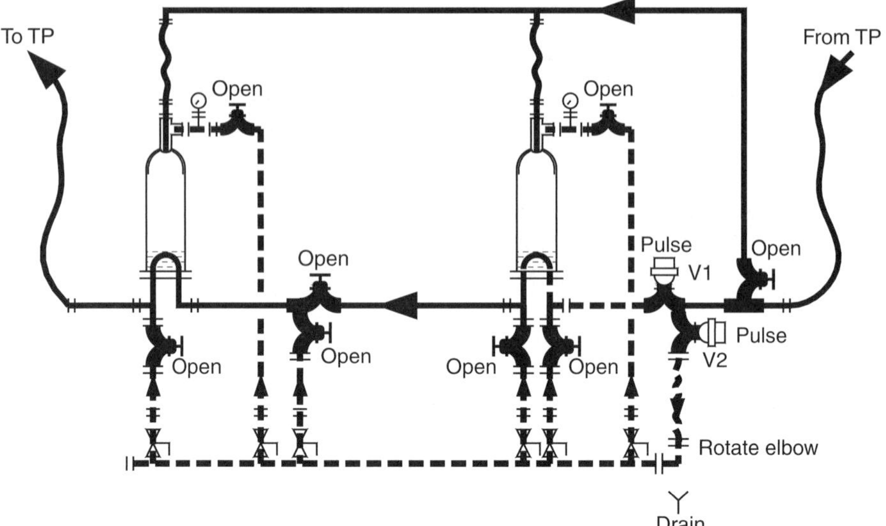

FIGURE 18 Following clearing of the transfer line and filter housings, the filter cartridges are removed and CIP of the housings and transfer line is accomplished in reverse flow as described in the narrative.

Steaming a Vessel and Transfer Line

Figures 1, 9–11 include the supply of CS to the vessel and other parts of the circuit through the CIPS to the vessel legs and sprays would require the use of portable traps on the outlet transfer panel ports, or ports to a trapped header. Figure 1 recognizes this via the legend COND on the bottom of the panel. Figures 9 and 10 provide the CS supply to the vessel and transfer line, and also a condensate trap at the low point of the outlet/transfer line. Figure 9 also provides CS to the vessel outlet valve body (closed to tank) to permit steaming the transfer line forward to the destination vessel when steaming the destination vessel. Figure 10 could be modified to include this capability. Figure 11 includes all valves and traps to SIP the complete bioreactor and its legs following CIP and insertion of the cartridges in the filter housings. The mixing vessel and transfer line illustrated in Figure 14 would require the CS to the CIPS line and to the outlet valve, and a trap on the mixer pump drain valve, plus traps in the downstream continuation of the piping. SIP of the Filter Cart has been discussed above.

The reader may wish to review the subject *SIP of the Transfer Line with Filter Housing and Destination Vessel* as illustrated in Figure 9 of Chapter 1.

SUMMARY

The advantages of CIP far outnumber the disadvantages. Of greater importance, however, is the beneficial impact of a well-engineered CIP system, properly applied to a cleanable process. These benefits may include increased production capacity due to less down time for cleaning, improved product quality and reduced losses due to cross-contamination, less physical abuse of equipment, and, with automation, reduced labor for processing and cleaning. These benefits will accrue to the maximum levels only if CIP/SIP design integration is given as much consideration as the design of the process, from the very beginning of the project.

An effective CIP/SIP procedure is a prerequisite to the design, operation, and validation of multiproduct facility designed to accommodate the production of two or more products concurrently, or on a campaigned basis. The above narrative, figures and photographs provide an overview of how some proven design concepts have been applied to create a variety of different cleanable liquids processes in both the pharmaceutical and biopharmaceutical segments of the industry.

REFERENCES

1. Seiberling DA. Clean-in-place and sterilize-in-place applications in the parenteral solutions process. Pharm Eng 1986; 6(6):30–5.
2. Seiberling DA. Clean-in-place/sterilize-in-place (CIP/SIP). In: Olson WP, Groves MJ, eds. Aseptic Pharmaceutical Manufacturing. 1st ed. Prairie View, IL: Interpharm Press Inc., 1987:247–314.
3. Seiberling DA. Alternatives to conventional process/CIP design—for improved cleanability 1992. Pharm Eng 1992; 12(2):16–26.
4. Seiberling DA, Ratz AJ. Engineering considerations for CIP/SIP. In: Avis KE, ed. Sterile Pharmaceutical Products–Process Engineering Applications. 1st ed. Buffalo Grove, IL: Interpharm Press Inc., 1995:135–219.
5. Marks DM. An integrated approach to CIP/SIP design for bioprocess equipment. Pharm Eng 1999; 19(2):34–40.
6. Chrzanowski GA, Crissman PD, Odum JN. Valve assembly use: a case study in maximizing operational flexibility, cost management, and schedule benefits. Pharm Eng 1998; 18(6):70–7.
7. Shnayder I, Khanina M. Equipment cleaning-in-place in modern biopharmaceutical facilities: engineering concepts and challenges. Pharm Eng 2005; 25(1):1–10.
8. Domanico E. Tri-Clover Friction Loss Calculation. Personal correspondence, May 23, 1990.

14 Cleanable Solids Processing Equipment and Systems

Simon E. J. Forder
JM Hyde Consulting, Inc., San Francisco, California, U.S.A.

INTRODUCTION

Dry powder processing equipment can be cleaned through the use of clean-in-place (CIP) systems. As described in this book, there are a number of advantages to using CIP systems for cleaning many types of biopharmaceutical equipment including: controls of cleaning parameters, control of variability in the cleaning process (consistency), and the ability to more easily validate the cleaning process. The ability to establish controlled conditions of time, concentration, and temperature via application of a CIP system has been demonstrated to be superior to manual cleaning under any conditions.

Dry powder equipment and piping that have been designed to be totally disassembled for manual cleaning, as in the pharmaceutical industry until the recent past, are not suitable for application of automated CIP cleaning. Similar to tank and vessel design, there are a number of design criteria that need to be met to successfully use CIP with dry powder processing equipment. The general design criteria for processes that handle dry or powdered products, which must be maintained in a very clean or sterile condition will include:

1. All equipment that will be in contact with cleaning solutions must be of stainless steel, glass-lined construction or equally corrosion-resistant and CIP-cleanable materials, sealed and closed with elastomers that are Food and Drug Administration (FDA) approved for the intended application.
2. Traditionally, 316-L stainless steel has been specified for bioprocess equipment and should be used in the construction of dry powder equipment. Additionally, attention to surface finish on product contact surface should be considered.
3. All welds need to be fully penetrated and sanitary. Edges should be rounded and sanitary as well to minimize powder collection and holdup. A surface finish of less than 25 rouchness average (Ra) is sufficient to ensure residue removal. The use of electroplating will also facilitate residue removal.
4. The equipment must be designed to confine the solutions used for flushing, washing, and rinsing. Recirculation is essential for economical and effective CIP operations.
5. The design or modification of the process equipment should allow cleaning solutions to freely drain from all equipment surfaces to one or more return collection points.
6. All parts of the piping or ductwork should be pitched at $\frac{1}{6}$ in. (5 mm/m) to $\frac{1}{8}$ in. (10 mm/m) per ft to drain points. Pitch must be continuous.
7. A minimum radius of 1 in. (25.4 mm) is desirable at all corners, whether vertical or horizontal.

8. Projectile-type thermometer sensors are acceptable for use with filled tube or resistance temperature detector (RTD)-based temperature indicating and recording systems. Thermocouple(s) or RTD(s) installed so as to sense only the temperature of the tank surface provide an even more satisfactory installation from the standpoint of cleanability.
9. Clamp-type joints of CIP design are acceptable for semipermanent connections. An acceptable CIP design infers (*i*) a joint and gasket assembly which will maintain the alignment of the interconnecting fittings, (*ii*) position the gasket so as to maintain a flush interior surface, and (*iii*) assure pressure on each side of the gasket at the interior surface to avoid product buildup in crevices that might exist in joints which are otherwise "watertight."
10. The design of the equipment should allow drying of the equipment after cleaning.

A major portion of all CIP cleaning of the dry granulated material processes will be via spray operations.

SPRAY DRYING APPLICATION

Spray drying applications are being used in more and more pharmaceutical production operations than ever before. The use of spray drying technology allows the manufacturer to produce a dry, stable, and homogeneous product. Applications are being developed to use sterile spray drying processes to facilitate final formulation and in some cases, bypass traditional lypholization. Spray drying operations are also being used for encapsulation and the development of extended release products.

Traditional spray drying operations involve spraying an aqueous or nonaqueous solution into the dryer body via an atomizing device. The solution is sprayed into a heated gas stream and the solution evaporates leaving a powdered product behind. Depending upon the method of atomization, powders can be generated in the 2- to 100-μm size or larger. Powder is carried out of the dryer body and collected in a collection device, usually a cyclone or a baghouse. Dried powder is then collected at the bottom of the collection device and stored for final processing.

During the operation of spray drying equipment, residue and product buildup are common on the processing surfaces. Despite the use of vibrators and rappers, the spray dryer will be covered with a fine powder at the conclusion of processing.

The use of CIP spray devices (Fig. 1) either permanently installed or installed at the time of cleaning will facilitate the distribution of cleaning solutions to these processing areas. Many types of spray devices can be employed in CIP cleaning including balls, wands, and rings.

Atomizers

Atomizers (Fig. 2) used in forming the solution droplets, which are dried, come in a variety of types. Rotary atomizers are usually used in the production of larger particle size powders and single- and two-fluid atomizers are used for producing smaller size particles.

Cleaning of the atomizer is probably the most problematic cleaning task. Typically, the small-diameter product fluid path through multifluid atomizers is difficult to clean and CIP flush, wash, and rinse solutions must be directed through

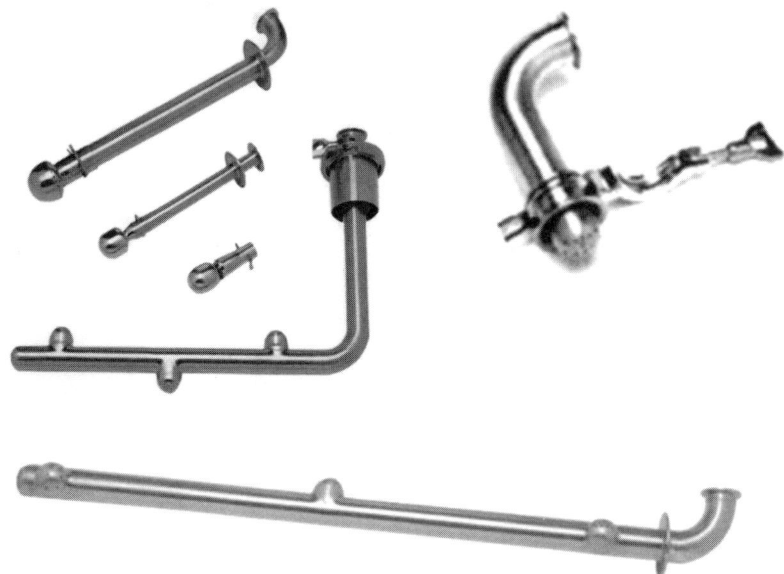

FIGURE 1 Typical spray balls. *Source*: Courtesy of Electrol Specialities Company.

the full product path to ensure cleanliness. A part washing cabinet may be set up and used for cleaning the fluid atomizers. Rotary atomizers generally require disassembly and manual cleaning or cleaning in a parts washer.

Specific Spray Dryer Design Considerations

Headplate/Air Inlet

Spray dryers use an air inlet distribution system installed in the dryer headplate to control hot air distribution into the dryer body. This is sometimes referred to as an air swirler. The air swirlers come in a variety of sizes and configuration to allow

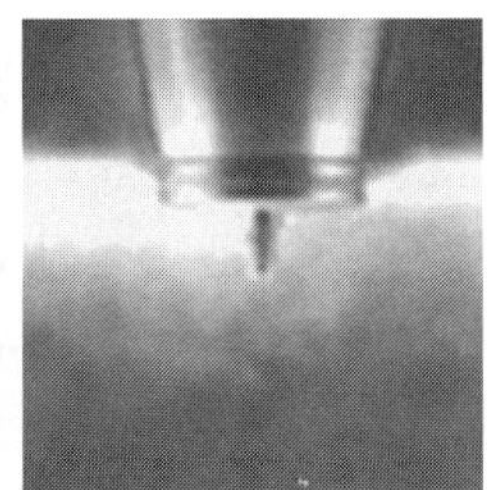

Rotary atomizer

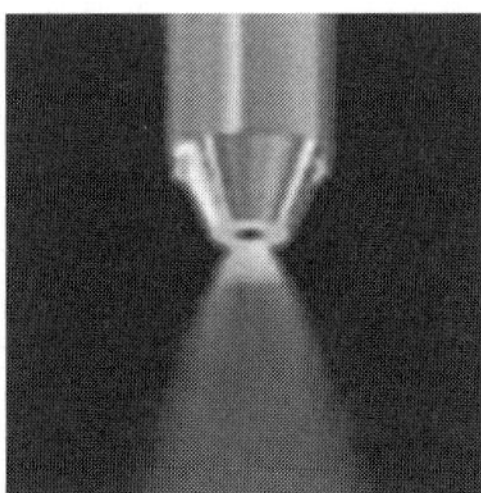

Two-fluid nozzle co-current mode

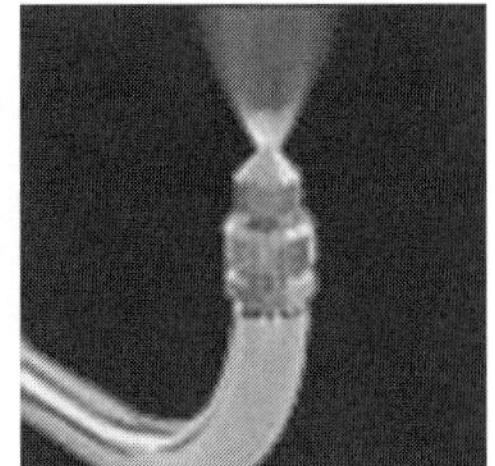

Two-fluid nozzle fountain mode

FIGURE 2 Atomizers (including rotary and two-fluid). *Source*: Courtesy of GEA Niro Inc.

control of the drying process. Hot, drying gas passes through the air swirler on its way into the drying chamber on a dryer. Despite the high air velocities through these air inlet devices, powders can "creep" into the air inlet pathway during operation. Attention needs to be paid to ensure that the air swirler is adequately cleaned during CIP. This can be achieved by installing spray ball or spray rings into the headplate and allowing cleaning fluids to contact and clean the air swirler.

Some dryers may incorporate a removable air swirler or diffusion plate in the drying airflow inlet. The swirler devices in these applications use a pin to locate and hold them in place in the headplate. There is a possibility of powder to eventually accumulate in the gaps between the swirler and the inlet path. This powder is difficult to clean unless the swirler is removed and the powder manually scrubbed off. In commercial equipment used for routine good manufacturing practices, the final design of the swirler should be fixed and the air swirler should be permanently installed and welded in place to eliminate any gaps or crevices.

Spray dryer installations will require isolation valves and proper pipe slope to facilitate CIP removal of dry powder from all product contact surfaces. An isolation valve should be located as close to the top of the spray dryer as possible to prevent cleaning solutions upstream of the dryer body. This valve will prevent solutions from coming into contact with upstream heaters or air filters.

Additionally, downstream isolation valves should be incorporated to protect final filters. The valves are closed during cleaning and prevent cleaning solutions from entering the final filter housing or upstream air filter systems. Isolation valves should be of sanitary design and typically of a butterfly design to minimize drying gas flow interruption.

Dryer Body Explosion Equipment

The actual dryer body will be of sufficient length to ensure complete drying of all liquid droplets before they reach the bottom area of the dryer. Most aqueous dryers will require some type of explosion suppression/containment system (Fig. 3) to prevent loss of product or life in the event of a conflagration. In cases where these suppression systems are added to the walls of the dryer body ensure that a sanitary fitting is used with a minimum amount of offset from the dryer wall. Fike uses a suitable sanitary design to attach both the suppression system and the pressure sensor to the wall of the dryer (Fig. 4).

Dryer Gaskets

To facilitate fabrication, the dryer body will typically come in two parts, an upper and lower half. Most nonpharmaceutical fabrications will use a common flange and gasket to combine the two halves of the dryer. While simple and relatively cheap, these flange gaskets are difficult to clean and accumulate residues. GEA Niro offers an alternate sanitary O-ring gasket to combine the two halves of a dryer which eliminates the flange and provides a sanitary seal between the dryer parts. The O-ring seals the two halves of the dryer and eliminates the gaps of the standard flange/gasket configuration.

GEA Niro Sanitary Seal

Additional area of design concern to ensure proper cleaning includes the use of sanitary instrumentation. Temperature sensors (e.g., dryer outlet temperature) should incorporate a sanitary well-type design allowing cleaning solutions to

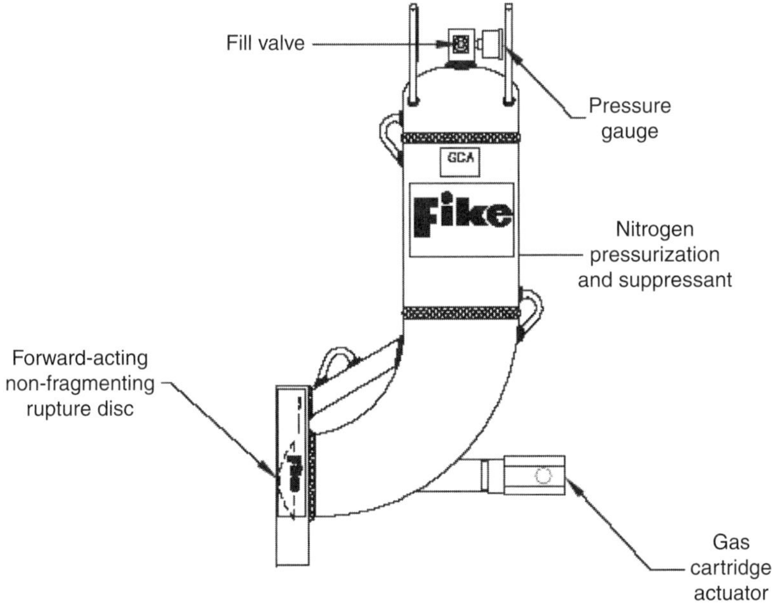

FIGURE 3 Explosion suppression system. *Source*: Courtesy of Fike Corporation.

clean the instrument protrusion and rinse free. In-process pressure gauges should incorporate a design with a positive pressure air sweep or a sanitary pressure-sensing surface. Positive pressure on the pressure sensors will ensure that powders do not accumulate in the pressure lines.

Spray Dryer Powder Collection

Dried powder is conducted out of the drying chamber by the drying gas. The powder exits the dryer and is carried to a collection device. Most spray dryer

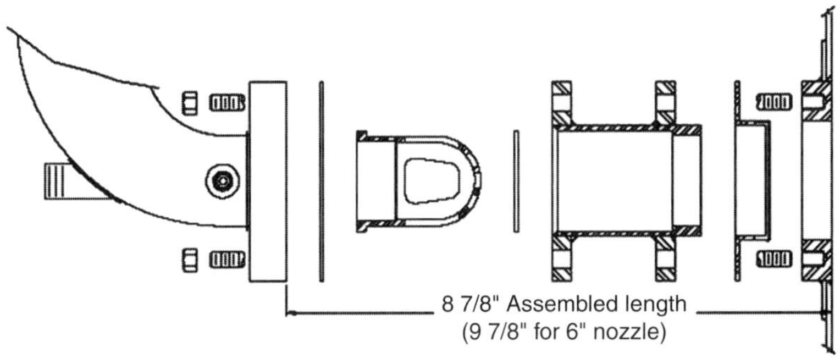

FIGURE 4 Flush mount telescoping nozzle detail and dimensions. *Source*: Courtesy of Fike Corporation.

FIGURE 5 Pharmaceutical cyclone assembly. *Source*: Courtesy of Fisher-Klosterman, Inc.

applications will use either a cyclone(s) or baghouse to collect the dried powder. Some dryer arrangement may use both cyclones and a final baghouses. In either application, baghouse or cyclones, the piping (referred to as the crossover piping) between the dryer body and the collection device needs some attention to ensure proper cleaning. The piping immediately after the dryer and before the collection device offers some interesting cleaning challenges. The crossover piping between the dryer and the cyclones/baghouses should be of 316-L stainless steel (Fig. 5).

Depending upon the size and length of the crossover piping, retractable or insertable CIP spray devices may be needed to clean this section of piping. Ensure that this section of pipe is sloped to the collection device. Proper slope will not allow cleaning solution holdup.

Cyclones

Dried powder conducted out of the dryer and connecting piping can be collected in a cyclone. The cyclone relies upon centrifugal forces to separate the dried powder from the gas stream. Dried powder is spun out of the gas stream and drops down the sidewalls of the cyclone.

During recent years, cleaning of cyclones has advanced. Again, depending upon the size, automated CIP cleaning of the cyclones is feasible assuming the cyclones have been designed with this in mind. At a minimum, sanitary stainless

FIGURE 6 Retractable washing nozzles. *Source*: Courtesy of Glatt GmbH.

steel should be used in the construction of product contacting cyclones. Any two-piece designs (two-part cyclones, etc.) should incorporate sanitary gasketing. Again, a sanitary design using O-rings is preferable to gasket and flange. Retractable spray devices can be installed to facilitate cleaning of the underside surfaces and cyclone walls (Fig. 6). Additionally, specially designed spray lances may be inserted into the cyclones to distribute cleaning solutions to the walls and underside of the cyclone.

Baghouses

A number of powder processing applications incorporate a baghouse for powder collection (Fig. 7).

Additionally, baghouses are used for process gas cleanup before final exhausting. The baghouse is installed in line and captures particulates and or product by way of filter elements. Product-laden airflow is directed at the elements and filtered. The cleaned processed gas is then exhausted out of the baghouse. At defined intervals, dry gas is pulsed onto the inside surface of the baghouse filters to blow product off the filter element surfaces. This is done to minimize powder buildup on the filter elements and ensure filter efficiency.

Baghouse Design

There are a number of design and application considerations when using baghouse for powder collection. For primary product contacting surface within the baghouse, sanitary materials of construction should be used and attention should be paid to welds and surface finish. A type 304-L stainless steel is acceptable for nonproduct contacting applications while 316-L stainless steel should be used for product contacting surfaces within the baghouse.

There are some commercially available CIPable baghouse systems (MikroPul, Inc.) which incorporate permanent spray device applications and washable filters. CIP solutions are applied to the inside body of the baghouse, the outside of the filters, and the headplate. Spray lances or spray balls can be removed without vessel breakdown. The spray balls allow complete coverage of all internal surfaces. Pharmaceutical units are suitable for FDA validation. Figure 8 shows a MikroPul pharmaceutical baghouse.

The headplate/plenum should incorporate a free draining design to avoid solution pooling during CIP. Some design of plenums may incorporate a filtered drain plug.

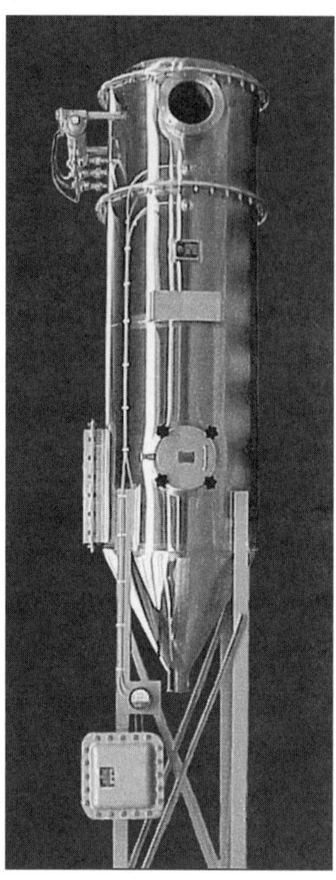

FIGURE 7 Pharmaceutical baghouse. *Source*: Courtesy of Fisher-Klosterman, Inc.

Final design considerations are to ensure that the baghouse can be dried post-cleaning. Hot drying gas can be supplied to the baghouse by way of upstream equipment (spray dryer, fluid bed). In baghouse application using filter elements which are not routinely removed, a bypass is sometimes requirement to allow filters to dry while avoiding excessive surface tension on the filters.

Baghouse Filter Material Considerations

Baghouse filter material selection can affect both powder recovery cost and cleanability. Filter materials come in a variety of shapes, sizes, and types but can be narrowed down to three general categories: cloth and fabric, cartridges, and stainless steel sintered metal material.

Traditional cloth and fabric filters are routinely used in baghouse operations. The cloth filters are robust, fairly efficient, and by far, the cheapest solution to filtering operations. The cloth filter elements are supported by a wire and mesh cage within the baghouse. The cloth filter and cage, also called a sock, is installed in the headplate or plenum of the baghouse. The biggest disadvantage of the cloth filter is that they need to be removed prior to cleaning or the baghouse. After removal, cloth filters may be reused or discarded after use. Some clothes elements may even be washed in washing machines and reused!

FIGURE 8 Pharmaceutical CIPable baghouse. *Source*: Courtesy of MikroPul, Inc.

Cartridges filter elements (Fig. 9) are more robust and cleanable than the traditional cloth elements.

An element is spun around a metal frame and the filter surface is pleated to increase surface area. The cartridge element has excellent strength and most are a one-piece design. Cartridges come in a variety of material but most pharmaceutical applications would use polytetrafluoroethylene or Teflon® (Fig. 10). The cartridge design offers excellent powder recovery and, in some cases, is cleanable in place with cleaning chemicals without filter removal. The cost is more than that of the fabric elements.

Stainless steel sintered metal filter technology has been used in military and nuclear applications for many years. The use of sintered metal filters for pharmaceutical applications is relatively new. Stainless filters have been developed to address issues with conventional fabric and Teflon cartridges. Conventional fabric filters cannot be cleaned while in the baghouse. Even newer Teflon cartridge filters, which can be cleaned in the place without removal, usually have their problem areas. Cartridge filters are also mechanically delicate and have short service lives.

Stainless steel filter elements by contrast have a number of advantages. During cleaning operations, the stainless steel filters do not need to be removed to clean. CIP cleaning chemicals are not reactive with the stainless metal of the filters. Additional advantages include the ability to clean the baghouse and ensure total containment for operator and product protection (Figs. 11 and 12).

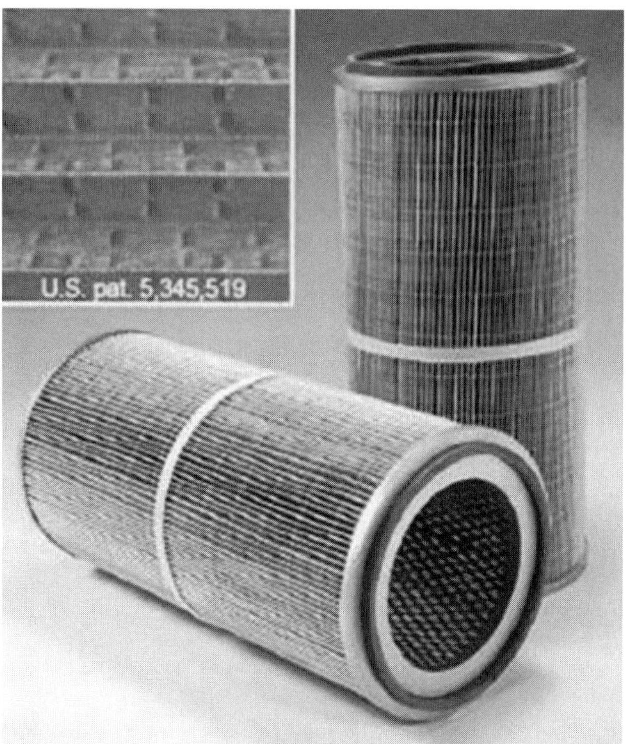

FIGURE 9 Pleated filter cartridges. *Source*: Courtesy of MikroPul, Inc.

Fluid Bed Dryer

Fluid bed dryer and granulators use hot gases to dry and/or agglomerate dry powders. Wet product is sprayed into the gas stream through an atomizer or spray arm (similar to a spray dryer). Product is dried and collected in a hopper at the bottom of the dry chamber. After all liquid is sprayed into the drying chamber, the dryer is shut down and the hopper removed.

Incorporation of automated cleaning can be done with fluid bed dryers and granulators. There are many advantages in being able to CIP a fluid bed dryer including ensuring cleaning repeatability and consistency of the cleaning operation. Additionally, CIPable fluid bed dryers can significantly reduce exposure risk associated with operation and cleaning.

Fluid Bed Dryer Design Considerations

Many of the design considerations applicable to spray dryers are applicable to the fluid bed dryer. Several specific challenges exist in cleaning the fluid bed dryer. Filters are installed in the top of the drying chamber to collect exhaust particles in the drying stream. These filters are removed and washed or if of a suitable material (sintered metal, Teflon cartridge) may be cleaned in place.

Again, care needs to be focused on the gasket and flange design as with spray dryers. Similar to the spray dryer, sanitary flange connections minimize the

FIGURE 10 Polytetrafluoroethylene filter cartridges. *Source*: Courtesy of MikroPul, Inc.

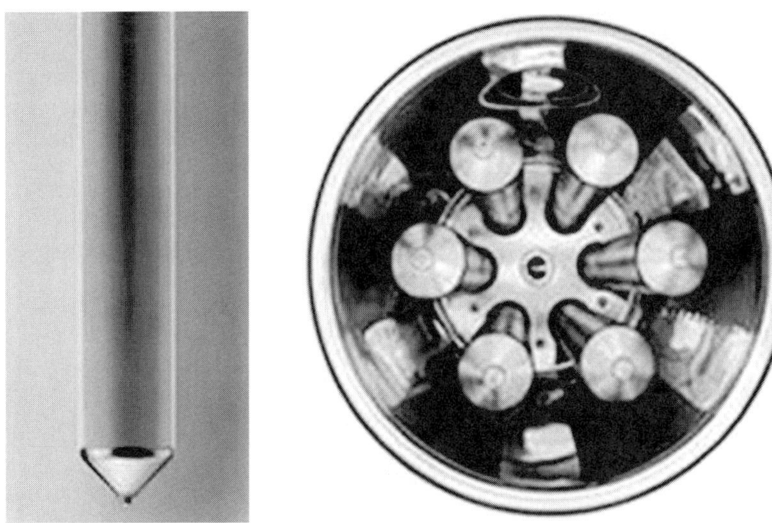

FIGURE 11 SC superclean stainless filters assembly. *Source*: Courtesy of Glatt GmbH.

FIGURE 12 Disassembled stainless filters in baghouse. *Source*: Courtesy of Fisher-Klosterman, Inc.

likelihood of product carryover and cleaning problems. Instrumentation and explosion suppression equipment need to be of a sanitary design. Again, attention needs to be spent to ensure sanitary design of temperature/pressure equipment as well as flanges and detectors.

Fluid bed dryers should incorporate spray devices for CIP. New systems include removable spray devices which are inserted and removed. Figure 13 shows a Glatt fluid bed dryer with spray balls incorporated into the dryer.

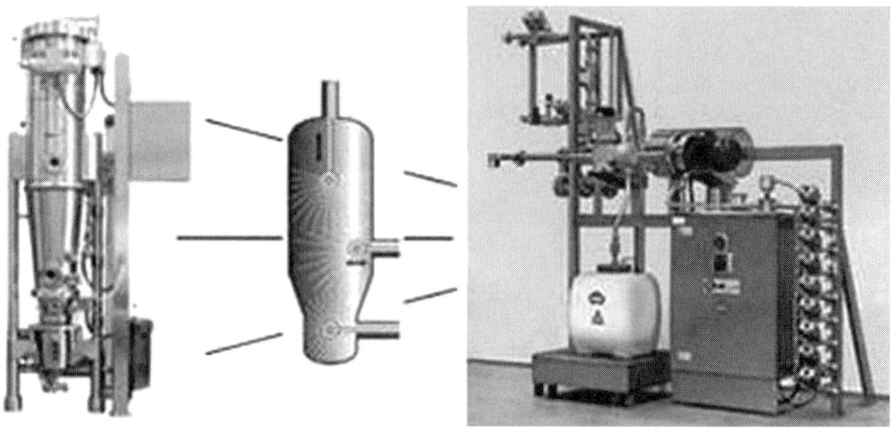

FIGURE 13 Sanitary fluid bed dryer and CIP system. *Source*: Courtesy of Glatt GmbH.

Additional areas of concern are the isolation valve in the bottom of the fluid bed dryer body and the diffusion screens. These pieces of equipment may be cleaned in place but will need to be inspected during start-up to ensure they are adequately cleaned.

Cleaning of the atomizer for the fluid bed dryer can be approached in a similar manner to the atomizer cleaning used in spray dryer operations.

The actual integration of the automated cleaning cycle and equipment can be approached in much the same manner as other tanks and vessels. Figure 14 illustrates an approach to incorporating the many different components of a dry powder process into a CIPable process, and though published in 1995 (1), it is still representative of an approach in current use.

Typical Spray Dryer CIP Cycle

Separate cleaning circuits may be developed and used to ensure adequate flow to sections of the dryer and ancillary equipment like the baghouse. A set of circuits may clean the dryer body, the cyclone, and then a final collection device or baghouse. The CIP sequence should incorporate the same steps as cleaning a fermentor or tank, including rinse, caustic wash, acid wash, final rinse, and dry.

The prerinse will be used to wash off powder contacting services and depending upon application may be sent to waste or pretreatment prior to waste. An alkaline wash is used for primary cleaning of the dryer.

The wash solution should be at temperatures 65°C to 70°C and solution makeup should incorporate a formulated cleaner or sodium hydroxide or potassium hydroxide. A water rinse and airblow can be used after the alkaline wash.

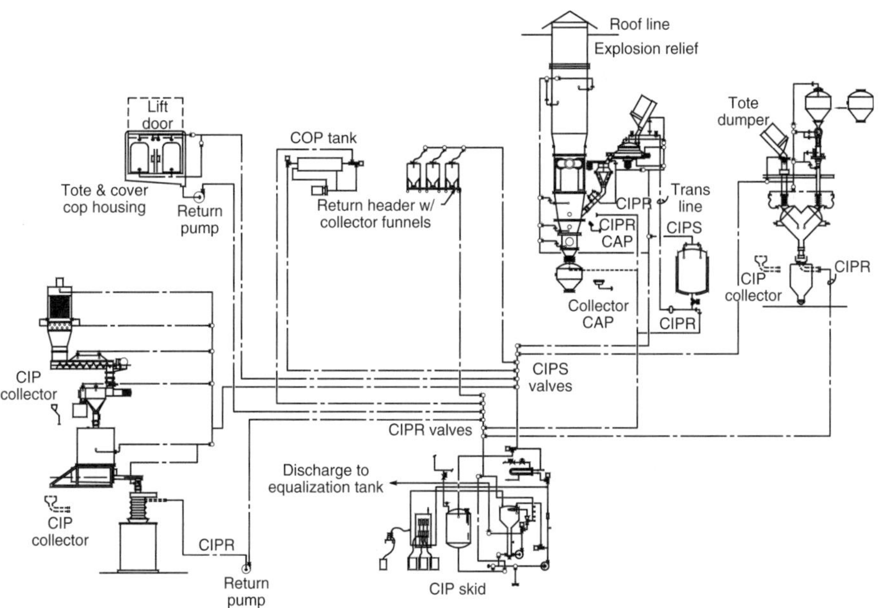

FIGURE 14 Schematic diagram of a CIPable dry products process. *Abbreviations*: CIPR, clean-in-place return; COP, cleaned-out-of-place; RTD, resistance temperature detector.

The acid wash of 85% phosphoric acid or equivalent is then used for neutralization and mineral removal. The acid wash is followed by a rinse and airblow followed by a final water-for-injection rinse.

Where this would normally be the end of the washing cycle for tanks and vessels, the configuration of the spray dryer allows for quick and efficient drying of the dryer after CIP. A final drying step should be incorporated into the dryer CIP sequence. This can usually be nothing more than passing heated process gas through the dryer body and collection equipment.

GLOVE BOXES AND BARRIER EQUIPMENT

Barrier equipment or glove boxes are used in dry powder operations for containment and product/operator protection. The glove box provides a physical barrier between the processing operations and the outside environment. Some examples of application include protecting operations personnel during cytotoxic product processing or keeping hydrophilic powders dry during processing (Fig. 15).

Design Considerations for Glove Boxes Incorporating CIP

The glove box uses a high-efficiency particulate air (HEPA) or similar filter to filter air into or out of the glove box. Glove boxes may be purged with clean dry air or

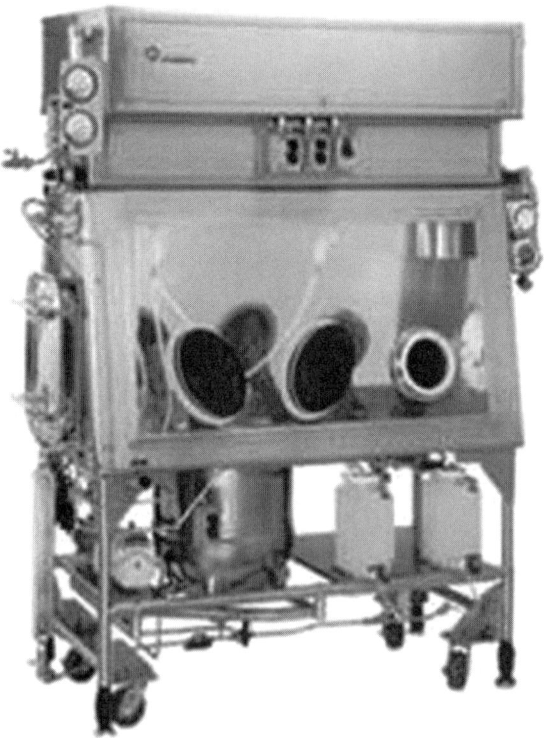

FIGURE 15 Sanitary glove box. *Source*: Courtesy of La Calhene, Inc.

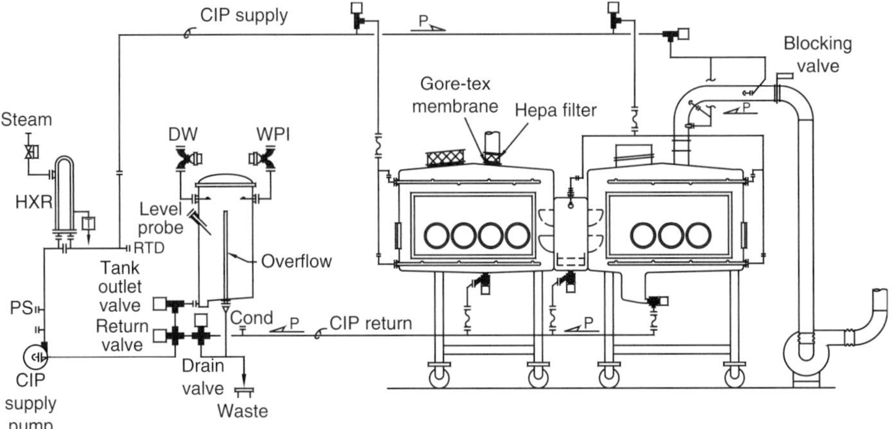

FIGURE 16 CIPable barrier equipment with dedicated CIP unit. *Abbreviations*: DW, de-ionized water; HEPA, high-efficiency particulate air; HXR, heat exchanger; PS, pressure switch; RTD, resistance temperature detector; WPI, purified water.

nitrogen to control humidity. In specifying and designing the glove box, the design needs to either keep product in (negative to the environment) or keep the environment out (positive to the environment). Each design will affect the cleaning of the glove box.

Material of construction should incorporate 316-L stainless steel for product contacting or washable surfaces. Additionally, attention should be spent on minimizing intrusion and irregular surfaces within the glove box. CIPable glove boxes may incorporate spray devices and wands for distribution of cleaning solutions. The glove box floor should pitch to a common drain. The drain allows cleaning solution to return and recirculate. The HEPA filters should be protected with covers during CIP. Use of a tri-clover fitting on the removable cover will ensure a tight and sanitary design. Alternately, a Teflon cover may be used to protect the HEPA filter and allow a washable surface.

Figure 16 shows a CIPable glove box. In the illustration, a dedicated CIP unit is used to clean the glove box. The glove box is of a two-piece design incorporating a pass-through, which is also cleaned by CIP. Cleaning solutions are delivered to the two processing boxes, the pass-through, and a section of sloped exhaust piping by way of the clean-in-place supply (CIPS) piping. A blocking valve on the exhaust piping ensures that no cleaning solutions end up downstream in the suction fan. Solutions are delivered to spray devices and spray balls for consistent coverage in the process areas. In this application, CIP supply is toggled repetitively through the CIPS paths. After being sprayed into the product contacting areas, cleaning solutions then proceed through drain valves installed on the gloves boxes and return to the CIP skid. After cleaning, the glove box fan is turned on to allow drying of the process surfaces.

BLENDERS, INTERMEDIATE BULK CONTAINERS, AND TOTE BINS

V Benders, traditional pharmaceutical bending equipment, and intermediate bulk containers (totes) can be designed to incorporate automated cleaning operations

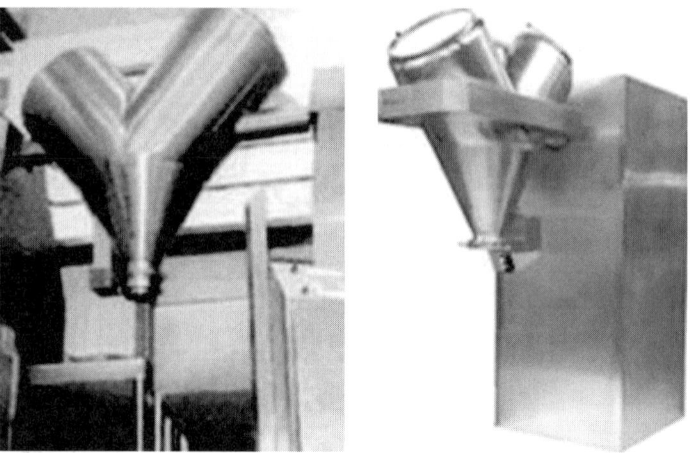

FIGURE 17 Sanitary V benders. *Source*: Courtesy of Pharmatech Ltd.

(Fig. 17). Materials of construction need to be compatible with cleaning chemicals. The type 316-L stainless steel is used for these applications. The use of clean welds and rounded corners in the construction of the equipment will facilitate the cleaning operations. More exotic metals including Hastelloy® may be used for highly reactive products.

Equipment can be cleaned using similar principles of tank cleaning. The required flow of flush, wash, and rinse solutions needs to be used to ensure coverage of all equipment surfaces. Spray balls (insertable) are used to distribute the cleaning solutions. In larger bender and tote operations, automated CIP systems can be incorporated into the design of the cleaning system. Areas of concern in cleaning these types of equipment include ensuring adequate coverage on the equipment (verified by coverage testing) and cleaning of valve assemblies (typically butterfly valves).

SOLIDS TRANSFER EQUIPMENT

Solids conveying equipment come in a variety of configurations. Pneumatic conveyers typically operate on a push or pull principle. Powders are either pushed through the conveying piping by gas pressure or pulled by a vacuum source. Conveying systems can be found on dryer applications or other powder processing systems.

The use of a double valve, sometimes called a flapper valve system, is used to maintain the pressure in the conveying line. These valve systems can be cleaned in line if both valves can be opened at the same time (not the way they are used in operation—one open, one closed, and vice versa).

Again, design considerations include material of construction and sanitary welding. Retractable or insertable spray ball systems can be incorporated into the design of these systems. Provision for slope and drain must also be incorporated during design and fabrication to ensure adequate drainage. Finally, the process equipment can be used to dry the conveying equipment and minimize bioburden.

Pharmaceutical Sifter/Screeners

Sifters/screeners are used in a number of applications for sizing dry intermediates and bulk products. The sifter/screener use a motor to impart vibration to spring-mounted screening decks. Powdered product is moved over multiple-sized screens. The vibration of the screen causes oversize particles to vibrate across the screen surface to the screen's periphery where they are collected. Undersized particles pass rapidly through the screens and are discharged through a spout at the periphery of the bottom collection chamber.

Multiscreen sifters may be used to separate several desirable sizes from a bulk intermediate. At the conclusion of the processing, fines and small product/intermediate particulates are often trapped in the screens and gaps and require cleaning. Traditionally, the sifters and screens are disassembled and cleaned manually. Advances in design are now allowing manufactures of sifting/screening equipment to incorporate CIP technology.

New sanitary sifters for pharmaceutical applications allow for hands-free wash down by using CIP spray heads (Fig. 18).

In Figure 18, the screener is configured with a gap-free screening deck. Unlike traditional circular screen separators, the pharmaceutical sifter has no gaps between the mounted screen and the frame wall. This helps eliminate powder buildup on the outside of the screens. Three CIP spray balls are used with the multilayer screens:

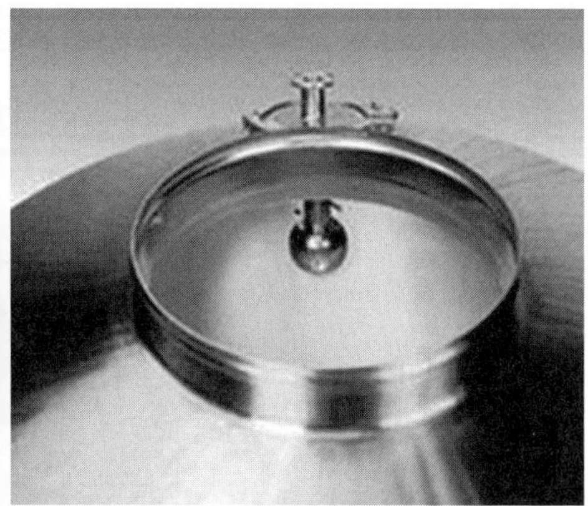

FIGURE 18 Screener incorporating CIP and installed spray balls. *Source*: Courtesy of Kason Industries, Inc.

one each for the upper, central, and lower chambers. Tri-clover clamps allow easy disassembly of all CIP supplies, CIP spray balls, and separator body sections for inspection or screen changes. The construction is of 316-L stainless steel.

REFERENCE

1. Seiberling DA, Ratz AJ. Engineering considerations for CIP/SIP. In: Avis KE, ed. Sterile Pharmaceutical Products—Process Engineering Applications. 1st ed. Buffalo Grove, IL: Interpharm Press Inc., 1995:135–219.

15 Cleanable API Processing Equipment and Systems

Gerald J. Cerulli
Integrated Project Services, Somerset, New Jersey, U.S.A.

INTRODUCTION

Active pharmaceutical ingredients (APIs) are the drug substances generally produced by chemical synthesis or chemical manipulation of materials derived from biological origins. The most important trend in APIs is that these materials are becoming more potent or more specific in their biological target. Another trend has been the design of facilities for multiproduct or multipurpose production. These two trends have ignited interest in incorporating the concept of clean-in-place (CIP) technology into bulk API facilities.

This chapter focuses on the design considerations relative to the process equipment, piping, and spray devices specifically for API systems. It will illustrate a comparison of CIP versus traditional boil-up methods, discuss design for cleanability issues relative to a range of process equipment, spray design approaches, results measurement, safety issues, and regulatory positions. The goal is to give the designer or operator of bulk API facilities a working knowledge of how these systems can be designed to be CIP cleanable.

DEFINITION AND HISTORY

CIP is a repeatable, automated process whereby cleaning solutions are properly brought into contact with all product contact surfaces of process systems to effectively remove a soil or contaminant by chemical action based on controlled time, temperature, and concentration of a cleaning agent without disassembling the process system. CIP methods have been used since the late 1940s to clean process equipment and piping. The initial applications were in the dairy industry, but CIP quickly spread to other fluid applications, primarily food and beverage processes. CIP methodology later evolved into the pharmaceutical and biotech industries where it is the accepted standard. The API synthesis industry was faced with a few more design challenges relative to CIP application, and has very slowly started to move toward CIP in the last 10 years. This may be because of the increased potency of drug substances. Some of the reluctance to embracing CIP technologies in bulk pharmaceutical facilities has been due to the perception of relatively high initial cost of implementing CIP in a new facility or in revamping existing plants. The complexity of developing flexible software to handle the changing cleaning requirements of new products is also a perceived concern, as are the safety issues involved in spraying solvents. Possibly the greatest factor-limiting change toward CIP is the perception that existing approaches are sufficient to clean bulk reactor systems, which are involved in producing active pharmaceutical chemicals, to an appropriate level of cleanliness.

CIP VS. TRADITIONAL BOIL-UP METHODS

For years, in API chemical facilities, the typical method for cleaning batch reactors was to perform "boil-ups." This consists of introducing a solvent in which the APIs and/or the soils are soluble. The solvent is heated to reflux. Solvent vapor condenses in the overhead system and on the top head of the reactor vessel. The hope is that, given enough time, the condensing vapor will dissolve any chemical contaminants. The boil-up is also augmented by pumped flushing of associated transfer lines. This is quite effective in cleaning lines 3 in. or smaller (larger lines require inordinate amounts of fluid to produce the required velocities). The boil-up method is straightforward and requires no additional piping or spray devices beyond those required for the process. Hence, it does not involve additional capital expenses. There are, however, several important drawbacks. The process is time and energy intensive, aqueous cleaning solutions cannot be used in the most effective way, the process is very difficult to make repeatable, and, most importantly, the result may be unsatisfactory.

In the past, API producers have accepted the time required to clean their process trains. Now many companies desire to decrease the time spent on cleaning, which may be as much as 30% for multiproduct or multipurpose plants facilities that change products frequently.

Possibly the largest problem with boil-ups is that the refluxing vapor only acts where it condenses. The actual rate of vapor condensing is actually quite small.

Aqueous cleaning solution applied via spray devices, in many cases, can be used to supplement organic solvent rinses. These solutions, which are primarily caustic in nature but can also be acidic, are applied in low concentrations (0.15–1.0% typically, but as high as 5.0% for heavily fouled heat exchange surfaces), and are cost-effective in cleaning many types of soils. These cleaning additives are not generally volatile and therefore cannot be distributed via a reflux method.

To illustrate the boil-up method, consider a 1000-gallon reactor undergoing an atmospheric reflux with acetone, a common cleaning solvent. Assume that the reactor heat transfer area is limited to the bottom head. After achieving steady state, the fluid can only condense on the reactor top head and vapor riser at the rate of heat loss to the environment. Therefore, assuming an uninsulated reactor top, head/vapor riser, and a generous value for natural convection, the amount of solvent vaporized and condensed is illustrated in Table 1 for a vigorous boil-up.

These rates compare unfavorably with typical wash rates that might be applied during a CIP cycle, as illustrated in Table 2.

Clearly, the CIP method will introduce many times more cleaning fluid per unit time than will a boil-up. This intense rate of solution application can shorten cleaning cycles. Highly soiled areas can have more solution applied while less soiled areas can be treated less aggressively.

TABLE 1 Boil-Up Rates

Vapor boil-up rate	13.0 lb/min 2.1 gpm
Vapor condensing in top head	0.6 lb/min 0.1 gpm
Vapor condensing in the riser	0.7 lb/min 0.12 gpm

TABLE 2 Typical Wash Rates for Clean-In-Place (1000-Gallon Reactor)

Top head	434 lb/min 70 gpm
Vapor riser	312 lb/min 50 gpm
Condensers	370 lb/min 60 gpm

Another important factor is that the fluid can be distributed evenly using the CIP approach, whereas, in the case of a boil-up, the vapor can only clean where it condenses. Soil actually inhibits the condensation of vapor and, therefore, cleaning. Most of the vapor is condensed in the overhead condenser, while dead legs, like valved or blinded nozzles, receive little cleaning solution. This also applies to ducts and chutes, as the rate of condensing does not provide adequate rundown over vertical surfaces.

In spite of the limitations of boil-ups, many companies have developed protocols, in which reactor systems are cleaned to a level deemed acceptable with this traditional method. This, however, may take hours and be subject to failure because of the nonuniform nature of soil distribution, batch to batch.

After a traditional boil-up or a CIP rinse and flush sequence, it is a common practice to take a sample aliquot of solution and subject it to testing to confirm that the level of specific and/or total contaminants have been reduced to a predetermined value. A test failure at this point requires additional cleaning and should initiate an investigation of what caused the failure. However, many cleaning procedures that the authors have seen rely on a "test until clean" philosophy. That is, the boil-up or flush solution is analyzed and the equipment is subjected to additional cleaning cycles if required.

FOOD AND DRUG ADMINISTRATION OBSERVATIONS ON TEST UNTIL CLEAN

The Food and Drug Administration has commented in the "Inspection Guide on Cleaning Validation" that testing for cleanliness in this manner may not be acceptable. Retesting until an acceptable residue level is attained is "acceptable only in rare cases."

What is required is a "validated" cleaning method, or in the case of a multipurpose or development facility, a system which is "qualified" for its ability to wet a process surface. This ability to wet all process surfaces can be confirmed by a riboflavin test similar to those performed in the biotech industry.

TRADITIONAL REACTOR SYSTEMS—CLEANING CHALLENGES

Glass-lined reactor systems are an essential element of most multipurpose bulk pharmaceutical production facilities. They are favored because of their wide range of resistance to corrosion and buildup of products/surface contamination, as well as for their comparatively low cost.

The basic design of the glass-lined vessel started to evolve in the 1880s. Up until quite recently, these glass-lined reactors and Teflon®-lined piped systems were not designed to be cleaned in place. However, process designers and glass-lined vessel manufacturers have made great strides in increasing the cleanability of these systems. Many of the strategies employed on reactor systems can be used to clean other pieces of equipment common to API production. Therefore, this chapter will address these systems in detail and then discuss other systems (isolation, drying, etc.).

It is instructive to compare a traditional design of a glass-lined reactor system not designed for CIP with the one incorporating features that will enhance CIP effectiveness. Examination of a typical glass-lined reactor system, as illustrated in Figure 1, will permit identification of those areas that are difficult to clean/wet using a boil-up or simple flushing approach. While the examples shown are of glass-lined systems, the findings can also be applied to alloy reactor systems.

Most readers will recognize this illustration as a simplified diagram of a glass-lined vessel system. There are provisions for liquid and solids charges, nitrogen inertion, reflux/distillation, emergency relief, and process transfer. Those areas in this example that constitute potential cleaning challenges, or that need to be properly designed for CIP, include the reactor vessel (A), the condenser (B), the emergency vent line and relief device (C), the vapor riser and instrumentation (D), the baffle and unflushed nozzles (E), the solids charge port (F), the dip tube (G), the agitator (H), and the bottom outlet (I).

This traditional glass-lined reactor system was never intended or designed to be cleaned in place. Some of the major deficiencies of this system are as follows:

- Condensers mounted horizontally or with a slight pitch are difficult to clean efficiently. Condensing boil-up vapors will not uniformly clean the tubes, and little cleaning action will be directed to any soil located on the bonnets and tube sheets. Flushing with solvent would be an effective way to clean, but this would require hundreds or thousands of gallons per minute to achieve. One approach that has been used for this condenser orientation is a liquid fill-and-soak with an effective cleaning solution, but the lack of solution movement will lead to inconsistent results. An alternative (not shown) is the addition of a recirculation pump to maintain a low rate of recycle through the flooded tubes.
- Another area of concern is the vapor riser, emergency vent piping, and all the instrumentation incorporated in this system. In operation, these lines become soiled due to a reactor foam-over and the carryover of dust from solids charging operations. As was pointed out previously, boil-ups produce only small amounts of condensing liquids. In the case of the emergency vent line, it is improbable that vapor will rise into this space, effectively displace the inert atmosphere, and efficiently wet the surface uniformly at a rate sufficient to remove all soil consistently.
- Nozzles designed for liquid charges can be cleaned easily, but those that are not externally flushed, such as the baffle, level transmitter, dip tube, and solids charge port, may be difficult to clean effectively and consistently with a solvent boil-up method.

The reactor vessel itself has much to do with the ultimate success of cleaning. Older style glass-lined vessels were designed for simplicity of fabrication and ease of maintenance with little thought to cleaning. These vessels are often characterized

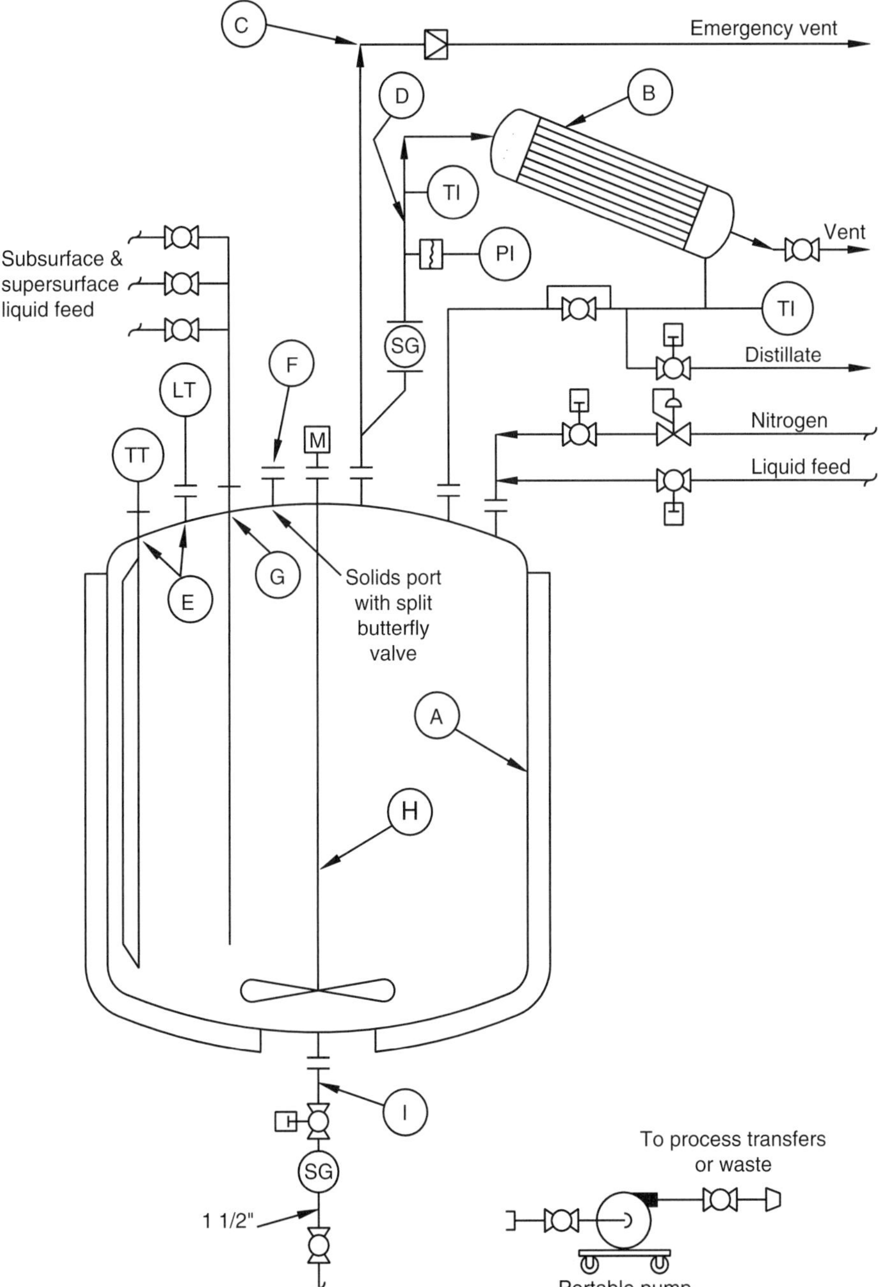

FIGURE 1 Typical traditional glass-lined reactor. *Abbreviations*: LT, level transmitter; TT, temperature transmitter; SG, sight glass.

by large flanged connections for the removal of one-piece agitators and small nozzles with long, extended necks positioned close to the centerline.

CIP SPRAY DEVICES

Clearly, to obtain an effective repeatable method of cleaning this system, it is necessary to wet all surfaces with sufficient quantities of fluids, which are directed precisely against the areas where soil is likely to accumulate. Spray devices (which include spray balls, spray bubbles, and bubble spray arms) are commonly used to distribute cleaning solutions on large surfaces such as the vessel and chutes, and pipes larger than 3 in. in diameter. Small lines less than or equal to 3 in. can be successfully flushed at velocities of approximately 5 ft/sec.

Examples of the most frequently used types of devices are provided in Chapter 9. Traditional spray balls are used for cleaning vertical vessels. If two/three nozzles can be allocated for spray devices on a reactor head, the traditional spray ball is a good choice, especially if it can be positioned deep enough into a tank so that spray streams can be directed to unflushed nozzles. Such nozzles must be the target of two streams of fluid, which can "splatter or ricochet" and effectively wet these areas.

Some important considerations when using traditional spray balls for cleaning API vessels are as follows:

- There is a preference for using minimum spray ball installation nozzles of 3 in. in glass-lined vessels, to allow insertion of 2½ in. diameter spray balls.
- Two-inch nozzles can be used on alloy vessels, and 1¾ in. sprays, for lower flow rates.
- All sizes of spray installation nozzles can be effectively cleaned by drilling three holes into the spray supply tube, a short distance below the installation flange.

Often, especially when confronted with a revamp situation, an adequate number of nozzles with the appropriate position cannot be obtained. In these cases, spray tubes fitted with multiple bubbles are often used to get the required coverage. Some important design considerations to be observed when using spray tubes are as follows:

- Avoid positioning below solids charge points.
- Avoid obscuring site glasses or shadowing lights.
- The minimum horizontal clearance to the agitator should be 3 in.
- Ensure that if the device cannot be removed from the insertion nozzle that it can be removed by two men working at the insertion nozzle and manway.

Finally, one of the most versatile types of spray devices is the spray bubble. This type of spray is especially effective in cleaning piping, both horizontal and vertical, where the diameter is greater than or equal to 4 in. The great benefit of this kind of device is that it does not insert more than ½ to ¾ in. into the pipe annulus, making it perfect for solids charge chutes or for cleaning the undersides of rupture disks, and instrument probes or sensor elements.

The spray flow rate required to clean simple vessels is based on the diameter. Good results have been obtained using 2.5 gpm for each foot of vessel periphery. Therefore, a 5-ft diameter vessel should be effectively cleaned with

40 gpm of cleaning solution, most of which would be directed upward to wet the top head and nozzles and then to cascade down the sidewall in a uniform sheeting action.

As vessels become more complex with the addition of agitators, dip pipes, and baffles, the requirement for CIP solution rates increases. While the precise amount required to clean the reactor depends on the specific layout of the top head, a value of 50 to 60 gpm or more for a 5-ft diameter (1000 gal) reactor would not be unusual. A 78-in. diameter vessel (2000 gal) may require 60 to 75 gpm. The depth and position of the sprays in the vessel also have an impact on spray efficiency and the subsequent flow rate. After a minimum flow rate is established for a piece of equipment, that flow rate must be looked at relative to the entire CIP process and adjusted upward, if necessary, to balance circuit flow rates, and achieve the required velocity in all associated piping in the circuit.

REACTOR SYSTEM DESIGNED FOR CIP

The modern designer has better glass-lined vessels and Teflon-lined components to employ. Given the opportunity to redesign a reactor system using modern components, best practices, and a little imagination, the system may look like that portrayed in Figure 2.

This figure illustrates an approach where both the process and CIP requirements are given equal consideration. Process piping and equipment is used as much as possible to achieve CIP. The process transfer pump is utilized for both removing process waste and for circulating CIP solutions at rates required for effective cleaning. A CIP header (A) has been added to supply solution to all subcircuits. Spray devices of ball (B) and bubble (C) types have been added to wet surfaces where line flushing is not practical. Valves have been included so that the sequence in which each circuit is brought into service can be controlled. This use of automated valves also serves to limit the total volumetric amount of wash solution required to a rate that can be handled by the reactor transfer pump.

This system allows for both once-through and recirculating modes of operation. Most of the cleaning will be achieved by recirculating CIP solutions from the vessel through the process transfer piping to the sprays or inlet tubes, or directly from the CIP header to supplementary sprays. An aliquot of CIP solution can be prepared, warmed, and circulated through each leg of the CIP header and the sequence repeated as required. The solution acts over time to solubilize and remove soil. The aliquot can then be directed to waste.

Once-through flush solutions can be supplied by an outside system (D), which delivers CIP solutions at the required temperature, pressure, and concentration. These flushes are important for initial removal of solids that may be present on surfaces but not tightly adhered. This avoids the "bathtub ring effect," i.e., what is observed when a kitchen sink is drained. Oil and particles remain in the sink after the liquid has drained. Refilling the sink does little to remove this soil. The solution is to use the hand spray nozzle and continuously sheet the material down the drain. This effect is completely analogous to what occurs in a glass-lined reactor. The bottom head is often coated with soil after a recirculating aliquot is drained, requiring a once-through flush to completely remove the last of the soil or cleaning solution.

To assure a successful result, the designer also needs to provide for the ability to drain liquid from the vessel as quickly as it enters. Outlet piping has been sized to minimize vortexing and minimum holdup by provision of bottom outlet lines larger

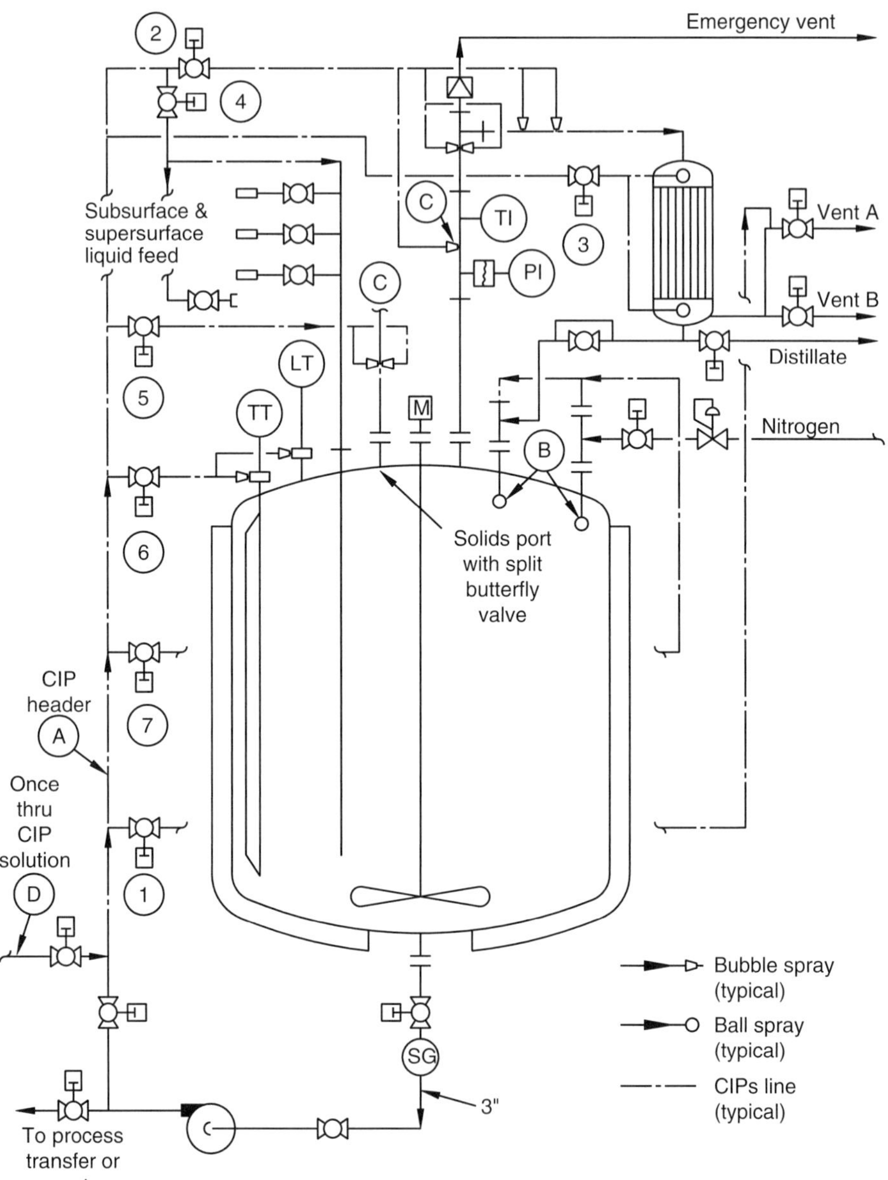

FIGURE 2 Glass-lined reactor designed for CIP. *Abbreviations*: T1, temperature indicator; P1, pressure indicator.

(3 in.) than that would otherwise be required for the process. And, the overhead system has been reconfigured. This arrangement comprised of primarily vertical ductwork and chutes, and a vertical condenser allows for the charge chute (E) piping and instrumentation (C) to be completely wetted by CIP solutions.

The areas of cleaning concerns can be addressed as follows.

Reactor Vessel

Areas of concern regarding reactor vessels include the vessel configuration, the manway, baffles and dip pipes, and unflushed nozzles. In the past, reactors were designed with a top head, which had a large agitator removal cover. The large agitator opening suffers from an additional large diameter gasket, and the fact that the agitator cover does not allow for the CIP liquid sprayed into the top head to sheet uniformly down the vessel head and sidewalls. These reactors are not as cleanable as the more modern one-piece top reactor that is available from the major manufacturers of glass-lined vessels. These models not only offer better CIP solution flow but also allow for more and larger nozzles that are important from the standpoint of supplying spray devices.

In larger sized vessels (>48 in. diameter, >500 gal), welded top vessels are clearly preferred. These reactor types do not have a body flange and rely upon a special removable agitator type. The advantage is in the elimination of the very large gasket, which may interrupt flow on the vessel sidewall and may also hold up solid materials.

If there is no way to avoid the use of a split flange, then one of the more hygienic gasket types should be used. The manufacturers of glass-lined equipment offer several types of gaskets. These should be evaluated. Gaskets that are cut back and do not have any ledges which could hold up liquid or soil should be selected.

There are examples where vessels failed a riboflavin test using a standard envelope gasket but passed the test easily after the top charge port, manway, and head gaskets were replaced with a hygienic design. Apparently the ledge of the envelope gasket held up to riboflavin which slowly bled off, causing the test failure. It is reasonable to assume that this ledge would hold up API product in actual operation leading to residue test failures.

The manways of glass-lined reactors are much further from the head surface than those of an alloy vessel, because of the methods required to fabricate glass-lined equipment. Because the interior surfaces are far from the head plane, special attention is required to distribute sufficient CIP solution into this area to effect cleaning.

The additions of baffles and dip pipes further complicate cleaning within reactors. These vertical structures create shadows, which block cleaning fluid from striking the far side of the element and the vessel wall beyond. Multiple sprays are required to eliminate shadowing.

The number of nozzles that can be fitted onto a glass-lined reactor is limited due to several constraints related to vessel fabrication. It is critical that the designer and the owner work together to optimize the position and the number of spray devices without compromising those nozzles required for processing reasons.

Figure 3 illustrates a 2000-gallon (78 in. diameter) vessel. Three spray balls were employed. The manway has a spray device on either side to assure full coverage of all manway surfaces. The dip pipe and baffle each have a spray device immediately adjacent and another spray device about 90° away on the other side. The spray balls are positioned approximately 120° apart. Note that two of the spray devices (inserted into nozzles G and K) share the nozzles with a clean fluid. These fluids enter from the side through full-size instrument tees.

The occasional lack of space for nozzles for multiple sprays can often be overcome by use of specially designed assemblies consisting of a spray ball, arm, and bubble, thus providing multiple spray sources while utilizing only one nozzle. Though properly designed spray devices can spray enough liquid into each nozzle

Nozzle	Use
A	Manway
B	Agitator
C	CIP/Liquid feed
D	Sample dip pipe
E	Vapor rizer
F	Liquid inlet
G	CIP
H	Solid charge
I	Baffle
J	Liquid inlet
K	CIP/Liquid feed

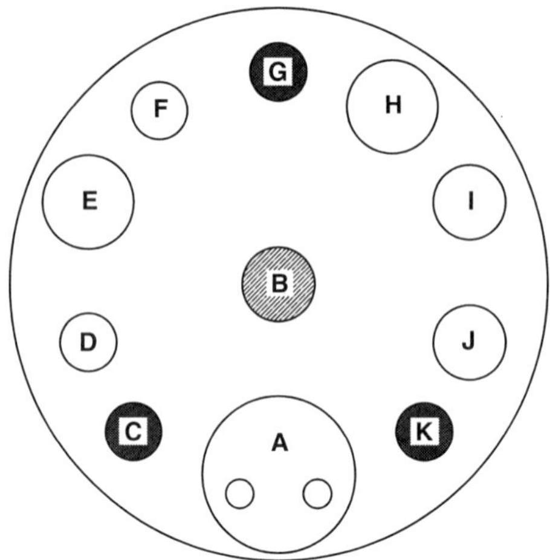

FIGURE 3 The 2000-gallon vessel top head.

to assure full coverage, the need can be alleviated if the nozzle is externally flushed as part of the line cleaning. Examples of externally flushed nozzles include all liquid feeds, reflux, and the vapor riser. Shorter neck nozzles are easier to spray and should be favored in the design stage if they can be fabricated by the vessel vendor.

In the past, the baffle nozzle has been a problem. Typically, this nozzle must be cleaned by spray directed from below. This has been addressed by extending the baffle and inserting a glass-lined instrument tee to allow for CIP fluid to be injected into the annulus of the baffle/temperature transmitter nozzle temperature transmitter (TT) and level transmitter (LT) in Figure 2. This approach could be used for any unflushed nozzle with a vertical member inserted into the vessel, such as a dip pipe.

Condensers

A vertical condenser is shown in Figure 2. Assuming the need for a 200 ft^2 condenser with 174, $\frac{3}{4}$ in., tubes to achieve a flushing rate of 5 ft/sec would have required a flow of over 1200 gpm. Clearly, this is not feasible. Even with a rate of 1 ft/sec, the flow is much greater than that which could be provided by the process pump. However, if a single-pass (i.e., process on tube side) condenser is mounted vertically, it is possible to introduce CIP liquid at such a rate that a falling film is established in each tube. This film will consistently wet the tube surfaces and sweep away soil continuously. A second smaller spray ball can be used to wet the bottom tube sheet. This method has proven effective in cleaning reactor condensers. The individual tubes are treated as small tanks, which need not be filled, to be effectively wetted and cleaned.

Overhead Vapor Lines

Overhead vapor risers and reactor emergency vent piping can be the largest bore process piping in a bulk pharmaceutical facility. Typical sizes may range from 4 to 6 in. for a 300-gallon vessel to 8 to 10 in. for 2000-gallon reactors. It is not possible to pressure flush such large diameter lines at rates that would achieve 3 to 5 ft/sec.

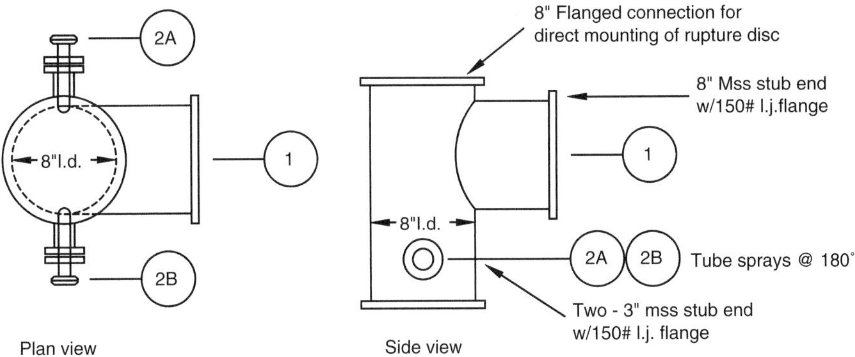

FIGURE 4 Rupture disk holder/vapor sidearm.

The only real alternative is direct application of CIP solution from spray devices permanently installed in the vapor riser and the overhead piping. A solution is to install a pair of bubble spray devices (see Chapter 9, Figure 6) positioned as shown in Figure 2 at the high point of the vapor riser vertical section, positioned at 180° relative to each other, so that they can spray upward and across to form a falling film, which will wet the sides of the riser and sweep away soil. Figure 4 illustrates such an arrangement via reference to sprays 2A and 2B. Because the bubble-type sprays were inserted only ½ inch into the pipe annulus, less than 2% of the space was impacted. This is important when dealing with relief systems.

In designing this type of overhead system, it is important to install a vertical riser without bends, and to aggressively minimize horizontal runs, which cannot take advantage of a falling film to decrease the number of sprays required. Horizontal runs must be fitted with multiple bubble sprays to spray the top of the pipe which must pitch to a drain point for process and CIP fluids.

Special Fittings

The insertion of bubble-type sprays into small diameter lines can sometimes be handled with available Teflon fittings, but as the line sizes increase, specialized pieces are required for the best cleaning geometry. Instrument tees may be used or special fittings can be fabricated from Teflon-coated steel, Hastelloy-, or glass-lined carbon steel pipe. The objective of the designer should be to provide a simple-to-fabricate piece that positions spray devices, so that they can direct cleaning solution directly at the point where soil may accumulate.

An example of such a special fitting is illustrated on Figure 5, a drawing of an Instrument holder consisting of a special piece that employs the use of a single bubble spray to clean three instrument connections (two pressure traps and a temperature probe) mounted opposite the spray.

A clever designer can use bubble-type sprays and specially fabricated pieces to solve many cleaning problems, including those related to solids charge and solids discharge chutes utilized in API isolation and drying systems.

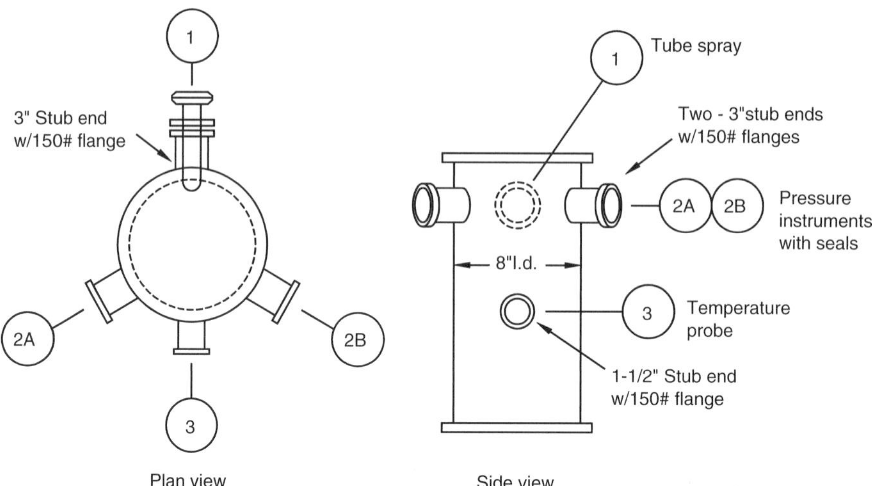

FIGURE 5 Instrument holder connections.

Solids Charge Nozzles/Chutes

Solids charge nozzles and chutes can be a challenge to clean. By their nature they are always heavily soiled after use. They can, however, be cleaned like vertical tanks by applying falling films of CIP solution, initiated from properly positioned bubble-type sprays.

The use of a split butterfly valve for contained solids transfers has become quite popular in bulk pharmaceutical processing. Vendors of these devices offer special adaptors fitted with spray devices that allow for CIP of their solids handling valves.

REACTOR CLEANING CIRCUITS

As previously stated, in the design of API CIP systems, it should be a goal of the designer to use as much of the process piping and equipment as possible to distribute the cleaning solution.

- There is a long-standing "rule of thumb" that lines should be flushed at a rate of 5 ft/sec. This rule of thumb was developed by early workers with CIP systems comprised primarily of glass pipe. Simple observations of flow in lines of 1½ to 3 in. in diameter, in extensive circuits, indicated the need for a 5 ft/sec velocity to ensure that the flow is highly turbulent and will sweep air from long horizontal runs, that it will flood vertical runs in a downflow regime, and that it will sweep gas pockets out of vertical tees.

Obtaining 5 ft/sec is not always possible, but this is a good starting point for evaluating the system. The designer is usually confronted with the fact that the system uses Teflon-lined pipe of four different sizes, as shown in Table 3.

In the design of reactor systems between 300 and 4000 gal, the most typical liquid transfer line is generally 2 in. Consequently, there is an upper limit to how much liquid can be distributed at any one time because of the limitations of the piping components and the characteristics of the process transfer pump.

TABLE 3 Inside Diameter, Area, and Flow Rate for 5 Feet/Sec in Condenser

Nominal diameter (in.)	Internal diameter (in.)	Area (ft^2)	gpm at Q (5 ft/sec)
1	13/16	0.0036	8
11/2	1 5/16	0.0094	21
2	1 13/16	0.0179	40
3	2 13/16	0.0431	97

The designer has to make some important decisions on how differing use points are combined to form a CIP circuit. Consider the system shown in Figure 2. That system has a 2-in. Teflon lined (TL) CIP header at the discharge of the process transfer pump. Assume that this pump has a nominal capacity of 50 gpm at 150 ft head. It is the designer's task to apportion adequate flow to various circuits and ensure that the system hydraulics can be met to ensure the proper flow to each branch of the circuit.

To do this, it is important to realize that there are three types of backpressure patterns which may be present in a circuit.

- *A spray device*: Generally, nonrotating sprays are designed for a pressure drop of 20 to 30 psi (46–69 ft of water).
- *An open line*: Pressure drop per 100 ft at 5 ft/sec is generally in the range of 1 to 3 psig per 100 ft.
- *A restricted flow*: Generally, a higher pressure drop than an open line, but less than a spray device.

It is important that a spray device is not included in a circuit in parallel with a circuit with an open line. The flow through an unrestricted (open) line would deprive the spray device of the appropriate flow of CIP fluid. Sometimes a restriction flow orifice can be used to backpressure an open or restricted line so that satisfactory hydraulic performance could be obtained with spray devices included in the same circuit.

The glass-lined reactor portrayed in Figure 2 utilizes the circuits summarized in Table 4.

The process transfer pump needs to be evaluated to ensure that it is able to provide adequate flows for both aqueous and organic fluids. It should be noted that the pressure drop through pipe and fitting is affected by the fluid specific gravity. In a similar way, the discharge pressure of a centrifugal pump varies with specific gravity. Typical solvent specific gravities vary between 0.6 and 1.3 SG.

TABLE 4 Example of Flow Rate Through Sub-Circuits in Reactor CIP

Circuit	Flow rate (gpm)
Reflux, distillate, or vent	>40 gpm
Overhead piping	5 bubble sprays at 10=50 gpm
Condenser	2 sprays=60 total gpm
Dip tube	>40 gpm
Charge chute	2 bubble sprays at 10=20 gpm
Instrument and baffle nozzles	20 gpm
Reactor vessel	2 sprays at 30 gpm=60 gpm

In most cases, the pressure drops and flows can be balanced without much trouble. Cases, however, do arise when the use of a pump discharge flow control valve or a motor with variable frequency drive is required to balance the circuits.

TYPICAL CLEANING CYCLES

A typical sequence of cleaning operations for a reactor system will include some or all of the following steps. For example, a given campaign might require the highly complex sequence (the order is not fixed and may change), as shown in Table 5.

Once-through rinsing is used for two purposes; the first to flush solids and contamination from the system and as a final rinse to remove the last traces of cleaning solution or dissolved soil from the system.

- As discussed earlier, once-through flushing is a method to avoid the bathtub ring effect. Some designers recommend cleaning the most soiled or most distant circuits first. In the case of the reactor in Figure 2, a possible once-through sequence could be (1) reflux return from condenser, (2) vapor riser, (3) condenser, (4) dip tube, (5) charge chute, (6) blind nozzles (via sprays), and (7) vessel via multiple sprays.

Obviously, when using a solvent, the time length of this flushing should be minimized for cost and waste minimization reasons. If the process allows for the use of a hot aqueous rinse, that avenue should be considered as a method of reducing cleaning expenses.

Recirculating flow is used to conserve solvent/CIP fluid and allows for extended period of contact. In the case of recirculating flow, the volume of solvent charged to the system has to be considered. Important considerations are:

- The initial fill must allow for the volume required to fill the CIP header.
- The liquid height during steady state must be sufficient to provide the net positive suction head required to allow the pump to operate without cavitation (with some allowance for nitrogen aeration of the suction line).
- The starting volume must be sufficient to allow for enough volume to sufficiently clean the waste (i.e., discharge lines) of the circuit to the limits of the reactor envelope.

Assuming that liquid CIP fluids and waste are set up with automated valves, a recirculation sequence may take the following single-pass cycle time (multiple passes required; Table 6).

It is important when designing a CIP system for API that the residual liquid can be purged from the system. Small quantities of pooled liquids, especially when

TABLE 5 Example of a Reactor CIP Program

Post campaign rinse (vessel to vessel to clear lines)
Solvent 1 once-through rinse
Solvent 1 recirculation wash
CIP 100 (basic cleaning solution) recirculation wash
CIP 220 (acidic cleaning solution) recirculation wash
Water for operation recirculation wash
Solvent 1 recirculation (sample for analytical test)
Solvent 2 recirculation (solvent for next campaign)

TABLE 6 Typical Duration Time of Reactor Sub-Circuit Steps

Step	Description	Duration (min)
1	Fill with 75–125 gal at 50 gpm	2
2	Recirculate through reflux/vent piping	1.0
3	Recirculate through overhead piping	1.5
4	Recirculate through condenser	1.5
5	Recirculate through dip tube	0.5
6	Recirculate through charge chute Instrument and baffle nozzles	1.0
7	Recirculate through reactor vessel sprays	3.0
8	Pump to waste at 50 gpm	2.0
	Subtotal	12.5

incorporated into a subsequent recirculating aliquot, will potentially contaminate any surface which it touches. Low points and reservoirs of residual liquids must be eliminated.

Methods to be considered include:

- *Gravity*: In modern, multipurpose API facilities, it is typical that the reactors are elevated in three- and four-level structures. Gravity can be used to drain fluids from low points.
- *Automated line blowing*: Nitrogen can be used to blow liquid from a local low point to waste.
- *Pump casings*: Pump casings form local low points. Designers have automated the draining or pressure purging of pump casings by utilizing a casing drain valve.

ISOLATION EQUIPMENT CIP

Isolation equipment in API manufacture is used to separate solid products from the saturated mother liquor. The types of equipment utilized include horizontal centrifuges, inverting basket centrifuges, and agitated Nutsche filters. The same processes and techniques used to clean reactor systems can be used to clean isolation systems.

Figure 6 illustrates an inverting basket centrifuge and associated equipment. CIP piping is shown as alternate short and long dashes. The basic aim is to introduce cleaning fluids into the header system and spray various portions of the system. The CIP solution then drains by gravity to the M/L tank. The fluid collected is either recirculated or pumped to waste.

It is important to note that in any isolation system, there are two major drainage paths for CIP fluid: the solids discharge port and the M/L drainage port. If the M/L tank is to be used as a collection point for the CIP fluid, it must be positioned below the solids discharge chute.

In the case illustrated in Figure 6, a temporary connection to the bottom of the chute has been made using a split butterfly valve with a CIP collector. These devices are available from the suppliers of split butterfly technology. The liquid collected is directed to the M/L tank by gravity.

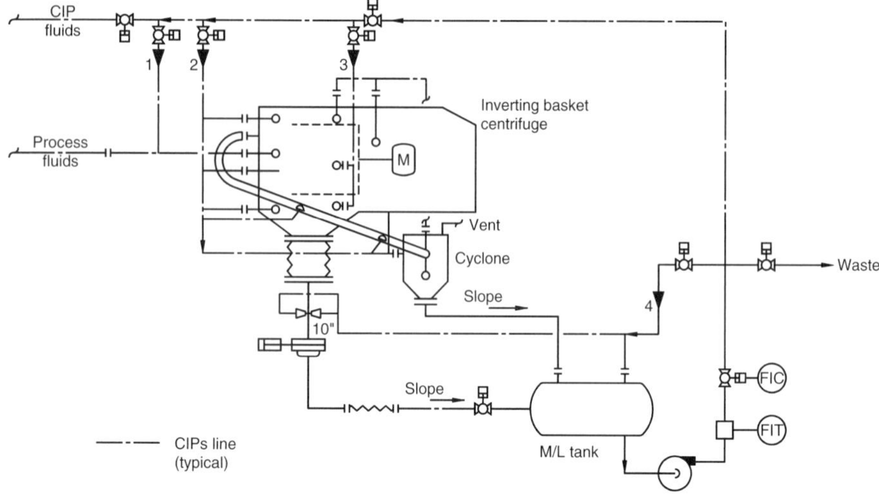

FIGURE 6 Typical centrifuge circuit. *Abbreviations*: FIC, flow indicating controller; FIT, flow indicating transmitter.

There are five major CIP concepts involved in this example including (1) the feed circuit, (2) the front housing and holding line, (3) the rear housing and vapor liquid separator, and (4) the M/L tank and solids chute.

Figure 6 only hints at the complexities of the internal sprays needed to clean the interior of a device as complicated as a centrifuge. Fortunately, in recent years, many manufacturers of isolation equipment have incorporated into their designs the spray device and appropriate internal design needed to make these devices suitable for CIP. This development includes procedures/software for use in cleaning these devices. It is usually a requirement that the machines are rotating and the equipment cycled through various ranges of motion while cleaning proceeds.

The flow rates for vendor-supplied spray devices generally exceed the process pump requirements. It is therefore not unusual to see flow control valves or variable frequency drive pumps in this application.

DRYER EQUIPMENT CIP

API facilities included equipment for drying the products. Materials isolated in centrifuges or filters contain 10% to 50% by weight solvent which must be driven off to produce a stable material that can be processed further to create the final drug substance.

Dryer types that are typically used include:

- Nutsche-type filter dryers
- Agitated pan dryers
- Conical dryers
- Rotary double cone dryers
- Vacuum tray dryers

The first three devices are generally chosen for a facility designed for CIP. The reasons are:

- The inlet/outlet connections are fixed
- The internal design can be adapted for ease of cleaning

Rotary double cone dryers can be cleaned in place, but the need to make and break large diameter connections makes them a less-than-ideal selection. Tray dryers are made less suitable for CIP due to the internal trays and structure.

Figure 7 depicts a conical dryer designed for CIP. The CIP piping is shown in alternating short/long dashes. A conical or pan dryer does not normally handle process liquids. In the example shown, a portable CIP skid is positioned below the dryer discharge chute. This skid includes a tank of 100 to 200 gal capacity, a circulating pump, and instrumentation. There are six major CIP circuits defined by the numbered arrow heads including (1) the dryer vessel circuit, (2) the dust filter housing, (3) the charge chute and emergency vent system, (4) the vent line, and (5) the discharge chute. Spent solutions are pumped to waste per path (6).

The vendors of dryer systems have recently begun to offer CIP spray devices and guarantee CIP performance.

PIPING DESIGN CONSIDERATIONS

To a large extent, the piping design of an API facility will determine the ultimate success of cleaning. The most important aspects include:

- L/D ratios of fittings
- Orientation of fittings
- Drainability

The L/D ratio can be defined as:

- L is the leg extension from the internal wall of the major flow axis.
- D is the internal diameter of the leg or extension.

Obviously, by making the L/D ratio as small as possible, the cleaning of dead legs is made easier. In recent years, the manufacturers of Teflon-lined pipe have started to make short stacked tees with greatly reduced L/D ratios. These should be used in place of standard tees.

Reducing tees should be avoided.

The orientation of piping branches and dead legs also should be carefully considered. Horizontal lines with a tee positioned in a vertical-up position will trap gas, while a tee positioned in the vertical-down orientation will trap solids. Tees positioned in a horizontal orientation will avoid both of these problems.

The piping designer should slope piping back to a vessel or local low point, which can be drained automatically at the end of each cycle.

CIP FLUID/WASTE DISTRIBUTION

The goal in the CIP of API systems is to use as much of the process system as possible to supply, condition, and distribute CIP fluids and direct cleaning waste to the appropriate final destination.

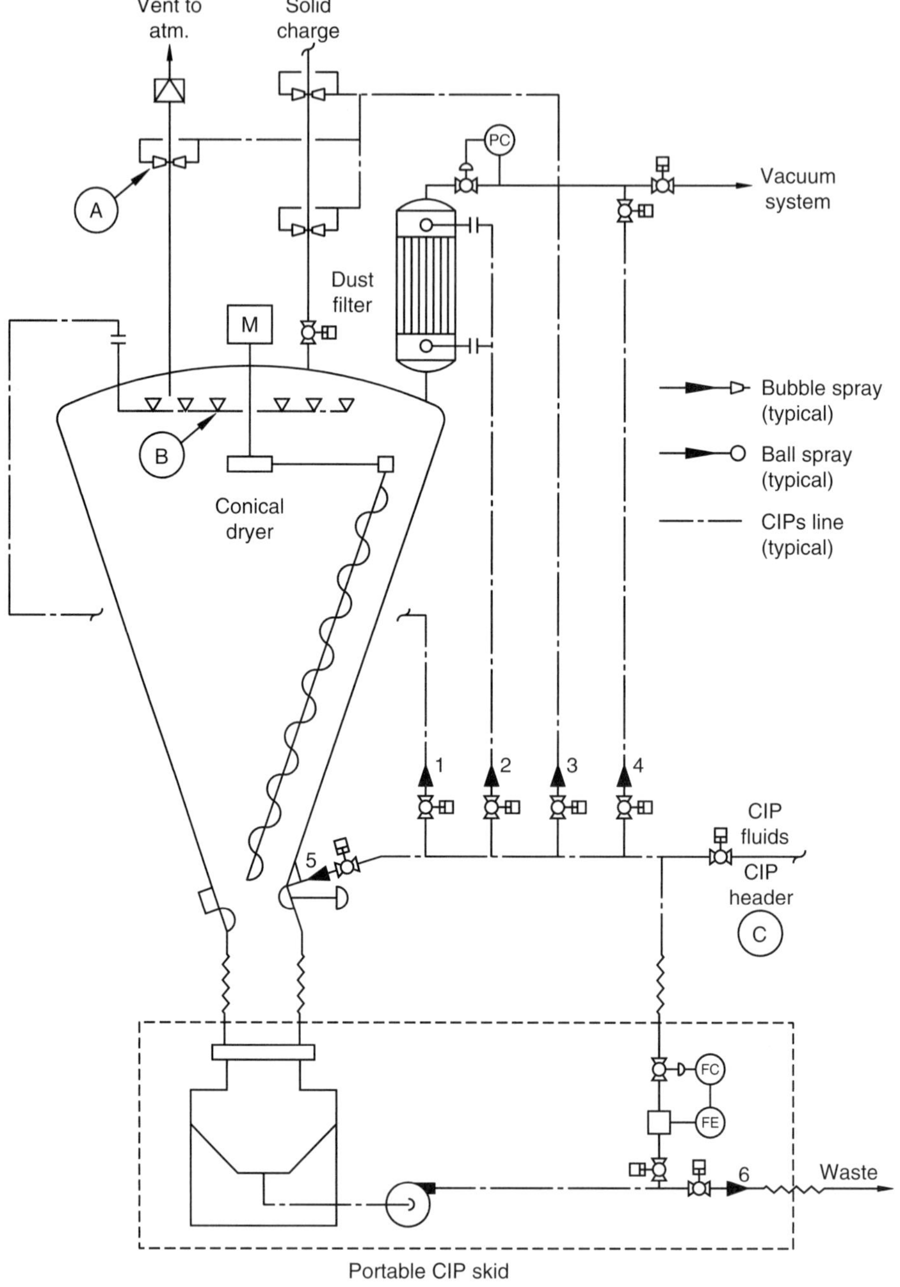

FIGURE 7 Typical conical dryer circuit. *Abbreviations*: FE, flow element; PC, pressure controller.

It is the usual practice that CIP solutions not be recycled to a central system. The concern is that potent compounds could be redistributed to other areas of the facility, greatly complicating the task of proving a system is clean.

There is also a safety concern of returning fluids that contain flammable solvents back to a centralized system located in an area not designed for solvents.

The other complicating factor is that most API facilities serve multiproduct or multipurpose functions. They are constantly being reconfigured to run new products. To accomplish this, most companies employ some sort of solvent, water, waste, and process manifold systems.

Experience has shown that these manifold systems can be designed to serve both the process and CIP functions, but careful attention should be paid during the design phases with regard to physical size, accommodation of automated valves, and drainability. A typical system is shown in Figure 8. The boxes represent arrays of valves to control flow.

In this system, organic solvents and process water is directed to process vessels through a manifold system for both process and CIP purposes. Acid–base wash solutions are prepared/conditioned and distributed in the same piping. The solutions can be applied to the process systems in the once-through mode or they can be recirculated.

Many API facilities utilize process manifolds to provide interconnectability between various pieces of equipment. These process manifolds can be set up with automated valves so that process transfers can be made from system to system and recirculation for cleaning purposes can be established.

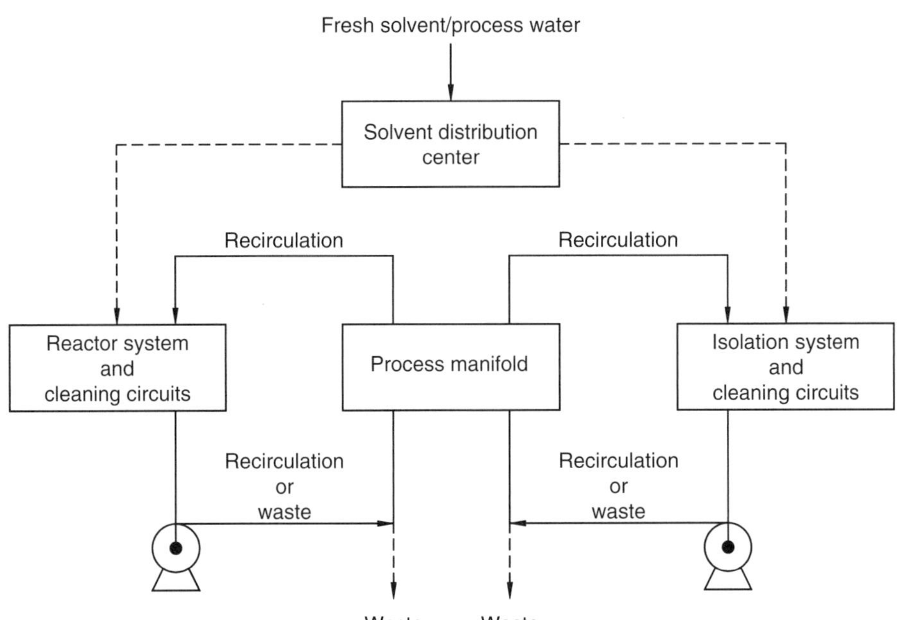

FIGURE 8 CIP fluid/waste distribution, active pharmaceutical ingredient facility.

One of the designer's first tasks when designing an API plant setup for CIP is to determine the locations, size, and configuration of the process manifolds' points.

TESTING TO PROVE CLEANING EFFECTIVENESS

Ultimately, cleaning effectiveness can only be verified by accepted recovery techniques and analytical methods of quantifying surface residuals. The ultimate end result is a function of how well all design parameters were incorporated into the design.

Riboflavin, more commonly known as vitamin B_2, is an orange-yellow needle-like solid that, when dissolved in water, produces a yellow liquid. This liquid, when exposed to black light, fluoresces with a greenish-yellow glow.

Riboflavin testing is frequently used to measure effectiveness of spray coverage. Riboflavin testing has no universally accepted standard, so there are a lot of variations of this test among end users. Generally, riboflavin is applied to the surfaces to be spray cleaned, and either a complete CIP cycle is run and the equipment is then inspected, using a black light, to verify that all riboflavin has been removed, or alternatively, a series of water rinse cycles are run and then success is based upon evidence that riboflavin dilution has occurred. Care must be exercised not to overinterpret the test results.

Riboflavin is usually applied at a concentration of 100 to 200 ppm; however, no recognized concentration standard currently exists.

SAFETY ISSUES

The use of solvents in the cleaning of API systems should be minimized for cost reasons, but it is very difficult to eliminate their use. Many API substances and soils have very little solubility in aqueous solutions. Therefore, the system must be designed to safely utilize flammable solvents. A fire or an explosion is possible only if three conditions are met simultaneously. They are:

- A flammable substance is present
- Sufficient oxygen is present to support combustion
- An energy source sufficient to ignite a fuel-air mixture is available

There is nothing that can be done to eliminate the flammable substance in all cases. Maintaining a safe working environment depends on inerting the equipment to be cleaned and eliminating ignition sources. All API facilities are designed to function with the levels of oxygen in systems handling flammable solvent at or less than 1% to 2% oxygen. This condition is achieved by the use of nitrogen sweeps or purges.

It is usually a simple engineering task to incorporate these systems into a CIP philosophy. The more difficult task is to make sure that the programming and processing scheme prevents any oxygen from getting into the system. It is imperative that the CIP sequence be subjected to a HAZard and OPerability analysis as rigorous as the process sequence itself.

An important energy source that could lead to a deflagration is static electrical charge.

Charge generation can occur at the interface of a solid and liquid. As the liquid flows, it generates a charge. The charge generation depends on the potential of the solid surface to accept or donate electrons, the speed of the flowing fluid, and the conductivity of the solid.

Little charge is generated by a polar liquid flowing through highly conductive pipes (i.e., water flowing in stainless steel). However, large potentials (thousands of volts) can be generated by nonpolar solvents flowing through insulated systems such as Teflon-lined pipe.

Charge can build up on the flowing liquid, insulating solids, and isolated conductors. This charge is sufficient to ignite flammable solvents if oxygen in sufficient quantities is present. It can also damage piping and glass-lined equipment if no steps are taken to dissipate this charge.

The charge buildup can be instigated by the proper grounding of conductive components and the use of conductive piping elements to dissipate charges from flowing liquids.

CONCLUSION

The recent trends toward more potent API compounds have spurred increased interest in the design of bulk chemical synthesis facilities designed for CIP. Numerous systems have been built or revamped, incorporating the philosophies addressed in this chapter.

The author believes that the need for automated, repeatable cleaning methods will only increase in the future driven by regulatory issues. The challenges for this improvement are currently being taken up by the manufacturers of equipment, piping, and the designers of API facilities. The API manufacturing process has needs for defining fabrication criteria for a wide variety of equipment including pumps, mixers, vessels, heat exchangers, etc.

Fortunately, even as this chapter is being written, the Steering Committee of Pharmaceutical 3-A Standards has begun development of P3-A Standards to allow equipment manufacturers to better understand the specialized needs of the API sector. This work has started with three coordinated efforts including the development of a document for pumps, and the development of a glossary document and another document for materials. The P3-A efforts are focused on supporting equipment designs for manufacturing of APIs. All three of these documents are being reviewed by a consensus body as part of the ANSI standards development procedures (1).

REFERENCE

1. 3-A Sanitary Standards Inc., 1451 Dolley Madison Boulevard, Suite 210, McLean, Virginia 22101–3850, U.S.A., Phone: 703-790-0295, Fax: 703-761-6284 (Accessed January 8, 2007 at www.3-a.org).

16 CIP System Troubleshooting Guide

Sally J. Rush
Seiberling Associates, Inc., Beloit, Wisconsin, U.S.A.

INTRODUCTION

Scope and Assumptions

Previous chapters of this book have identified the important variables that determine the efficacy of the clean-in-place (CIP) procedure, specifically time, temperature, concentration, and physical action. The first three are controllable and are subject to variations due to hardware and software failures. Physical action is initially determined by engineering design but may be affected by variations in flow and pressure caused by hardware or software failure. This chapter is intended to provide troubleshooting suggestions only for qualified CIP circuits, and generally on the basis of human machine interface (HMI) alarms to convey basic information regarding a CIP program failure. It is assumed that CIP circuit design issues have been resolved prior to or during commissioning and that the CIP system has been properly commissioned and qualified, meaning that, at one point in time prior to the CIP circuit failure; the CIP circuit had been known to run well with properly selected recipe parameters proven capable of removing the expected soil, and thus restore the equipment to a visually or swab analysis clean state.

TROUBLESHOOTING AIDS

Troubleshooting a CIPable process in modern biopharmaceutical facilities will generally require access to multiple areas, including the clean room in which the process is located, the control room where the HMI(s) will be found, and the mechanical space in which the CIP skid, supply/return valves and piping, and CIP supply/return distribution valves are located. As it is generally not possible to move freely from one area to another, troubleshooting may require the cooperative action of several people in these areas via electronic communication. The first, and most important requirement, is that all personnel involved have a common understanding of the problem and the troubleshooting approach to be followed.

The Human Senses

The most useful tools available to an individual experienced with the operation of CIP systems are the senses of sound, touch, and sight. Unfortunately, the CIPable process limits the use of these senses by a single individual, so teamwork and communication ability is required to verify the proper operation of CIP supply and return distribution valves, on the HMI(s), and in the field. When valves are not easily accessible, the result of their operation can be determined by touching the various downstream lines in sequence during heating and cooling of the flush, wash, and rinse solution. The use of hot water for injection (HWFI) for all phase of a program will make this approach less useful. Properly operating centrifugal pumps, the most common type used, can be verified by sound, in combination with observation of flow and pressure data, for pumps that are affected by air in the supply stream, or through failing seals, will cavitate, and perhaps cause substantial vibration in the attached piping. The appearance and sound of spray ricochet in a vessel being spray CIP'd will provide a reliable though not definitive indication of proper supply pressure, and uniform delivery of CIP fluids.

Historic Operating Trends

Much of the CIP control software in current use provides for the acquisition and storage of operating data for the important variables including level, pressure, flow, supply and return temperatures, and conductivity or resistivity. Prior to beginning field troubleshooting of a reported or confirmed operational problem, it can be very helpful to obtain a historical trend, if available, for the circuit or system involved. Trending data may be available for

- CIP supply flow rate
- CIP supply temperature
- Steam supply temperature control valve position
- Steam header pressure
- CIP supply temperature
- CIP return temperature
- Chemical supply source tank level
- Chemical feed header pressure (if applicable)
- Circuit system fills volume set point (volume of solution in circuit)
- CIP supply conductivity set point
- Conductivity return resistivity set point
- Water additions during the recirculated wash.

Interpretation of Trending Data

Some examples of operating problems that a review of trending data may be beneficial include

Repeated water addition during a cycle as the result of low level in a CIP recirculation tank (or air separation tank) will dilute the chemical solution and thus cause reduced conductivity.

Water addition may be the result of CIP circuit hydraulic imbalance, as the result of a failing return pump (generally seal or suction side leak). A U-Bend transfer leak, an improperly tightened or positioned return hose or spool piece installation may cause either loss of solution, and hence level, or incorporation of air in the supply to pumps.

Supply flow problems of a low or high flow nature and may be attributed to one of many issues. The alarm may be intermittent rather than continuous. The historical trend data can provide information as to when the intermittent alarm occurs, or at what point flow control has been lost.

PROGRAM FAILS TO START

Most CIP programs are initiated by a sequence of manual operations which include selection of (*i*) the equipment to be cleaned, (*ii*) the selection of the applicable CIP program, and (*iii*) the proper recipe. The program and recipe may be selected automatically or by the operator. After the menu driven HMI selections are made, the control system may provide a summary of the manual preparations required, and an HMI prompt may require confirmation to assure that the manual activity has been completed. After this confirmation, the CIP program will confirm all CIP permissives and check applicable interlocks. If all are in compliance, the program will be initiated. CIP program initiation failures are usually attributed to issues with the CIP "batch" assembly or CIP permissive and interlock failures.

Batch Recipe Failure

If operator selection of a CIP batch recipe is required, a failure to properly assemble or include required components will result in a CIP circuit initialization failure. For examples, consider

- If the operator selects a single pass rinse, when the operational sequence demands a recirculated chemical wash program, the CIP initialization will fail.
- The field devices (pumps, valves, agitation) for a CIP circuit are controlled by a CIP device sequence. If the CIP device sequence does not match up with the selected CIP program, the CIP circuit initialization will fail.
- The CIP device sequence also can define the circuit boundary and be used to acquire the equipment to be cleaned. If all devices and equipment within the CIP circuit boundary are not available for CIP operations, the CIP circuit initialization will fail.
- The CIP programs for unique CIP circuits are optimized by applying circuit specific recipe parameters, defining critical CIP parameters of exposure time, temperature, concentration, and physical action, as well as the noncritical parameters, like drain and air blow times. A CIP circuit boundary may have several recipes, and if the selected CIP recipe does not match up with the selected CIP program, the initialization will fail.

Permissive Failure

Permissives are established by the pre-checks performed by software prior to CIP circuit initialization used to confirm that the operating conditions will ensure the CIP programs success. Typical permissives include confirming the availability of all support utilities, process equipment, and the required physical setup to establish the CIP circuit.

The utility of greatest concern is the water supply. The minimum water volume required must be available, and the distribution loop must be operating at the required delivery pressure and temperature for CIP operation. Quality determination may involve an operator prompt to sample and test. Similar utility confirmation checks may be performed for plant steam, chilled water, clean air for air blows, cleaning chemical supply, and waste discharge capacity if pretreatment of CIP waste in needed.

The process equipment to be cleaned may be defined within the circuit boundary. As there may be overlapping process/CIP boundaries, an adjacent process operation may need to be concluded to permit automation acquisition of the devices for cleaning operations.

The manual preparation may be monitored by devices like proximity switches used to confirm the placement of transfer panel U-Bends. If the U-Bend is not properly positioned, a *U-Bend proximity switch failure* will result, preventing the CIP program from starting the cleaning activity. The confirmation of much of the other CIP circuit manual preparation is via an HMI query and an operator response.

Interlock Failures

CIP circuit interlock checks generally follow batch initialization but are completed prior to establishing an active cleaning program. After the CIP batch is loaded, valves may be commanded to establish the initial flow path. A device failure detected by a limit switch or proximity sensor on a valve stem at this point in a program must prevent continuance of the program to the first active phase.

Pre-checks may be performed on any or all analytical element or measuring instrument as the program begins, and if an instrument is noted outside of its' calibrated range, a CIP program interlock may terminate the program.

FLOW-RELATED OPERATING FAILURES

Pumps

Before a CIP supply pump starts a three to five second pump start delay may be provided to allow the CIP water tank outlet valve to open and the pump cavity to fill, excluding air to establish a primed condition. When the pump is active, a discrete pump "run" signal is sent back to the control system to confirm the pumps active operation. If the run or device position feedback signal is not received, an alarm noting the pump (or perhaps a valve failure) may be provided as the CIP program fails.

When a CIP pump fails to start or there is a run signal failure, the alarm provided to operations will typically identify the pump and the issue. The failure investigation should start with a review of the pump in the field, with the troubleshooter working back to the control system to determine the reason for the failure. The troubleshooter may observe

- The pump is running smoothly with proper flow being generated. This suggests the need to check the discrete *Run* input signal to the control system to determine if there is a failure with the I/O card monitoring the run status.
- The pump is running smoothly, but no flow is noted, indicating that there is a piping block, and the flow path must be investigated for a closed manual valve, missing U-Bend, or an automated valve failure.
- The pump is functional but operating at a reduced capacity. The pump should be investigated to ensure the pump seals are intact, and the pump suction conditions are proper. If the pump is variable speed drive (VSD) equipped, the output of the VSD should be checked to determine approximate pump RPM.

The investigation may reveal that the pump has indeed failed to run, and a few causes may include

- Mechanical obstruction which would prevent pump from rotating (and probably cause power interruption via supply breaker).
- Local electrical disconnect at pump motor is "off" (this frequently occurs following maintenance).

- Pump motor starter overload has "kicked out" the breaker or blown a fuse, disrupting power to the pump motor.
- I/O card failure results in no output signal.

If the pump is controlled by a variable speed drive, there may be a failure associated with the VSD. The initial assumption will be that the recipe parameter providing the drive with the required data for either a fixed speed or proportional, integral and derivative (PID) control has not been altered.

Valves

When a valve is energized to open to initiate CIP operations, or for sequential control during the program, the control system provides an output to energize the solenoid, and pneumatic pressure is provided to open the valve. When the valve is commanded closed, the output and solenoid is de-energized, blocking the air supply and venting the pressure off of the valve actuator, allowing the valve to close due to actuator spring pressure. A valve position or limit switch is monitored to confirm the valve is in the commanded open or closed position. However, not all valves may be equipped with position switches to confirm both the open and closed positions.

When a "valve failure to open" or "valve failure to close" *is* reported as an alarm, the first check is in the field to determine the actual state of the valve.

When a valve is reported failed closed, but by actuator indication, appears to open and close smoothly, the first check should be the functionality of the limit or mechanical proximity switch. This monitoring device may require adjustment to provide the necessary feedback signal to properly report the valves position. If the valve is a ball type, examine the linkage between the valve and the actuator.

A valve not fully open indication may be due to an obstruction within the valve body which may limit its' travel closed or open. Valve stem misalignment, which can cause the valve stem to bind, can result in the failure of a compression or mix-proof valve, often used for CIP supply and return flow control. The partial opening or closing of a valve may be related to an air leak, a leak, or crimp in the pneumatic line between the valve solenoid and valve actuator, or insufficient air header pressure.

A "valve FV-#### fails to open" message may be caused by any or all of the above, and in addition by failure of valve solenoid, or I/O card failure, with no signal provided to the valve.

Field investigation of the less frequent "valve FV-#### failed to close" indication may reveal the valve closes, but the proximity or limit switch fails to read closed valve position. Other cause may include a misaligned valve stem (in compression valve), which results in valve binding or failure of the actuator to valve stem linkage results in failure to close. A crimped pneumatic line between the valve actuator and solenoid and the subsequent slow bleed of air, or the failure of the solenoid to properly vent may also slow or prevent valve closure.

A valve may be commanded open, and by all appearance functions, but no flow occurs through the valve. For example, a tank outlet valve actuator may indicate the valve has opened, but the tank does not drain. This most commonly happens on diaphragm valves when the valve stem pulls free of the diaphragm material. The actuator functions and the proximity switch reports the valve as open, but the valve essentially remains closed (or partially closed) due to the failed diaphragm connection.

TEMPERATURE CONTROL PROBLEMS

The CIP supply temperature is a controlled recipe parameter, and the CIP return temperature is generally a monitored recipe parameter. There may be independent alarms related to CIP supply and return temperature, and troubleshooting the alarms requires an understanding of the desired control and monitoring functions.

CIP Supply Temperature

The CIP supply temperature is generally controlled by the CIP skid's heat exchanger and temperature control valve(s), operating under PID control based on a CIP supply temperature sensor. The heat exchanger may be used to heat or cool the CIP solution, adjusting the temperature to a circuit specific recipe set point. The key failures associated include *CIP supply over-temperature, excessive delay to CIP set-point temperature*, or *PID temperature control loop failure*.

Over-Temperature Alarm

This alarm occurs when the actual CIP supply temperature exceeds the maximum allowable CIP supply temperature and is provided to protect equipment and/or operating personnel from excessive temperatures. Possible causes include excessive steam header pressure (normally in the range of 50–90 psig) due to a failed pressure reducing valve generally set at 15 to 35 psig. A failed pressure reducing valve, and thus excessive supply pressure, may cause the temperature control valve to be unable to regulate steam flow in a proper manner.

If the plant steam supply is found to be in accord with specifications, the temperature control valve should be checked for a mechanical failure, which may involve a worn valve seat, failed actuator, or a pressure to current (P/I) transducer which may have failed or be out of calibration. Some CIP skids are fitted with a characterized ball valve to serve both shut-off and throttling functions. Alternatively, a stem-type throttling valve may be used in combination with a ball valve for positive shut-off. The ball valve is fairly reliable for positive steam shut-off but if it fails, steam may leak by the temperature control valve also, causing overheating of the cleaning solution, especially under low flow or long cycle conditions. The most common failures noted in ball valves are related to actuator or actuator linkage failures.

Leaky steam supply valves are most easily investigated when the CIP skid is idle by observing condensate discharge and/or hot condensate piping.

The temperature sensing element used as the basis of controlling the above valves is typically a reliable component, and if there is reason to suspect a device failure, the transmitter should be the starting point of investigation. The element itself or the well does not routinely fail, but element inaccuracies may be noted if the device was not installed with the proper thermal transfer liquid, if required. This failure may go undetected for some time period as the device will work, but not reliably within accepted instrument deviation.

Excessive Delay to Supply Set-Point Temperature

Most CIP programs establish recirculation through the circuit and then invoke a "heat and chemical feed" step as the first step of the chemical wash phase. A recipe value may establish a maximum time period for the circuit to achieve the CIP supply temperature set point. If the set point is not attained within the timed preset, the CIP circuit will fail and the subject alarm is provided. The investigation of the failure should start with evaluation of all of the steam supply and control described above. In addition, check to see if a manual isolation valve has been closed to permit safe maintenance on a temperature control component, and not reopened after the job was complete, and confirm full stroke operation of ball-type or stem-type shut-off and throttling valves. Then, follow with an evaluation of the condensate collection system and return system (if any). If condensate is not being effectively removed from the heat exchanger, it will impede effective heat exchanger operation. The condensate drainage problem may be related to a failed condensate return pump, a malfunctioning steam trap, or a closed manual valve in the condensate collection system.

If the investigation confirms the CIP skid's heating and temperature control components to be in order, attention should next be directed to the process vessel being cleaned. The presence of idle Cooling Media in a vessel jacket will not generally cause a CIP operation to fail, but will extend the heating time for the CIP circuit, and may cause the *excessive delay to set-point temperature* alarm. If the jacket cooling is active, the CIP solution temperature is reduced with every pass through the process vessel and will assure this alarm. And, if the alarm is not triggered, there may be other problems later as the mass of solution affects any effort to cool the circuit to prepare it for subsequent processing.

PID Loop Response

The throttling-type temperature control valve mentioned in the previous section will require PID-type control. The temperature control loop tuning should facilitate a steady rise of the CIP solution temperature to the set point, with minimal cycling around the set point prior to settling into a steady-state condition. The temperature control loop should be able to maintain the CIP solution at temperature set point, ± 1°C when properly tuned. The performance of this control loop is dependent on the proper functionality of the components in the system as previously described. If system response is not correct, and the steam supply and condensate discharge systems are satisfactory, the most probable cause will be an improperly adjusted PID loop or mechanical failure of the throttling-type valve. The temperature control valve operation should be tested to determine if the valve travels to its commanded position based on PID output. The control system should be used to transmit several signals to the valve, starting with full open, and gradually closing the valve using 75%, 50%, and 25% output, as well as the signal required to achieve a fully closed position. If the valve position does not reflect the output, the valve current to pressure transducer (I/P) setting should be evaluated and adjusted to ensure that the device is providing an appropriate response to the signal.

CIP Return Temperature Alarm

The CIP return temperature sensor is used to monitor overall CIP circuit temperature as it is the theoretical lowest temperature point in the circuit. The CIP chemical wash period (by time or volume measurement) should not begin until the CIP return temperature has achieved the recipe set point. If this temperature is not noted within a preset time period, the *excessive delay to CIP set-point temperature* alarm will occur. If during the chemical wash phase, the temperature is noted below the minimum acceptable set point for an excessive time period, most typically applied programs will produce a *CIP circuit temperature below set-point* alarm and fail.

CIP Return Temperature Below Set Point

There is always some temperature loss through a CIP circuit, usually 1°C to 5°C differential between CIP supply and CIP return. When the CIP return temperature is below set point, there are two primary areas to investigate, either insufficient CIP supply temperature or a process jacket cooling operation improperly active. If the CIP supply set point has been altered, or is lacking proper control, the CIP return may not meet the minimum temperature requirement. Cooling of CIP solution by an active process vessel jacket will assure such an alarm condition.

CHEMICAL CONCENTRATION CONTROL PROBLEMS

The target cleaning chemical concentration for a given CIP circuit is generally expressed using percentage by volume or normality. Diaphragm metering pumps are often used to deliver the target concentration, but in some cases, where the chemical supply is pressurized, a flow metering system may be utilized.

Common practice is to program for a chemical volume, delivered to achieve a chemical concentration within an established dead band based on the circuit's target concentration. The conductivities for the upper and lower concentration limits of this dead band are experimentally determined by collecting data using cite water at the chemical wash temperature using laboratory data. Samples of the cleaning solutions taken at a variety of conductivities are subjected to lab analysis to verify the chemical concentration/conductivity relationship.

During the CIP circuit's chemical wash step, the CIP supply conductivity sensor monitors the cleaning solution, and if the conductivity measurement is found to be outside of the established conductivity limits, the CIP circuit goes into alarm.

Conductivity Alarms

When the conductivity alarm is triggered, the initial investigation must determine whether the failure was from a low conductivity or high conductivity condition. After the alarm condition is recognized, the timing of the failure must also be taken into account to assess the possible cause.

During a recirculated chemical wash, the CIP circuit is filled with a specific volume of water. The chemical supply is then initiated for either a time period or a volume, depending on the chemical dosing system. Upon completion of the chemical feed operation, the circuit will be allowed to recirculate for a specified chemical mix delay time. This mixing step is essential as the chemical addition is not proportional to flow, and hence uniform at the end of the feed time. At the end of the mix time, the CIP solution conductivity is confirmed within range, and then monitored through the duration of the chemical wash.

CIP Solution Conductivity Below Set Point

There are four primary causes for a low conductivity alarm: inadequate chemical supply, chemical feed pump failure, inadequate mixing, and water additions to the circuit.

If the low conductivity alarm condition occurs during the initial check following mixing, there are three likely scenarios that may have caused the failure. The investigation should begin with a review of the chemical source. Generally, the chemicals are fed either from totes or tanks, and the tote/tank level should be visually confirmed. If the source is a pressurized chemical supply header, the header pressure at the delivery point should also be confirmed. If the chemical supply totes/tanks have suitable level and the delivery pressure is sufficient, all manual isolation valves located in the chemical supply to the CIP skid should be checked for proper position.

If the chemical supply is found adequate and delivery capability is confirmed, the chemical delivery or metering system should be evaluated. For chemical supply systems where a pump is used

- Confirm that the pump is actually running when commanded to turn on.
- For chemical pumps with a VSD or stroke control adjustment, confirm that the speed or stroke length is to specification.
- If the above investigation indicates the pump to be in working order from a control standpoint, it is necessary to focus on the internals of the pump, typically confirming the condition of the pump's diaphragm and suction and discharge check valves.

When considering a pressurized chemical delivery system that utilizes a flow meter and flow control valve (FCV), the action of the FCV must be reviewed. Confirm that

- The FCV positioner action is correct.
- The flow meter has been properly calibrated.
- The FCV I/P setting results in a proportionate valve response to an analog output.

The last check is to confirm the conductivity meter settings and calibration, as well as the temperature compensation, as this will affect the conductivity reading. Less elegant meter-based systems may utilize a simple shut-off valve in combination with a restrictor to control delivery flow through the meter at a constant pressure.

If the low conductivity is noted after the chemical concentration has been initially verified after proper mixing, there are a limited number of possibilities for the cause. This failure may be an indication of a CIP circuit problem, for example, a valve failure in a multiple path circuit. The improperly operating valve may be causing a restriction through one of the flow paths, hindering even chemical mixing, which may result in an early failure, or a failure after initial confirmation.

CIP Solution Conductivity Above Set Point

The two primary causes for a high conductivity alarm are failed chemical supply equipment and excessive soil load.

Most chemical feed systems which utilize a chemical supply pump, also employee a back-pressure/antisiphon valve to prevent chemical siphoning through the pump. A pressure regulating valve may be required to ensure even header pressure to feed the chemical pump. If there is a failure or improper adjustment of either device, the chemical feed may be excessive due to over-pressurization of the supply pump, or siphoning of chemical through the pump. For chemical feed systems which employ a FCV and meter, the concerns noted above for under feed apply equally to an overfeed situation.

CIP SUPPLY FLOW PROBLEMS

The CIP supply flow rate is a CIP circuit recipe parameter, typically using a flow element (FE) and FCV for PID control to a circuit specific set point. A CIP supply pump equipped with a variable speed drive may be used in place of, or in combination with, a FCV for CIP systems that support a wide range of flow rate set points. A VSD alone is not recommended on CIP circuits with device sequences which require quick transitions as the VSD is usually not quick enough to adjust flow control to ensure a hydraulically balanced circuit.

The most common flow control alarm failures are not necessarily found with the FE or the FCV, but appear as external factors that affect a properly tuned flow control loop.

CIP Supply Low Flow Rate

A CIP circuit obstruction in the CIP supply or process transfer line can result in a continuous or intermittent low flow or no flow failure. This obstruction may be a manual valve left closed after a maintenance procedure, blocking the primary CIP path and resulting in an immediate, continuous low flow failure. If the blockage is in a circuit sub-path, the result is an intermittent low flow alarm. Additional CIP supply obstructions which may result in a low flow failure include debris of various types in CIP spray device holes, restricting CIP supply flow, gaskets or other debris in pump impellers, impeding the effectiveness of the pump, and gaskets or debris impeding the action of the CIP supply FCV. The mechanical function of the FCV itself should be considered by evaluating the proper action of the throttling valve actuator, I/P, and positioner.

During rinse operations, the CIP Water Tank must provide the CIP supply pump with sufficient and relatively even net positive suction head (NPSH) conditions to support steady flow control. If the water supply tank at low level does not support the NPSH requirements of the pump, or if tank outlet vortexing is noted, the CIP supply flow control will be negatively impacted by unsteady operation of the CIP supply pump.

During recirculated cleaning, when the CIP Recirculation Tank is utilized for return flow air disengagement and as the supply puddle for CIP supply pump, recirculation tank vortexing, or repeated low-level conditions can impact reliable flow control.

During recirculated cleaning with a bypass-type CIP skid, the recirculation surge volume (puddle) is retained in the process vessel being cleaned. Therefore, process vessel outlet vortexing can result in an unsteady CIP solution return to the CIP skid; negatively impacting CIP supply pump and flow control operations.

CIP Supply High Flow

The alternative to a CIP circuit obstruction is that a device that would normally present a restriction in the CIP pump discharge path line is not installed. As a result, the PID loop tuning is inappropriate for the CIP circuit's downstream reduced pressure drop. Failure to properly install one or more sprays in the target process vessel can result in high flow failure. This can be intermittent or continuous, depending on whether all of the CIP supply passes through the vessel sprays all of the time.

The alarm message *flow control out of acceptable control range* indicates the flow rate is neither excessively high or low, but cycles around the set point, failing to properly control within an accepted deviation range under steady-state conditions. The CIP supply downstream pressure drop profile often presents a dynamic environment, i.e., when cleaning a bioreactor with multiple legs, and flow control through multiple paths may swing on either side of the set point every time the device sequence changes to an alternate path. Typically boundaries are placed on the acceptable deviation, and when the circuit repeatedly swings outside of this deviation an error may be noted.

RETURN FLOW PROBLEMS

A hydraulically balanced CIP circuit with an effective CIP return system will have a CIP return flow capability equal to or slightly greater than the CIP supply flow, resulting in no accumulation of solution in the equipment being cleaned. If CIP return flow does equal or exceed the supply flow, hydraulic instabilities result, and any number of CIP system alarms may be invoked.

No-Return Flow Alarm

During the first rinse phase of many CIP programs, the CIP return flow is confirmed with a simple return flow sensor or flow switch. The CIP return flow check is performed after a fixed volume of water has been discharged to the circuit, and CIP return flow is expected under normal operation conditions. If the return flow is not noted within a timed preset, an alarm will be provided and the CIP circuit will fail.

If the problem is from a CIP supply blockage, a CIP flow error would likely occur. Therefore, when a "CIP return no flow" alarm is noted, the initial effort should be to examine the circuit checking for solution buildup in one or more items of the equipment being cleaned. The CIP return no flow alarm may be the result of

- a manual isolation valve closed in the CIP return line;
- CIP flow blocked or misdirected by an improper CIP supply or return connection at a transfer panel;
- improperly installed CIP Return Hose sending CIP solution to an alternate process or waste destination.

Although most transfer panel connections are monitored via a proximity switch, not all manual connections are confirmed via a control system input acting as a permissive or interlock.

The CIP return pump *motor start failure—no run signal* has been covered previously under general pump failures. However, issues related to CIP return pump cavitation are not identical to those noted above for CIP supply pump cavitation. The CIP supply pump is designed to operate with steady NPSH condition for reliable CIP supply flow. When specifying a CIP return pump it is understood that the flow will be intermittent and the norm is that NPSH conditions vary. If the CIP return pump is not specified to be a self-priming, liquid ring pump, it must be a low speed pump (1750 rpm), and the return pump suction piping must pitch continuously upwards from the pump inlet to the vessel outlet. This will permit the pump to unload small quantities of air in the reverse direction (toward the vessel) and then move the air/water mixture through the return line to the CIP skid.

The CIP return pump may also mechanically fail; a failed pump seal allowing the pump to draw air through the seal, or a gasket partially plugging the impeller will result in insufficient discharge head for proper return flow.

Insufficient Return Flow

The failures noted above that result in the *No CIP return flow* alarm may be present but not severe enough to fail the first rinse. However, when present will result in a CIP program which "limps" along to eventual failure. For example, a minor return pump failure which causes the slow rise of level in the equipment being cleaned will eventually result in a CIP circuit instability, negatively impacting the CIP circuit hydraulic balance, CIP supply flow rate, chemical concentration, and temperature control.

The performance of an eductor-assisted or eductor-based CIP skid depends on a sealed CIP return system. Although the process equipment may be vented, there is a negative impact on return flow capability if there is an air leak in the CIP return between the process vessel and the CIP skid. Eductor performance is also impacted by excessive solution temperature, as the vacuum in the CIP return line will reduce the vapor pressure to a point where the returning solution essentially "boils" in the line, reducing CIP return flow. This problem is most commonly noted after circuit qualification when a facility increases the water-for-injection (WFI) loop temperature to over 80°C, often in an effort to control WFI loop quality.

Vortex Formation

Vortex formation, the result of vertical discharge through the center outlet of a dish bottom tank can result in a large puddle in the process equipment. As the vortex forms and builds, the projection of it's apex into the outlet opening restricts solution flow. The exiting solution will form a stable vortex of increasing volume, with reduced flow from the vessel for return to the CIP skid. The CIP return pump is starved and may become air bound by bubbles of air leaving the vortex and moving to the pump.

Vortexing can also occur in a CIP skid recirculation tank, and will have the same negative impact on pump performance and hence CIP performance. Vortex formation can be prevented by permanently installing a flat plate vortex breaker three times the outlet diameter 1 in. above the outlet. The alternative is to operate at higher solution levels, wasting water, time, and chemical, and perhaps causing sedimentation or bathtub ring formation associated with large relatively quiescent puddles.

CIP SKID WATER TANK LEVEL PROBLEMS

The sole function of the CIP skid water tank is to supply water to the CIP pump for prerinsing, wash fill, post-rinsing, and final rinsing of all circuits. If waters of two qualities are used for CIP, two tanks may be provided, or alternatively the means of draining and flushing the tank when changing from the low quality water to the high quality water. Fixed probes are the most common means of achieving on–off control of the water valve(s), though analog sensors operating via level set points meet the same needs and much more. Operation problems may involve high-level and low-level control points and alarm messages.

High Level

When the *water supply tank at high level* alarm is noted, it is not a "circuit" failure but an indication of an equipment maintenance or calibration issue, which may include

- a slow acting water supply shut-off valve;
- a calibration issue with the water supply tank level sensor (if analog, rather than a probe).

The analog level sensor provides a range of set points and is usually not responsible for reporting a high-level alarm; the duty of the high high-level sensor, usually a discrete level element from which the active input spurs the alarm. The shut-off valve is controlled by the level element, and a slow to close valve can result in a high high-level alarm, and the valve troubleshooting should be as previously described.

A field investigation at the water supply tank will aid in determining if there is a level element calibration issue, or if the valve simply failed to promptly close. The water supply valve is typically interlocked to close when the high-level probe input is noted, but if there is a mechanical failure of the valve, an overflow condition of short duration can result. This overflow condition is also an obvious indication of a level instrument calibration error if the valve promptly closed when commanded. If the tank is not overflowing, a visual check will enable the investigator to determine if perhaps the level sensor calibration is in error, and a high-level condition does not in fact exist. If the sensors are in order, the next consideration is the water supply valve.

If the field examination reveals the water supply valve actuator slow to respond, the problem may be with any number of valve components. As previously described, the troubleshooter should start at the valve, and work back to the solenoid and I/O to resolve the problem.

Low Level

When the *water supply tank—low level* alarm is provided, it may be considered a CIP circuit failure if the CIP supply operations must be suspended until a sufficient water volume can be accumulated. A low-level alarm is an indication of either an equipment or water supply issue, the most common being

- a slow to open water supply shut-off valve;
- a water supply tank level sensor calibration issue;
- insufficient water supply.

The suggested order of action includes the following field investigations

- the water supply valve should be fully open when a low-level condition is noted;
- investigate the actual water tank level to determine if there is a level element calibration issue;
- the water supply valve should be commanded open and closed. If the examination finds the water supply valve actuator is slow to open the valve,

it may cause a low-level condition and may be the result of a valve components failure.

If the valve is observed to work promptly, consideration of an insufficient water supply to support the circuit flow rate is warranted. During commissioning and qualification, the facility may produce sufficient water to support the CIP load. However, as the facility ramps up to manufacturing, the actual water usage may be greater than the design capacity, and the CIP skids are the common victim. Water capacity issues are not fixed overnight, although there are short-term remedies to improve water tank utilization, including

- extend the drain periods after flush, wash, and rinse operation to allow more time to refill the tank between and within CIP phases that draw water;
- confirm that the water surge tank is actively charging during all CIP program phases whether water is required or not.

IMPROPER CIRCUIT HYDRAULIC BALANCE

It is easiest to manage the volume of only one "puddle" in a CIP circuit, either in the equipment being cleaned or at the CIP skid. Managing the "puddle" at the CIP Unit is preferred as it minimizes the complications that accompany development of large process equipment puddles that retain insoluble soil in the equipment, create "bathtub rings," and require rinsing by dilution, thus increasing water requirements and time for rinsing and draining. As previously discussed, effective control requires a CIP return flow capability slightly in excess of the required supply flow, and control of vortexing.

CIP Recirculation Tank

A properly balanced CIP circuit, with a CIP skid managed puddle, will permit a rise and fall of process tank level, but within a recognized and controlled CIP recirculation tank level operating range. This varying level may be the result of some circuit sub-paths draining more freely than others, resulting in a surge or decline in the rate of CIP return flow, increasing or decreasing tank level.

The occurrence of a high- or low-level alarm for the CIP recirculation tank when running a CIP circuit which has been properly qualified is an indication of a CIP circuit failure which will require investigation. The CIP program may include the means of compensating for abnormal conditions described below, but these are not acceptable for normal operation, and when they occur, the reason must be found and corrected. Whereas the following comments are specific for an eductor-assisted CIP skid with an air separation tank, they apply to any skid mounted recirculation tank.

Recirculation Tank High Level

Under this condition the CIP program may respond to avoid overflow, or blinding a vent filter, by briefly opening the CIP skid drain valve for either: (*i*) a timed preset or (*ii*) until operating level is restored. As the cause is temporarily alleviated by dumping return flow to waste, the tank may then next reach a low-level condition.

The first point of investigation when there is a cycling high- and low-level condition on a qualified CIP circuit is confirmation of the level sensor calibration. CIP skid level sensors are prone to drift due to the widely varying temperatures they may be exposed to during the cleaning process. If the level sensor has been found to be properly calibrated, then a FE investigation may be in order; a FE out of calibration on the high end may overfill the CIP circuit, resulting in circuit recirculation instability. The troubleshooter must review the process equipment evaluate the CIP return conditions per previously presented methods to determine the cause for the temporary process vessel solution hold up.

Recirculation Tank Low Level

When the CIP recirculation Tank reaches low level, common practice is to open the water supply valve to charge a controlled volume of water into the CIP circuit to assist in bringing the CIP skid recirculation tank up to operating level again. If after the water addition, the tank is not within normal operating range, the water addition may be repeated a fixed number of times to achieve circuit stability during recirculation.

An unexpected loss of solution volume can result in a low-level condition at the CIP unit recirculating tank. This loss may be from a leaky U-Bend at a transfer panel, a poorly fitted supply hose connection, or an open manual low point drain valve. The key indicators for an unmonitored solution loss include the CIP skid recirculation tank is at low level, no rising level is observed the vessel being cleaned, one or more water additions are required to maintain circuit recirculation, and CIP Supply flow control and pressure levels may be low during the low-level excursion.

Impact of Water Addition

A water add action can assist in restoring circuit hydraulic stability. However, the water addition may have the following negative effects if this action is performed in excess:

- Reduction of cleaning chemical concentration
- Potential impact on CIP circuit temperature
- If low level is not due to circuit volume loss, may eventually result in a recirculation tank high-level condition and corrective action, or overflow.

It should be understood that software fixes, i.e., high- and low-level control actions, are not the ideal approach to balancing CIP circuits. Proper engineering design can assure proper circuit balance and make these hard to manage, and troubleshoot, functions unnecessary.

INADEQUATE FINAL RINSE

This section assumes that the piping installation has adhered to American Society of Mechanical Engineering Bioprocessing Equipment (ASME BPE) recommendations to keep process/CIP piping dead-legs less than two pipe diameters, and horizontal in position. It is also assumed that water of proper quality is being supplied for rinse purposes. The more common causes to be investigated are poorly performing CIP supply air blow and drain operations before the final rinse, instrument calibration, poor CIP return flow, and process vessel puddling.

An effective CIP supply air blow and drain prior to the final rinse is essential to rinse the circuit with the minimum water volume and maximum available volume. An ineffective air blow may be caused by:

- Inadequate air supply header pressure
- Restricted air flow from manual isolation valves in air supply header not fully open
- A partially plugged air supply filter
- Failure to sustain air header pressure during a long air blow

If the air blow drain period does not sufficiently drain the process equipment puddle, the following investigations should be performed:

- Confirm performance of CIP return pump
- Check CIP return line low point drain valves (if an) for proper operation during the post-rinse drain
- Confirm freedom of vortexing and excess tank puddle

Conductivity and resistivity sensors require fairly frequent calibration, and a calibration error may be (*i*) the element monitoring the water supply to the CIP Unit, resulting in water being delivered of insufficient quality or (*ii*) the result of a CIP return monitoring device out of calibration. As previously noted, it is desirable to clean with a minimum puddle in the equipment being cleaned. When puddles develop in the equipment being cleaned, the chemical is flushed to the puddle where it resides for an extended period of time during slow removal by dilution.

EQUIPMENT NOT CLEAN UPON COMPLETION OF PROGRAM

This chapter assumes that the CIP circuit operational and performance qualification has been completed, and the CIP circuit has been validated clean through visual, rinse, and swab checks. If the equipment fails to clean during subsequent use, consider.

Actual Soil Load vs. Expected Design Parameters

If the CIP circuit has run properly mechanically, and there have been no failures or alarms, the first consideration is to confirm that the process operations were performed according to the validated SOP and that there was not an excessive soil load.

Circuit Recipe Parameters

Next, consider the CIP program recipe parameters and confirm that they have not been altered since the CIP circuit validation. Although recipe parameter modifications are not to be performed on a qualified CIP circuit, substantial experience suggests the wrong cleaning recipe may be applied to a CIP circuit, especially if this is an operator decision.

Circuit Dirty Hold Time and Impact on Cleaning

If the soil is allowed to dry on the equipment, the cleaning burden is increased and the cleaning program may fail to remove all soil. The processing record or data historian should be checked to determine if the time between process conclusion and CIP operations should be examined to determine if the "hold dirty" time has been exceeded. If the dirty hold time is exceeded, it may take repeated executions of the CIP program to remove the dried on soil.

Evaluate Spray Coverage

Spray devices will be removed and reinstalled throughout the life of the facility, and if this operation is done by those unskilled in the process, one or more may be improperly positioned. The first field investigation is to ensure that there are no spray device installation or indexing errors, meaning the sprays destined for the vessel are properly installed in the correct nozzle and indexed according to its indexing pin or 0° notch.

If inadequate spray device coverage is suspected, the following investigations should be performed:

- The installation of the proper spray in its identified position should be confirmed.
- The spray device may have multiple pieces that require clamps or slip pins to assemble, and each individual component should be permanently identified with an identifying part number.
- If the sprays are not all welded, and are assembled pieces, confirm that the components are properly matched up and assembled.

As discussed in chapter 9, many sprays are directionally drilled and must be indexed for proper installation in the nozzle in which they are located to assure proper coverage. Directionally drilled sprays will have an indexing pin or 0° notch, and the spray device schedule may provide supportive data to confirm the index position within the nozzle for proper spray coverage.

Following confirmation of assembly, location, and indexing, it may be necessary to remove the spray to determine if there is debris in the spray, causing an obstruction to flow through the complete spray pattern. If several of the holes are obstructed with debris, most commonly a loose gasket, then the spray coverage will be flawed and soil may not be removed from the entire vessel.

An observation of the CIP supply pressure/flow relationship during an active cleaning cycle may provide an indication that the sprays may be plugged prior to having a failed CIP program. If a spray device is plugged with debris, the FCV may be operating within a different range to control flow, or the CIP supply pressure may be higher than typical to overcome the resistance caused by spray blockage. A simple visual check through a sight glass can sometimes reveal uneven spray coverage, suggesting a partially blocked spray.

SUMMARY

This chapter has presented what may seem to be overwhelming amounts of information to troubleshoot CIP systems, though the directives noted can be distilled into a few simple guidelines to troubleshoot any specific problem. A few quick rules for troubleshooting successfully include

1. When a failure is noted by the CIP program, obtain the best data possible before entering the field. This includes the pertinent data historian trends for flow, pressure, conductivity, and temperature.

2. Troubleshooting is performed in the field allows the personnel to use their senses in the debug process.
   a. Vision to confirm equipment states
   b. Hearing for detecting proper pump operation, active flow, or spray coverage
   c. Touch for detection of temperature or flow
3. Once you have entered the field, start at the source of the problem and methodically work back through the system from the field device to the control system hardware.
4. While troubleshooting requires field observation, a team approach involving an automation savvy helper at the HMI can support the field work to ensure prompt success.

As in all detective works, a thoughtful and methodical approach with proper background data and field observations can work to satisfactorily resolve any CIP system operational issue.

17 Waste Treatment for the CIP System

Linda Rauch
CH2M Hill, Boston, Massachusetts, U.S.A.

Jay C. Ankers
LifeTek Solutions, Inc., Blue Bell, Pennsylvania, U.S.A.

CIP EFFLUENT—WHAT TO DO WITH IT WHEN YOU ARE DONE WITH IT

A clean-in-place (CIP) system circuit is designed, in simple terms, to clean all process residues out of the equipment and piping included in the CIP circuit. With the help of water prerinses, alkali wash, acid wash, and final water rinses, all at controlled temperatures, the process residue is broken down and conducted to a drain. As a CIP system designer, what happens to the CIP effluent after it leaves the process area is not usually a concern. However, in most cases a waste treatment system of some type and size will be required to treat the waste from the CIP as well as the other building process effluents. CIP design decisions will have impacts downstream on waste treatment requirements.

CIP waste contributes a significant percentage of the overall waste effluent from a biopharmaceutical plant. Each circuit will typically send 1000 to 5000 L of wastewater to the drain depending on the tank and circuit size. If the designer has followed the guidelines and recommendations in this book that create efficient CIP circuits, the CIP effluent volume will be minimized.

This chapter will attempt to address the waste treatment strategies that are relevant to the CIP waste streams and why they are necessary. For the purposes of this book the boundaries of the facility's waste treatment system include the drain system (plumbing) through the actual waste treatment system, up to the municipal connection point.

HOW DO I KNOW WHAT I NEED TO TREAT FOR?

The typical biopharmaceutical facility is cleaned with the alkali and acid wash steps that are explained throughout this book. This cleaning system is effective for the media, cell cultures, harvest broths, and downstream protein purification steps and associated buffers. The effluent therefore contains trace amounts of the process fluids that alone do not require much treatment as a waste stream. The primary treatment strategy for a typical biopharmaceutical facility will need to be for high and low pH, and high waste stream temperature, at high flow rates. All of these attributes have discharge limits set by the appropriate permitting authority (local/regional/federal). Plumbing codes will also limit what can be sent through the facility's nonprocess plumbing system above ground, as well as, any plumbing or piping system that runs underground.

The receiving wastewater treatment plant will set these limits to protect their operators' health and safety, maintain their treatment plant processes, and ensure compliance with state and federal standards (1). There will be some variation in permit requirements from location to location depending on the size of the receiving facility, whether it serves primarily industrial sources or residential sources, and its design parameters.

A typical industrial discharge permit will regulate the following aspects of the CIP effluent:

1. Corrosive pollutants (acids and bases)
2. Temperature
3. Nutrients (nitrogen, phosphorus)
4. Organic loading or biochemical oxygen demand (BOD; measure of the readily decomposable organic content of a wastewater)

Why do these need to be limited before they are sent to a treatment facility? Acids and bases can cause corrosive damage to the transfer sewer piping as well as the equipment in the receiving *publicly operated treatment works* facility (POTW). Both temperature and pH spikes can inhibit microbial processes at the POTW that are critical to the treatment of organics. High temperatures can also pose a safety risk if there are materials with low flash points in the sewer systems. Nutrients and organic loading sent down the drains from the CIP systems can exceed the design of the POTW resulting in plant upsets and permit violations for the receiving plant. This can cause significant issues and penalty fines for the local authority. Phosphorus, in particular, is often the limiting nutrient for algae and plant growth. Phosphorus is tightly regulated depending on the receiving body of water (e.g., Chesapeake Bay, San Francisco Bay, local river, etc.) and the capacity of the receiving POTW for its treatment.

An example of typical U.S. facility discharge limits relating to CIP discharge is provided in Table 1 (2). It is expected that the requirements in Europe and Asia would be comparable. Some aspects of these permit limits can be negotiated with the local authorities whereas others are national standards and will not be negotiable.

It is important to recognize from these limits that the design of the CIP systems and the selection of the CIP chemicals used will impact the design of the on-site wastewater treatment system and the ability to meet the requirements of an industrial discharge permit.

WHERE SHOULD THE CIP EFFLUENT BE TREATED?

Most CIP system designers are installing their system in a facility that has or will have a facility-wide, centralized waste treatment system. Installation of the CIP

TABLE 1 Typical Discharge Limits Compared to CIP Effluent

Permitted aspect	Typical limit	How does it relate to CIP effluent?
pH	6.5–8.5	pH of untreated CIP solutions range from 2 to 12
Temperature (°C)	40°C (at POTW headworks), 60°C to the sewer	CIP solutions and water rinses are often 80°C or hotter
Total phosphorus	Varies by region and receiving POTW design	CIP cleaning agents can contain phosphates and often phosphoric acid
BOD (mg/L)	Varies based on receiving POTW design	Organic matter rinsed from process equipment (e.g., cell debris and proteins) will affect BOD loadings

Abbreviations: BOD, biochemical oxygen demand; POTW, publicly operated treatment works.

system requires only connecting the drains from the skid and low points of the circuit, to various floor drains.

A small quantity of new CIP systems are being installed in facilities that are scaling up from small processes that did not originally use CIP systems. These small process volumes may not have required treatment before discharging to the city (i.e., discharging within the "municipal discharge limits"). In this second scenario, it is tempting to try to treat the waste stream from the CIP circuit on the skid before discharging it down the drain. This approach is not recommended because it tends to deposit what you just cleaned from process surfaces back on the equipment and piping. The high pH that was so effective in breaking up the process residue loses its ability to hold the residues in solution when the pH is adjusted in the circuit back toward neutral.

In following the recommendation to send the CIP effluent to a facility-wide or on-site waste treatment system, it will be necessary to connect all CIP systems to a drain system. These process/CIP drain systems are similar to the gravity drain systems for building plumbing but there are some significant differences. The key aspects for consideration in the design of these process/CIP drain systems are as follows:

1. Appropriate materials of construction must be used for the high purity (read: "mineral hungry") water as well as the acid wash; both at elevated temperatures. The recommended materials include stainless steel or fiberglass-reinforced plastic for their corrosion resistance.
2. The equipment drains at the CIP skids should be appropriately oversized to prevent flooding the floors around the skids during the pump-to-drain sequences. It is also recommended to locate the skid in a diked containment area or pan (Fig. 1).
3. Instantaneous flow rates must be used when sizing drains, branches, and headers, not the average flow with usage factors, commonly used by plumbing designers. Special consideration should be given to routing these branches and headers directly to the waste treatment system without many turns.
4. As with the branches and headers, the vents from these high capacity drain systems need to be appropriately sized and located with consideration for plumbing codes as well as CIP flow rates.
5. Provision of an air break between the CIP skid and the floor drain to prevent backing up waste into process systems (Fig. 2).

The oversized drain at the CIP skid discharge point should give consideration to the following needs:

- A suggested size of 10 in. diameter and 24 in. deep
- Located directly below CIP system drain
- Appropriate air gap to prevent back siphoning
- Debris strainer to catch large objects in containment pan

HOW DO I TREAT THE WASTEWATER ON-SITE?

The on-site waste treatment system will need to address treatment of flows that are not within pH- and temperature-permitted ranges. The effluent from CIP systems can also impact a facility's ability to meet permit requirements for total phosphorus and BOD concentrations.

FIGURE 1 CIP skid in stainless steel containment pan. *Source*: Courtesy of Lonza Biologics, Portsmouth, New Hampshire, U.S.A.

Waste treatment systems can be designed for batch or continuous treatment modes. In the case of a batch system, the waste is collected in a tank, held and treated, and then discharged. In a continuous system, the wastewater continuously flows through a series of tanks where treatment takes place. A continuous flow system will be the preferred approach for most manufacturing facilities. A batch system may be appropriate for a small research or clinical-scale facility.

Treating for High/Low pH

Both the CIP designer and waste treatment system designer must first understand the chemistry basics with respect to pH. The pH scale is a logarithmic scale ranging from 0 to 14. A neutral pH is 7 on this scale. Solutions with a pH less than 7 are considered acids and with a pH of greater than 7 are considered bases. Acids and bases can be rated as strong or weak depending on how easily they dissociate in water to form ionic species (3). By combining acids and bases together, it is possible to neutralize the overall solution.

On-site wastewater treatment systems will typically use a strong acid and a strong base such as sulfuric acid and sodium hydroxide, respectively, for pH neutralization. By metering in and mixing relatively small quantities of these solutions with the CIP and process waste stream, the system can neutralize the wastewater pH to within discharge permit limits.

This process is usually accomplished in large tanks piped in series as shown in Figure 3. This common design, using three large tanks, follows a sequence of equalization, gross pH adjustment, and fine pH adjustment. The bulk acid and base are stored in vessels appropriately sized depending on the desired chemical

FIGURE 2 Oversized drain at CIP skid. *Source*: Courtesy of Lonza Biologics, Portsmouth, NH, U.S.A.

delivery frequency (Fig. 4). These acid and base solutions require appropriate operator safety precautions and special containment for their storage tanks. It is common to provide spill containment for 110% of maximum stored volume as well as complete segregation of the acid and base in case of a spill. The reaction caused by a spill of these stored quantities of acid and base could lead to dangerous heat levels generated from the exothermic reaction as well as noxious fumes.

The design and installation of equipment to handle the neutralization procedure must give consideration to the following factors:

- Spill containment
- Segregation of acid and base
- Color-coded and labeled piping and fill connections
- Metering pumps (with installed spares)

Because of the logarithmic nature of the pH scale, a different quantity of base is required to adjust the pH from 2 to 3 as compared to adjusting the pH from 5 to 6. The quantity of chemical required for neutralization is also affected by: (*i*) incoming solution pH, (*ii*) type and concentration of CIP chemical used, (*iii*) type and concentration of neutralization chemical used, (*iv*) overall alkalinity of the wastewater stream, (*v*) buffering capacity of the wastewater stream, and (*vi*) dilution of

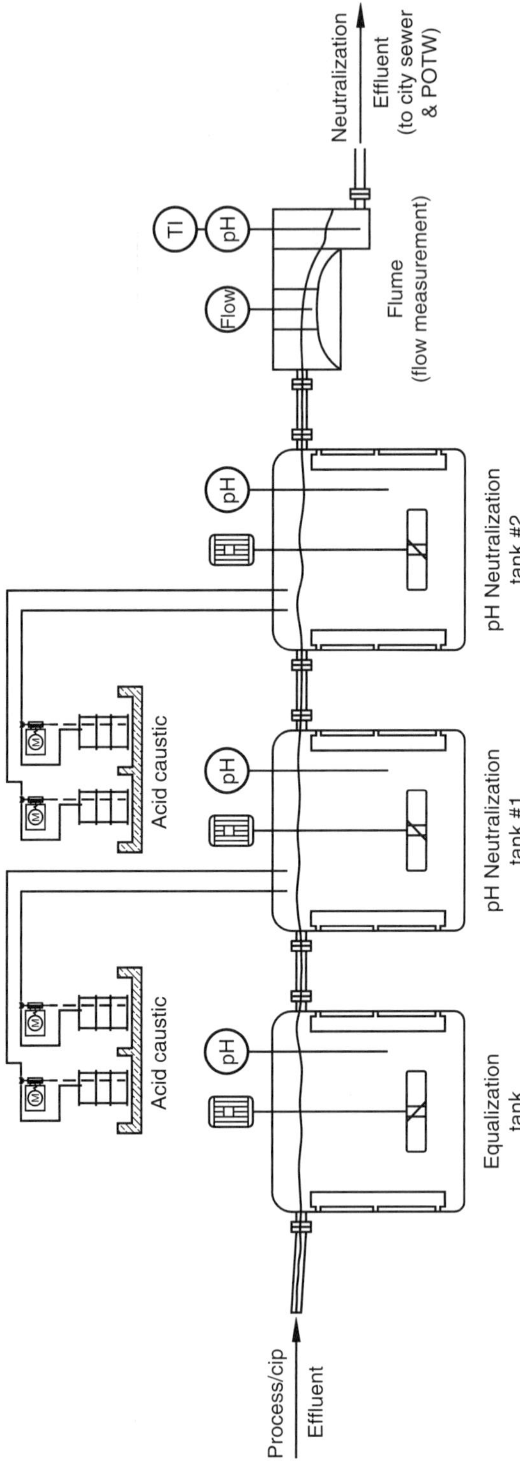

FIGURE 3 Three-tank waste neutralization system. *Abbreviations*: POTW, publicity operated treatment works.

FIGURE 4 Acid and base storage and pump system. *Source*: Courtesy of Human Genome Sciences Inc., Rockville, Maryland, U.S.A.

the CIP wastewater stream by other simultaneous processes in the manufacturing facility.

This neutralization concept is demonstrated in Figures 5 and 6. These figures were generated using Visual MINTEQ (version 2.32) to model the effect on pH of acid and base addition to typical CIP wastewater solutions (4).

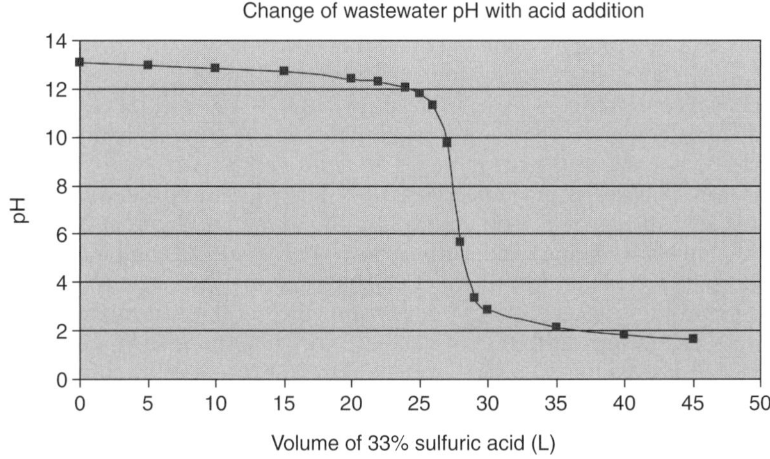

FIGURE 5 Computer model output of the pH resulting from an incremental addition of neutralizing acid to a typical basic CIP wastewater (containing 1% by weight Steris CIP 100).

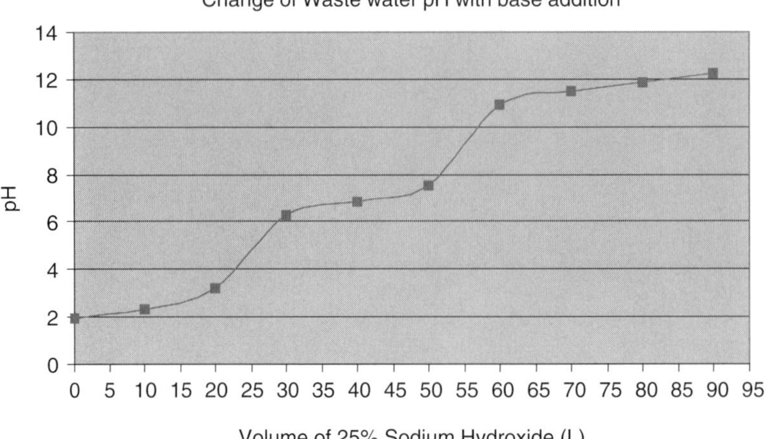

FIGURE 6 Computer model output of the pH resulting from an incremental addition of neutralizing base to a typical acidic CIP waste water (containing 1% by weight Steris CIP 200).

This first graph, Figure 5, demonstrates the change in pH for a basic CIP waste solution with each incremental volumetric addition of 33% sulfuric acid per 1000 L of CIP wastewater. The starting conditions were taken as 1% by weight solution of a CIP caustic chemical containing 25% potassium hydroxide (e.g., Steris CIP100). The starting pH of this solution would be approximately 13. The key item of note is that it takes 25 L of acid to change the pH by 1 unit from 13 to 12 and then only 1 L of acid to drop from a pH of 10 to below 6. This highlights the neutralization system control challenges of hitting a discharge permit limit range of 6.5 to 8.5. Waste treatment systems designed to treat large volumes of high or low pH effluent tend to overshoot neutral and waste both acid and base as they overcompensate in both directions.

The second graph, Figure 6, repeats the same exercise starting with an acidic CIP waste solution and adding incremental quantities of 25% sodium hydroxide per 1000 L of CIP wastewater. The starting conditions were taken as 1% by weight of CIP acidic chemical containing 50% phosphoric acid and 10% by weight citric acid (e.g., Steris CIP200). The starting pH of this solution would be approximately 2. Again, it should be noted that the pH increase does not vary linearly with sodium hydroxide addition. The reader may observe that the shape of the curve in Figure 6 is different from that in Figure 5. The neutralization curves are different because different acid chemicals are being used, sulfuric acid compared to a combination of phosphoric and citric acids, respectively. The shape of the curve is related to whether there are multiple acids or bases present and whether they are monoprotic, diprotic, triprotic, etc. (referring to the number of hydrogen atoms to be given up per molecule) (3). It is possible for a facility to determine plant-specific curves like these by conducting lab bench tests with a CIP wastewater sample and the neutralization chemicals to be used. This can provide some understanding of the sensitivity of acid and base additions to the resulting pH.

Mixing volumes of these acidic and caustic CIP chemicals will bring the pH closer to neutral. This demonstrates the potential benefit of mixing these CIP chemicals together in a large equalization tank prior to the addition of neutralization chemicals in subsequent mix tanks. In practice, this result may be difficult to achieve in a large facility with multiple process and CIP systems operating simultaneously. Often the wastewater flow rate carries the alkaline wash through the system before the acid wash can blend in to equalize it 20 to 30 minutes later.

High Temperature

Wastewater from CIP circuits is often hot (65–85°C) and can cause the facility's waste stream to exceed allowable discharge limits for temperature. The heat of the CIP effluent is combined with blowdown streams from the large distillation systems and other steam condensate sources to significantly raise the temperature of the waste stream leaving the facility. Provisions should be made for cooling this wastewater to below permit requirements. The waste stream can be cooled down to 35°C to 40°C with a plate and frame type heat exchanger and plant cooling water of appropriate temperature. The opportunity exists for energy savings if the CIP circuits are designed and validated to operate at lower temperatures.

Total Phosphates

Many cleaning chemicals contain phosphates or phosphoric acid both of which contribute to the total phosphorus in the wastewater stream. As was noted previously, this nutrient is often tightly regulated to prevent excessive algae and plant growth in the nation's lakes and rivers. Not all POTW facilities are designed for nutrient removal (e.g., phosphurus, nitrogen) which can result in a requirement for the biopharmaceutical facility to meet the final nutrient discharge requirements for the receiving body of water. The on-site treatment of phosphorus is not preferred as it typically involves either a chemical precipitation reaction or a complex biological process. The best approach is to avoid CIP chemicals that contain phosphates or phosphoric acid if at all possible. One alternative to phosphoric acid for CIP cleaning is hydroxyacetic acid (also called glycolic acid). The use of these alternate cleaners needs to be considered early in the process development and cleaning validation.

Biochemical Oxygen Demand (BOD)

Finally, CIP wastewater may contain high levels of biochemical oxygen demand (BOD) depending on the area of the biopharmaceutical facility being cleaned. In particular, circuits that clean cell debris from fermentation and harvest areas result in high BOD loads in the CIP effluent, especially the initial rinse steps. BOD is a measurement of the organic loading in a waste water stream. BOD is typically measured with a 5-day analytical test that determines the amount of dissolved oxygen used by microorganisms in the biochemical oxidation of organic matter. Because most POTW facilities use microbial processes to treat their incoming waste stream, this measure is used to size and operate their treatment plants. High or variable levels of incoming BOD will directly impact their ability to operate.

Similar to phosphorus, on-site treatment of BOD is not preferred due to the complexity of operating these systems. However, BOD equalization may be necessary depending on the facility permit requirements. In this case, high BOD waste streams are segregated within the plant and collected in a large vessel.

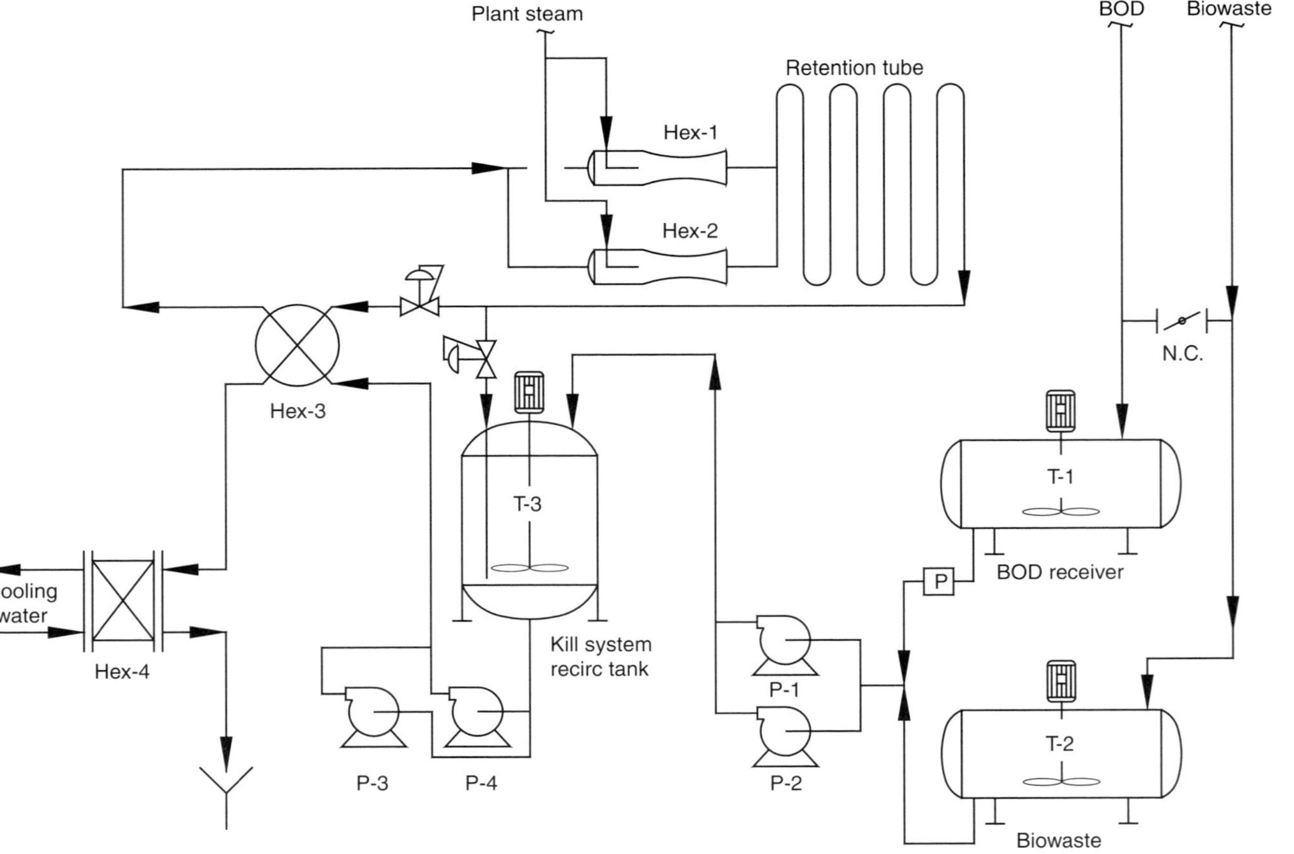

FIGURE 7 Typical continuous biowaste inactivation system. *Abbreviations*: BOD, biochemical oxygen demand, NC, normaly closed.

From this tank, they are metered in with the plant effluent so that large surges of concentrated BOD waste streams are not sent to the receiving POTW at one time.

SEGREGATING AND TREATING THE VIABLE ORGANISMS

When a CIP system is used to clean equipment and piping that has been exposed to living cells, the effluent may need to be inactivated using a specialized waste treatment system before it is sent to waste neutralization for pH treatment. These biowaste systems are sometimes referred to as "kill systems." They use heat or chemicals to kill the active cell cultures in either batch mode or continuous mode.

Steam heat is the preferred method of inactivating any viable organisms in the waste stream. The goal of 6-log reduction in the number of microorganisms is easily obtainable using steam directly injected into the waste stream or using a heat exchanger. Most designers use direct steam because of its heating efficiency as well as it being less likely to foul from the waste stream getting "baked on." The steam injector is well suited for both the batch systems as well as the continuous flow systems (Fig. 7).

A continuous biowaste inactivation system will include, but not be limited to, the following components:

- Steam eductor(s) for direct steam heating
- Receiver tanks for both biowaste and high BOD waste streams
- Heat recovery exchanger on discharge line
- Additional cooler for reducing discharge temperature

How does biowaste inactivation affect the CIP system designer? First, the designer needs to be aware of the need to treat the CIP waste streams that *potentially contain viable organisms from the process*. Then system drains at the skid and each low point will need to be grouped so that they can be sent in the dedicated biowaste drain system to the inactivation system. In large CIP systems, it may be beneficial to install a divert valve on the common drain of the CIP skid that sends the pre-rinse and caustic wash, which will carry the biological material to the biowaste system and then diverts the remaining cycles directly to the pH neutralization system. The divert valve is an inexpensive way to save money on the energy required to heat a large quantity of biowaste. If a batch system is used for inactivation, a divert valve can save a significant amount of the capital cost for larger equipment.

CONCLUSION: IN THE FUTURE, HOW MIGHT CIP SYSTEMS BE IMPROVED WITH RESPECT TO WASTE TREATMENT?

The design features that improve a CIP system's ability to efficiently clean a circuit with less water and chemicals will also reduce the waste effluent load from each CIP circuit. Minimizing the amount of water used is the first step in reducing the waste volume that a facility sends to the POTW. The reduced volume of water directly corresponds to a reduced chemical and heat load from each circuit. As phosphate discharge becomes more restricted in the United States and around the world, other cleaners with little or no phosphates should be chosen early in the process cleaning development. Choosing alkali and acids that both clean well and neutralize easily in the waste treatment system will save money and benefit the environment. CIP is one of the areas in a facility where incremental improvements directly affect the facility's overall impact on the municipality and environment.

REFERENCES

1. United States. Office of Wastewater Management 4203. EPA Local Limits Development Guidance. EPA 833-R-04-002A, July 2004.
2. United States. Office of Wastewater Management 4203. EPA Local Limits Development Guidance Appendices. EPA 833-R-04-002B, July 2004.
3. Benjamin M. Water Chemistry. New York: The McGraw-Hill Companies, Inc., 2002.
4. Software used to generate Figures 17.4 & 17.5: Visual MINTEQ, ver. 2.32, Compiled in Visual Basic 6.0 on 7 April 2005 by Jon Petter Gustafsson; Website link: www.lwr.kth.se/English/OurSoftware/vminteq/

18 Commissioning and Qualification

James P. Norton
Eli Lilly and Company, Indianapolis, Indiana, U.S.A.

INTRODUCTION

One of the comments made to me when I was considering writing this chapter was that "many people think that clean-in-place (CIP) is something that you buy out of a catalog." I know some of those people. My reply to that statement is that to those same people this chapter in its entirety could be titled, "Hit the start button."

Now that you have gotten this far in the book you should understand that you cannot buy a CIP system out of a catalog. Designing a well-integrated CIP system is a complex task. Similarly, the commissioning of CIP systems can be complex. Operational qualification (OQ) can be made simpler and straightforward with a well-developed and executed commissioning program.

DEFINITIONS

Installation qualification (IQ) is "Documented verification that the equipment or systems, as installed or modified, comply with the approved design, the manufacturer's recommendations and/or user requirements." OQ is "Documented verification that the equipment or systems, as installed or modified, perform as intended throughout the anticipated operating ranges."

Performance qualification (PQ) is "Documented verification that the equipment and ancillary systems, as connected together, perform effectively and reproducibly based on the approved process method and specifications in a setting representative of routine commercial processing." Commissioning is "A planned, documented, and managed engineering approach to the start-up and turnover of facilities, systems, and equipment to the end user that results in a safe and functional environment that meets established design requirements and stakeholder expectations."

Cleaning validation is discussed in great detail in Chapter 19 and will not be defined or discussed here.

For CIP systems, the OQ and PQ are virtually the same thing. Both can be performed by operating a full circuit CIP run and verifying that all acceptance criteria are met. OQ will be discussed only briefly in this chapter. IQ for CIP systems is performed in a manner similar to process systems. Most companies have a well-developed program for IQ, so IQ will not be covered in detail in this chapter.

The most interesting and complex part of commissioning CIP systems is the functional testing portion. Functional testing is the development activity that takes an installed system that has undergone the preliminary elements of IQ and makes the system ready to successfully complete OQ. CIP functional testing is far more complex than qualification and will be covered in greater detail in this chapter.

WHO SHOULD PLAN COMMISSIONING AND QUALIFICATION?

The ideal individual to plan for and lead the execution of the commissioning and qualification (C&Q) of CIP systems is the person who designed it. Understanding the design is essential to properly plan the C&Q. Commissioning a CIP system will also make an individual a much better designer. Unfortunately, sometimes the system designer is not available or not knowledgeable of C&Q requirements and methods. The next preferred choice for planning and executing C&Q is the engineer who will become responsible for supporting the system in operation (owner). A joint execution by the two individuals above is ideal. The least attractive alternative is to have neither the designer nor the owner responsible for C&Q. The functional testing execution allows an individual to learn more about the system than they can possibly pass on. Third party executions leave that valuable experience and knowledge with the third party only.

For completeness, this chapter is written for an individual charged with a C&Q planning responsibility without having the benefit of early design involvement.

HOW DO I GET STARTED?

Understand the Requirements

You need to begin planning with the end in mind (OQ and cleaning validation). OQ and cleaning validation planning require an understanding of the requirements. Hopefully, these are the same requirements that were the basis of the design. If not, expect a lot of rework.

Cleaning Validation Requirements

It is your task to make a designed and installed system ready so that cleaning validation can begin and be successful. Therefore you must have a good understanding of the cleaning validation requirements. The cleaning validation requirements should include:

1. Tested attributes (residual product, residual wash reagent removal, total organic carbon, total viable organisms, etc.)
2. Test methods (swabbing, rinse samples, in-line monitor)
3. Acceptance criteria.

Product residue can typically be determined by taking a swab sample of the process equipment and submitting the sample for appropriate assay. It is necessary for you to know if process piping, CIP piping or the CIP system components will be sampled. Pipe sampling will require sections to be installed for easy removal.

Residual wash reagent removal is typically determined via an in-line conductivity or resistivity sensor during the final rinse. Total organic carbon may be similarly measured in-line. Rinse sampling could be used to validate either attribute as an alternative. Total viable organisms are typically determined by rinse sampling. During commissioning you will need to ensure that in-line sensors are working properly and sample points are in place. Sampling from a CIP return line can be difficult. The return line typically discharges into a waste hub at a low level. Sample locations located just upstream of the point of discharge typically do not work without special design (due to the siphon effect in the return line and problems keeping the return line full). You should ensure during functional testing that your sample location actually can be successfully sampled. If your return line does not remain full this can also cause problems with in-line sensors. During functional testing you need to ensure that in-line sensors function properly for the operating period.

Critical Operating Parameters

The requirements should also identify the critical CIP operating parameters necessary to clean the residual product. Critical operating parameters typically include:

1. Wash temperature
2. Wash reagent concentration
3. Contact time
4. Flow rates

During OQ the CIP system operation should be documented to properly control these critical operating parameters for each circuit in the intended ranges. This can typically be achieved by running a complete circuit CIP, without alarm, while verifying acceptance criteria have been met. Where total viable organisms may be a concern, the effectiveness of the final drain step should also be tested in the OQ. Complete drainage can be subjective. After an air blow and drain, your equipment will not be 100% visually dry. A practical criterion to use is that the low points are drained to no more than a drip.

OQ acceptance criteria for wash temperature, reagent concentration and contact time can be determined in the laboratory or can be based on previous experience cleaning the residue. The effectiveness of the chosen criteria will not be proven until cleaning validation, however.

Although flow rate is a critical parameter (and the ability to control flow should be part of the OQ), flow set points should not be included as a requirement. Flow rates are determined by design and in some cases the design flow rate is confirmed during functional testing.

OQ should include verification that the cleaning sequence performs as designed. Development of the sequence is performed in functional testing and is discussed in more detail later.

Understand the Schematic Design

Effective C&Q planning requires a thorough understanding of the CIP system design. I was recently assigned to plan and execute C&Q of a large CIP system having no previous involvement in the design of the system. This project will be used throughout the remainder of the chapter for examples and to explain key concepts.

Design Scope

The assigned project had two circulating CIP supply skids and approximately 35 total CIP circuits in the facility. A portion of the process was to be brought on-line for early validation work. This portion of the process required 10 of the designed CIP circuits. I will refer to this project as Project A. During the commissioning for Project A, several major CIP system issues were discovered. In Project A, work-arounds were successfully implemented for these issues to qualify the CIP systems in order to complete the intended study. However, more appropriate solutions were desired for long-term production. A CIP consultant was hired to redesign and commission the modified system prior to the next production phase.

The next production phase was for process validation for a different portion of the process. It involved 10 different CIP circuits. I will refer to this as Project B. The remaining production process steps and circuits would be validated later. Cleaning validation activities were performed for both Projects A and B. For Project B, I was involved with the preparation of software functional requirements, but was not involved for the majority of the commissioning. I was consulted to help troubleshoot issues with the system after the consultant completed its commissioning.

Skid and Circuit Schematic Design

A typical CIP circuit is shown in Figure 1. A schematic of a CIP skid is shown in Figure 2. Following is the description of a typical CIP operation. (This is the way the systems for Project B work):

1. A CIP circuit consists of the CIP skid, the process vessel and process piping, and the CIP supply and return piping.
2. The CIP system and process systems are controlled by the same distributed control system (DCS).
3. There are a number of transfer panels and valve clusters used throughout the distribution system.

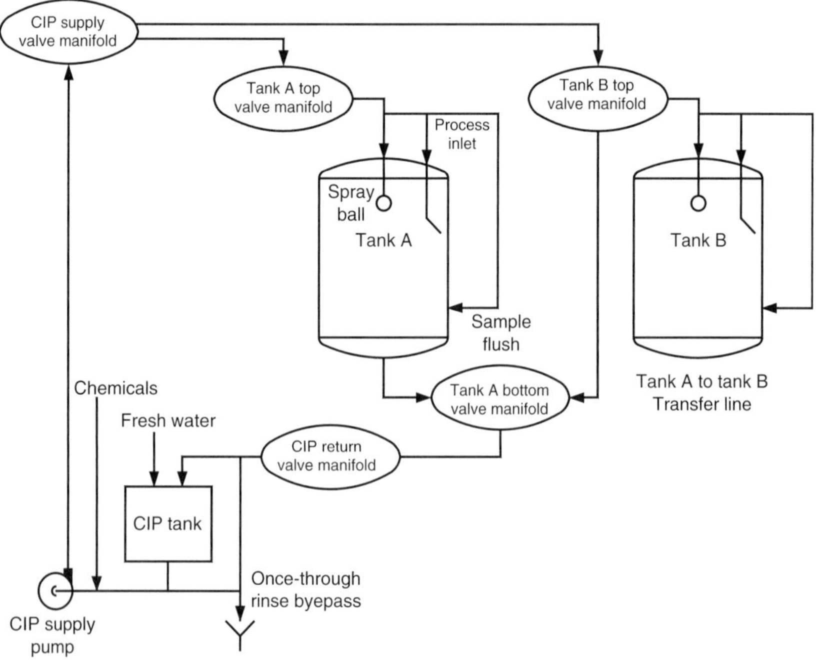

FIGURE 1 A typical CIP circuit.

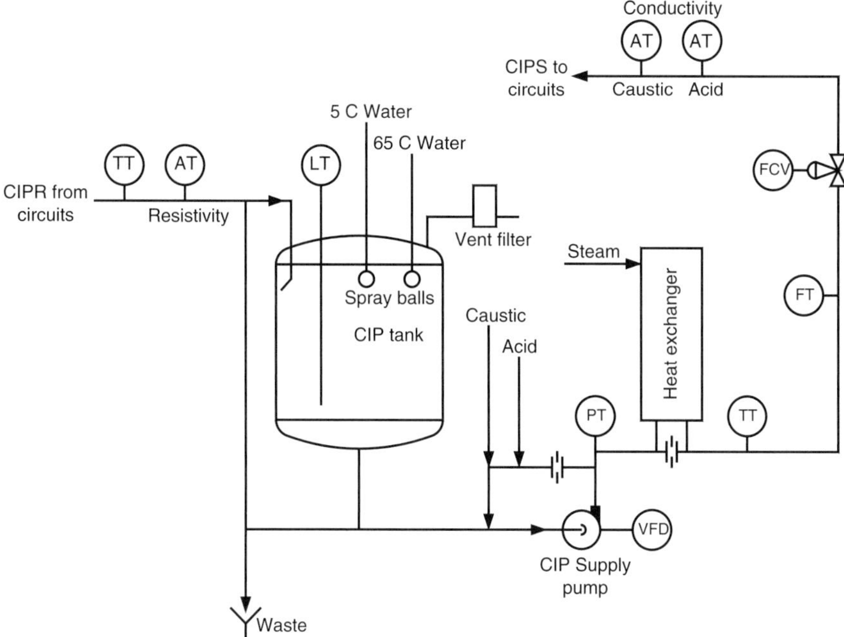

FIGURE 2 Schematic of a CIP skid.

4. Solutions are pumped through the CIP supply piping with a supply pump. Return flow is via gravity or a return pump.
5. Cleaning of the process equipment is via stationary spray balls. Product contacting piping will be cleaned via appropriate cleaning velocity through the piping.
6. Cleaning cycles are: water prerinse, alkaline wash, water post-rinse, acid wash, final water rinse.
7. Water rinses are once through followed by a short drain of the process equipment and return piping.
8. The alkaline wash and acid wash are circulated via the CIP skid solution tank, and are followed by an air blow of the supply piping, drainage of the complete CIP circuit, and a water rinse of the CIP tank.
9. During each cleaning cycle and air blow, fluids are passed through the CIP circuit in a repetitive pattern. The wash sequence is called the normal device sequence (NDS). The air blow is in accord with an air blow sequence (ABS).

Device Sequences

A simplified NDS can be found in Table 1. The ABS sequence looks similar but is normally less complicated than the NDS. These sequences are prepared for each CIP circuit. Sequence steps are similar for similar process equipment. However, step times and specific valve and equipment identities will vary for each circuit.

The sequencing of the CIP skid program must be equally well understood. The CIP skid prepares the solution required for various CIP cycles to be delivered to

TABLE 1 Normal Device Sequence

Tag no.	Description	Spray ball	Dip tube	Sample valve	Transition	Transfer line	Transition
	Step time seconds	60	10	10	1	35	1
	Sequenced valves						
FV-AA	Spray ball supply	O	O	O	O	C	O
FV-AB	Process dip tube CIP supply	C	O	C	C	C	C
FV-AC	Sample CIP supply	C	C	O	C	C	C
FV-AD	Tank bottom valve	O	O	O	O	C	O
FV-AE	Tank A to B transfer CIP supply	C	C	C	O	O	O
	Maintained open valves						
FV-AS	Tank-A top manifold feed to tank-A valve	O	O	O	O	O	O
FV-AR	Tank-A bottom manifold CIP return valve	O	O	O	O	O	O
FV-SA	CIPS manifold tank A CIPS valve	O	O	O	O	O	O
FV-SB	CIPS manifold tank B CIPS valve	O	O	O	O	O	O
	Maintained closed valves						
FV-BS	Tank-B top manifold feed to tank-B valve	C	C	C	C	C	C

Abbreviations: CIP, clean-in-place; CIPS, clean-in-place supply.

a circuit. The skid sequence typically has little variation from circuit to circuit (other than set points and time). There may be some variation between CIP cycles; however, the circulated wash steps are very similar to each other (as are the once-through rinses).

Design Basis

The NDS and ABS design and determination of some of the critical operating parameter values require the engineering and commissioning function to be somewhat iterative. The NDS and ABS and associated times are typically determined using a design basis from rules of thumb (or designer experience) and assumed flow rates.

Flow rates are initially estimated during the design phase. Flow rates for cleaning via spray balls (and the spray ball design) are usually initially determined based on the tank geometry. Minimum flow rate requirements for pipe filling and cleaning can be calculated by engineering during design based on achieving appropriate velocity (generally a minimum of 5 ft/sec). Sequence time for line cleaning is typically based on volume turnovers.

Sequence times for flow through spray balls are based on the number of circuit volume turnovers as well as a minimum tank contact time. The total number of NDS sequence iterations planned per cycle is based on the anticipated needed contact time for the alkaline and acid cycles and on the anticipated number of volume turnovers needed for the rinses. Adjustments are made as data is acquired via functional testing (such as actual process transfer line flow rates).

It is important to understand and document the basis for design so that adjustments made during functional testing are within the capabilities of the design. Flow rates and times affect virtually all aspects of the CIP design. Affected areas include:

1. CIP tank sizing
2. Feed water flow requirements
3. CIP tank vent sizing
4. CIP circuit supply and return pump sizing
5. Circuit supply and return piping sizes
6. Heat exchanger sizing and associated utility needs

IQ AND FACTORY COMMISSIONING ACTIVITIES

Equipment is purchased and systems are fabricated and installed from the design documents. The functional testing portion of commissioning is typically performed on site after a completed and verified installation. Functional testing does not usually proceed until IQ activities have been completed for the CIP skid, supply and return piping, and all of the process equipment system components used for (or immediately adjacent to) the CIP circuits. Prior to starting the functional testing on site it is prudent to ensure that all components associated with the CIP circuit operations are per specification and are installed and working properly to avoid rework.

Installation Qualification

Every company has their methods and definitions in place for IQ and the associated elements of commissioning that may support IQ. Those distinctions will not be made here. Instead, all of the functional testing prerequisite activities will be lumped together for purposes of discussion.

Prior to functional testing ensure:

1. Procured components and piping of the CIP circuits meet the intended specifications.
2. Components and piping of the CIP circuits are installed correctly.
3. Components of the CIP circuits have been individually functionally tested for performance and proper hookup (wiring, pneumatic, etc.).
4. Components requiring calibration have been properly calibrated and are transmitting properly to the DCS system.
5. Piping has been constructed of the right materials, and constructed as specified (e.g., proper slope).
6. Appropriate safety inspections have been completed.

Many times components of CIP circuits are assembled off-site in skids or modules. Most common is the purchase of a CIP skid containing components required to make up and deliver CIP solutions. The process systems can also be assembled off-site as modules. Valve manifolds and transfer panels used for supply and return piping flow path changes are generally assembled off-site. It is common and prudent when this approach is taken to perform as much of the functional testing prerequisite activities prior to shipment to ensure specifications are met. However, because of factors such as: disassembly for shipping, damage during shipping, and changes after shipping, much of this activity will need to be rechecked on site. In addition, module fabricators may not be accustomed to creating documents that meet your needs for IQ. Therefore, you should use great caution in expecting factory testing to meet your qualification needs. You should also ensure that your specifications clearly state what documentation is required from the vendor.

Coverage Testing

One functional testing activity that is possible to perform at the factory is coverage testing. As discussed previously, flow rate via spray balls is initially estimated. The spray balls are strategically positioned with holes drilled to contact all nozzles in the top head of the vessel, agitator shafts, etc. The effectiveness of the design can be tested by covering the vessel with a water soluble material and flowing water through the spray balls for the intended contact time at the intended flow rate or inlet pressure. Riboflavin is commonly used for coverage testing because of its high water solubility, low toxicity and it is easily seen under ultraviolet light. Adjustments may be required for flow rate (or pressure), contact time and hole position based on coverage testing results. Management of this activity is important to an effective project. Mismanagement can have disastrous results.

For Project A, the CIP skid design was completed early in the design phase. The project team decided to have the vessel manufacturers be responsible for the design of the spray balls and for performing a factory coverage test. By the time the coverage testing was done, CIP designers were on other projects and not available. The vessel manufacturers used on the project had little experience in spray ball design or coverage testing. Their solution for unsatisfactory coverage test results was to increase the CIP flow rate (often significantly).

When Project A C&Q planning was initiated, the discrepancy between design flow assumptions and coverage testing results became apparent. Equipment and piping were already installed. A problem with supply and return pump sizing was suspected and their operating capabilities were checked. The engineer checked the supply pump sizing based on the circuits with the highest anticipated flow rates and recommended larger impellers and motors (which we changed). In anticipation of undersized return piping, a drain step was added within the NDS to keep the process tank liquid levels low (if necessary). Later other related issues were discovered:

1. The worst case for CIP supply pump sizing needs was actually low flow, high head circuits. This resulted in further pump upsizing.
2. The combination of the CIP tank size and fresh water supply design rate would not allow us to perform once-through rinses at the higher CIP flow rate without running out of water. This was a real cycle time issue.
3. For Project B, the capacity was increased for the water supply loop, but the CIP tank vent was too small for the higher fill rates.

Allowing the flow rates through the spray balls to be increased significantly above the design basis was a mistake. In retrospect we should have

1. initially had the spray balls designed and fabricated by someone with the proper experience;
2. had the spray ball designers present during coverage testing for necessary modifications;
3. had a better documented design basis; and
4. replaced the spray balls as per number 1 above.

Coverage testing performed at the factory does not necessarily eliminate the need for coverage testing on site with the actual CIP system. Effectiveness of the coverage testing can be affected by:

1. Flow rate or spray ball supply pressure
2. Contact time
3. Pulsing sequence
4. Equipment configuration

It is important to document the conditions used for factory coverage testing. It is difficult to duplicate at the factory all the conditions that will be present in your plant. Repeat coverage testing at your site may be necessary.

FUNCTIONAL TESTING PLANNING

As you might expect, the complexity of functional testing grows as your system complexity grows. Functional testing planning requires an in depth understanding of the detailed design. As discussed before, knowledge of the design basis is crucial to good planning and execution. Just as important is an understanding of the software design.

Understand the Software Design

A typical project software construction method for a CIP system will be described. This typical software design will serve as an example to aid in understanding functional testing planning described later. The intended software design is documented in software functional requirements. The functional requirements describe in text form the exact sequence to be programmed. The functional requirements identify the variables that need to be quantified (via commissioning or process design) in order to clean specific CIP circuits. The NDS and ABS described above are the starting point of the functional requirements.

In this example, the CIP and process systems are controlled by the same DCS. Terminology that is used in the following description is the terminology specific for this DCS system. Figure 3 shows the described software architecture used for this CIP system.

The CIP skid and the process equipment being cleaned are controlled by a CIP "procedure" using two separate "operations" that occur simultaneously (one for the CIP skid and one for the process equipment). The CIP skid "operation" controls the sequencing at the skid required to makeup and send CIP solutions out of the skid to the various CIP circuits. The operation for the process equipment positions the CIP supply and returns valves to send flow to the appropriate circuit and cycles the valves at the process equipment to run the NDS and ABS.

Operations are broken up into "phases." A "phase" is the lowest level of sequencing. Phases are planned in order to minimize the actual software code. As an example, alkaline washes for all circuits are run using the same phase logic or code. Variation from circuit to circuit is achieved via variables or "phase parameters." The CIP skid sequences program steps in the same order for an alkaline wash cycle regardless of the circuit being cleaned. The wash solution is prepared and sent by the skid in the same order with circuit specific phase parameters such as: flow rate; pump speed, and reagent addition time. In the above system there is enough similarity so that a single phase is used for the acid and alkaline wash and a single phase is used for each once-through water rinse.

Each process equipment circuit has its own phase to describe its NDS and ABS. Although these are distinct phases for each circuit, they are very similar. The phase parameters associated with the process equipment are primarily step times.

Identify Functional Testing Variables

It can appear to be a daunting task to develop a functional testing plan for a complex CIP design. The suggested method below will help you identify activities and determine a logical test order. To determine the total scope of functional testing it is necessary to get a complete listing of variables requiring determination. Those variables can then be organized into a logical order for testing activities. A good starting point is to develop a spreadsheet with a complete listing of the software phase parameters. These parameters will form the rows in your spreadsheet. Next you want to identify activities that may be necessary as a prerequisite to determining or testing a specific "phase parameter" value. As an example, there should be a circuit dependent phase parameter for flow set point. The ability to maintain flow set point will be tested in functional testing and OQ. The final flow set point used for each circuit will be confirmed via coverage testing. Before you test the ability to maintain the flow set point, you need to tune the flow control loop. Therefore, list the activity "flow loop tuning" as a row in your spreadsheet along with the phase parameters.

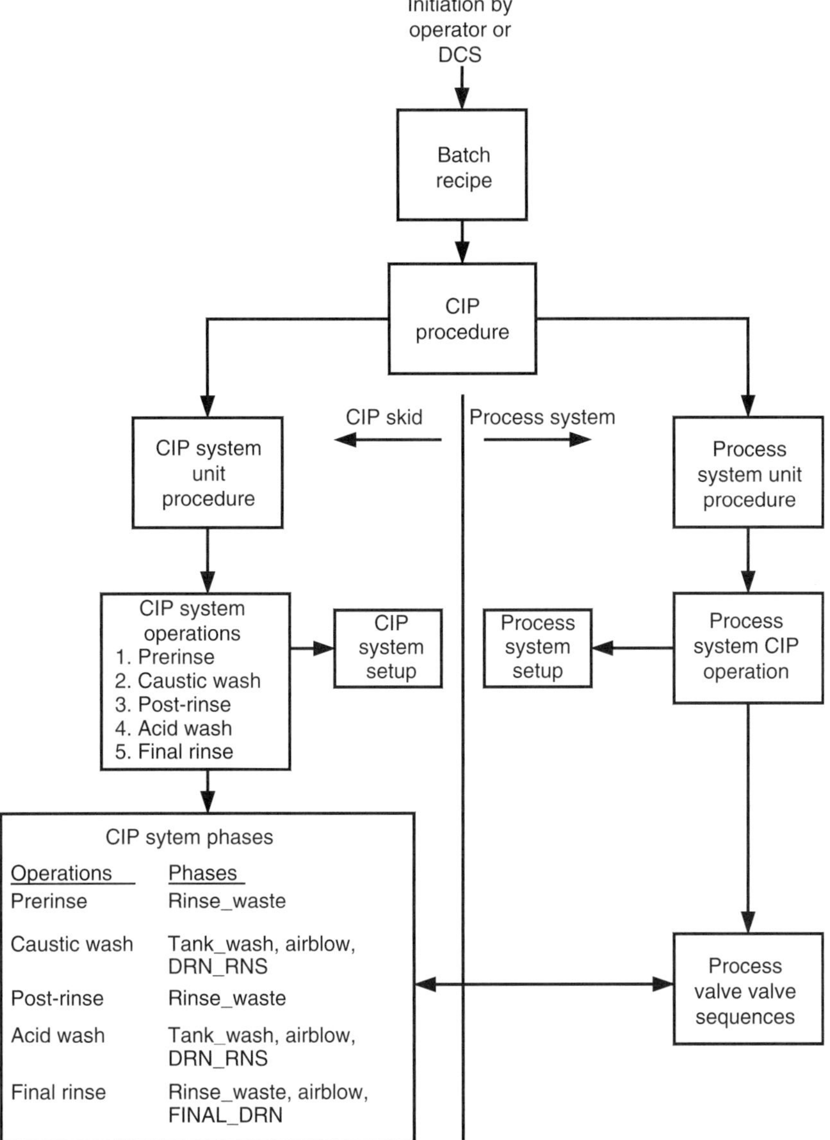

FIGURE 3 Software architecture for a typical CIP application.

Characterize Your Variables

It is useful to characterize the listing of variables to help determine what type of testing or determination is required, what equipment or systems will be needed for testing, what test method you may apply, and what order you want to test. Table 2 provides attributes you can use to characterize your variables or parameters to help in determining a logical order for testing. These attributes are discussed in detail below. Your spreadsheet should have a column for the four attributes listed.

TABLE 2 Phase Parameter Characterization Attributes

Priority	Type	Equipment needed	Test method	Test order
1	Requirement	Skid only	Flushing	Prerinse
2	Engineering	Circuit	Manual	Caustic
3	Coverage	All circuits	Auto	Post-rinse
4	Field		Full cycle	Acid
5				Final rinse
NA				

The "priority" column is added for each attribute. Only one priority column is shown in Table 2, for brevity. Numerical priority values (rather than the description) are used in the spreadsheet so that the spreadsheet rows can be easily sorted to help determine a logical order for testing.

Variable Type

The attribute "type" is based on whether the parameter is determined by design, functional testing, or if it is just tested in qualification. An attribute type identified as "requirement" is as defined previously. A requirement will not be determined in functional testing. However, the ability of the CIP system to control a requirement type parameter is typically confirmed in OQ. An example is the supply temperature set point. The set point is a phase parameter to allow flexibility for differing CIP cleaning recipes. In our system all set points were 65°C for the alkaline and acid cycles.

An "engineering" parameter is determined via calculation or engineering judgment and its value is not functionally tested. An example of this type parameter is maximum CIP supply pressure. If exceeded, the skid would go to hold. The maximum supply pressure is determined by engineering based on the lowest pressure rated device in the supply system.

Many parameters can and should be calculated where possible even though final confirmation via functional testing will be performed. These are not categorized as engineering-type parameters here since they will also be field-tested.

A "coverage" parameter is determined based on coverage testing. Coverage testing may be performed with a separate, preapproved test protocol, making it a special form of functional testing. Flow rate through the spray balls is an example of a coverage type parameter. Another example could result from an expectation that coverage should be demonstrable with the lowest contact time used for any of the CIP cycles. This expectation exists at our plant. We performed coverage testing in the field using the prerinse cycle; therefore, the coverage test typically determined the prerinse contact time. All other CIP cycles had a contact time greater than or equal to the prerinse.

A "field" parameter is determined via site functional testing. Drain time is an example of a field parameter. Field parameters may be initially estimated by engineering but require field confirmation.

Equipment Needed

Parameters can be sorted by what equipment is required for functional testing. Functional testing is performed after IQ-type activities. The CIP circuits involve the CIP skid plus all of the process systems. Testing requiring the use of the entire circuit is done after the IQ-type activities for the process system(s) associated with the CIP circuit(s). It is convenient to begin functional testing of parameters that require the use of the CIP skid only. An example of a "skid only" parameter is CIP tank drain time. CIP skid only testing can be setup and executed manually using the CIP skid only.

Most parameters typically are circuit specific and are determined while using the whole circuit for functional testing. In some cases, there may be a universal parameter that cannot be determined until all circuits (or most circuits) have been commissioned. For Project A, the CIP system controlled flow by varying pump speed. Control problems occurred because the controller would occasionally drop the pump speed too low. A minimum speed setting was added to the flow controller to keep this from occurring. Since there was only one flow controller per CIP skid, all of the circuits in the skid first were run to determine the appropriate minimum setting.

Special challenges can exist when utility systems are being installed simultaneous with the CIP systems. In this case it will be useful to list the utility as one of the "equipment needed" values.

Test Method

The parameters that require flow through circuits for testing are sorted based on the method used to perform the functional test. The methods used are dependent on what options are available within your control system to operate the CIP. The methods described are: circuit flushing, manual circuit testing, automatic circuit testing and full circuit testing.

Circuit Flushing

Parameters can be determined while performing a thorough, planned manual water flushing of the CIP circuit. Circuit flushing should be done first to remove construction debris from the fully assembled circuit. Experience on large projects has shown that regardless of the hydrotesting, cleaning, and passivation performed by the installing contractors, circuit flushing after complete assembly is beneficial. A good flush of CIP circuits can be achieved by manually mimicking the prerinse sequence. A flush of the complete circuit to waste treatment is performed without returning solution to the CIP tank until the circuit has been determined to be clean and leak-free. When the first circuit is determined to be free of debris, the CIP skid is free of debris as well.

Flushing can be followed by debris inspection of spray balls, pump screens or any other natural debris collection points. The flushing and inspection are continued until no sign of debris appears. A debris screen can be installed at the CIP return discharge point to waste to ensure that the entire circuit flows through a final inspection point. Cheesecloth can be used in lieu of the debris screen to detect potential debris that could pass through the screens undetected. This proves valuable in finding residue such as layout dyes (used by vessel manufacturers) that can pass right through debris screens undetected. If cheesecloth is used splashing will occur and may require some additional provisions for capturing the flush water.

A number of parameters can be determined during the circuit flush. For example, the Project B CIP design used a variable speed pump. The pump was run at a fixed speed per circuit and used a control valve for fine flow control. Pump speeds were determined (while flushing) by setting the control valve at 50%. The pump speed that resulted in the intended flow set point through the spray balls was used as the circuit pump speed. Flushing also can be used to determine what the flow rate through the transfer lines will be at steady state.

Manual Testing

Manual functional testing for determination of parameters is preferred (where possible) over automatic sequencing. Manual sequencing allows you to control the test conditions for long periods and determine one variable at a time. Determine all phase parameters possible manually before running the CIP automatically. The number of parameters that can be determined manually depends on how the sequencing is programmed. Typically flow and temperature loop tuning are easiest to perform using manual sequencing.

On Project A, flow control was designed to be controlled (while cleaning via spray balls during the NDS) by varying the supply pump speed (no control valve). The circuit piping was filled by a sequence that was separate from the NDS. During the fill sequence the pump speed was fixed. The CIP supply piping was filled while flowing through the spray balls followed by flowing through the process transfer line. The fill step completion was based on totalized supply flow volume. The circuit fill pump speed was set (at the speed determined in the circuit flushing) and the equipment was manually sequenced to fill the circuit through the spray balls. Flow rate was monitored on the DCS screen to determine at what totalized volume the piping became full and the air was gone. When the flow rate reaches steady state, the system is full. A similar approach was used to determine the totalized volume needed to fill the CIP circuit supply through the process transfer line.

Automatic Testing

Other parameters are easier to determine with some form of automatic operation for testing. In our DCS system, the phase parameters can be adjusted (for testing) when a batch is initiated. This allowed some parameters (such as circuit drain times or blow times) to be determined by running with an excessively conservative parameter value and observing what the actual parameter value should be. Some parameters may need to have an estimated value from engineering that can be optimized during automatic operation. One example of this is the CIP tank level set point used for the wash or rinse steps.

Full Cycle

Some parameters cannot be effectively determined until the circuit is ready to be run completely automatically with all other parameters at their determined value. This is because they depend on normal operation. Checking that the desired contact time is achieved via the spray balls, checking flow control, and checking that the number of NDS iterations required are completed requires full cycle testing. If volume is the parameter used to end the cycle, it must be determined after all other parameters are determined and a complete normal cycle can be run.

Test Order

Once the determination of test method is done, the order of testing should be determined. The cycle where the parameter is first used in the CIP is listed in the spreadsheet (in the "test order" column). One can further sort for the exact sequential order if desired.

Some phase parameters may be repeated in different cycles. We had a CIP tank level set point that could be set differently for each once-through rinse. However, we found that the same setting worked for all three once-through rinses.

It is also important to note dependencies of related parameters. In the circulating CIP systems, for reagent washes, the circuit is filled with a quantity of water first. A timed shot of reagent is then added and circulated to mix. The reagent pump has a manually settable stroke adjustment. A stroke adjustment is performed followed by a manual flow test to set the flow at the desired flow rate. A test is performed to determine what the volume of the filled circuit is. This allows the reagent shot time to be calculated. Functional testing then confirms the solution is made up in the proper range using the calculated parameter value.

Table 3 is an example of a sorted parameter table. A few examples of parameters are included to aid in understanding. There are typically far more parameters than are shown in Table 3.

PREPARING TEST DOCUMENTS

Once you have determined all the required testing, the method, and the order of testing, the next step is preparing test documents. Several factors must be considered before preparing test documents; the appropriate format of the data to be collected, the appropriate levels of approval prior to and after execution, and the level of detail required.

TABLE 3 Sorted Parameter Table

Parameter description	Type	Equipment	Method	Test order
Flow-deviation percentage	1	NA	NA	NA
Supply temperature set point	1	NA	NA	NA
Final resistivity limit	1	NA	NA	NA
Agitator speed	2	NA	NA	1
NDS time spray ball	2	NA	NA	1
NDS time dip tube	2	NA	NA	1
Supply flow set point	3	2	3	1
Caustic pump stroke adjustment	4	1	2	2
High caustic conductivity	4	1	2	2
Low caustic conductivity	4	1	2	2
Water rinse drain time	4	1	2	2
Pump flood time	4	2	1	1
Pump speed	4	2	1	1
Flow loop tuning	4	2	2	1
Wash fill volume determination	4	2	2	2
Caustic feed time	4	2	2	2
Caustic shot time	4	2	2	2
Prerinse drain time	4	2	3	1
Prerinse fill level	4	2	3	1
Prerinse level set point	4	2	3	1
Acid/caustic chemical mix time	4	2	3	2
Wash drain time	4	2	3	2

Abbreviation: NDS, normal device sequence.

Approval Levels

Approval levels should be based on the overall purpose of the test and its potential to affect safety, identity, strength, purity, or quality (SISPQ). The majority of the functional testing is a cycle development–type exercise that will be proven in OQ and in cleaning validation. This type of testing has little potential to affect SISPQ for systems that are not yet in use for production operations. There are some exceptions. The circuit flushing process may be used to demonstrate that the functional testing phase was begun with a circuit that is clear of construction debris. Documentation for the circuit flushing might be preapproved and post-approved by the quality unit. Conductivity is used to verify cleaning reagent concentration. Concentration is a requirement and a critical parameter. Before functional testing of our system, no data was available for the conductivity of reagents at the acceptable concentration limits and temperature. A functional test was developed to determine the appropriate conductivity limits. Because the results of this test were used to determine OQ acceptance criteria, the quality unit preapproved and post-approved the testing documents. Coverage testing documents might also be preapproved and post-approved by the quality unit (depending on how your company views coverage testing).

Level of Detail

The level of detail to be used in preparing test documents depends on the necessity for repeatability, the expected approval level, and the qualifications of those executing the test. Most of the commissioning testing should be witnessed, if not performed, by engineering. However, some of the testing may be highly repetitive and labor intensive and may warrant having operations personnel perform the testing. For our circuit flushing, a testing template was developed that was very detailed and instructive and was easily customized for each circuit. Documents that describe activities that have the potential to affect SISPQ warrant the addition of detail.

For functional testing supervised by appropriate knowledgeable resources, detailed procedural descriptions are less important. The most important part of the functional testing documentation can be the form and format of the data to be collected. Consider whether DCS trends should be set up and/or forms or spreadsheets constructed. It is important to have enough description so that another knowledgeable resource could replicate the testing later and can understand how the data was collected.

Here are some other items that should be considered when preparing test documentation and planning functional testing:

1. What prerequisites should be completed before beginning each test?
2. What miscellaneous equipment, utilities, tools, or resources will you need?
3. Will the control system allow you to execute manually, adjust parameters as needed, get the needed trends, etc. in order to carry out testing?
4. Will data compression in the historian necessitate live trending or live observation?

FUNCTIONAL TESTING EXECUTION

Now you have prepared commissioning test documentation and are ready for commissioning. There are some principles to keep in mind during the execution:

1. Think about what you expect to see before testing. If you do not see what you expect, investigate and understand why.
2. Do not tweak, changes should be based on engineering principles. As described in point 1, you must understand what is happening and the root cause of any problem before implementing a change or modification.
3. Test your data for consistency.
4. Avoid using software to correct hardware mistakes.

Investigate and Understand Why

The first example below of something gone wrong actually violates the first two principles above. But the root cause (for the far reaching effects it had) was the violation of the first principle.

Some background information on the CIP system for Project B is necessary first. The CIP tank and process vessel both have 0.2 μm hydrophobic vent filters. The process vessel vent is closed during most of the CIP to protect the filter from moisture (and plugging). The prerinse and final rinse are performed once through with water from the CIP tank (fed from a 5°C water supply). The post-rinse is once through and fed using a 65°C water supply. The supply-water feed rate was increased during the upgrade of the system between Project A and B. An air blow of the supply piping, a circuit drain, and a rinse of the CIP supply tank follow alkaline and acid washes. The supply pump has a variable speed drive, but speed is fixed for a given circuit. Flow control is via a control valve. A flow-deviation alarm of 5% will shut down the CIP (after a time delay). The supply-pump discharge-pressure transmitter is used to detect an empty CIP tank and shutdown the CIP (after a time delay).

Proper pump speeds were determined for each circuit during circuit flushing (as described previously). The engineer performed functional testing by running the circuit automatically. The engineer noticed that with the previously determined pump speeds, flow rates were higher than anticipated and flow-deviation failures occurred. At that point he should have investigated why the predetermined pump speeds were not working as expected. Instead he lowered the pump speeds and was able to finish all functional testing successfully for most of the circuits before leaving the project.

Just as OQ was about to start, low-flow-deviation failures started to occur in circuits cleaned by one of the two CIP skids. Trends of previous (apparently successful) CIP runs were reviewed. The pump discharge pressure showed dramatic pressure swings corresponding to CIP tank level changes. This indicated that the CIP tank vent filter was likely plugged. Current failed runs did not exhibit this trend. Inspection of the CIP tank manway revealed that it was not bolted down. This was allowing the tank to vent via the manway.

Further investigation revealed that the CIP tank vent was too small (and in a bad location) and that when 65°C water was added to the tank (via spray balls), flashing occurred and the vent filter was plugged with wet steam. The CIP tank was being pressurized when it was filled adding pressure to the CIP supply pump suction. A redesign of the CIP vent line corrected the venting problem. Pump speeds were returned to the values determined during the manual flushing. However, other issues began to occur.

1. CIPs started to shutdown due to low pump discharge pressure at the beginning of the post-rinse and final rinse.
2. Much larger final rinse volumes became necessary to meet return resistivity criteria during the final rinse.

The shutdowns were occurring in circuits with return pumps. The reason is that the process vessel became slightly pressurized during the alkaline wash due to the heat up of the solution. The process vessel was never fully vented during the blow and drain steps. The process vessel pressure was at approximately 3 psig when the post-rinse started. This was enough pressure to keep the CIP supply pump from filling so the pump was not able to fill and develop pressure. A revised circuit venting strategy was needed to correct the problem. This did not happen with a plugged vent filter since the CIP tank pressurized while filling.

The larger final rinse volumes were occurring because the CIP tank was no longer completely draining during the rinse after the acid wash. The purpose of the CIP tank rinse was to clear out reagent in the tank so that clean water would be used to flush the remainder of the circuit. When the CIP tank had a plugged vent, it pressurized during filling and drained quickly. After fixing the vent, the tank was not fully draining. Fresh water was simply being mixed with reagent and diluted. This greatly increased the rinse water required to reach final rinse resistivity criteria. A separate CIP tank drain step (after the CIP tank rinse) was added. Final rinse volumes have decreased substantially since this change.

Had the commissioning engineer investigated why the predetermined pump speeds would not work, the problem would have been discovered quickly and fixed early. Instead he tweaked (violating principle 2) weaving a tangle web that caused a delay in the startup.

Do Not Tweak

In the above CIP systems for alkaline and acid washes the CIP circuit is first filled with water. Alkaline or acid is added using a pump with an adjustable stroke length to set flow rates. A timed reagent addition is used. For Project A, the CIP circuit piping volumes were calculated prior to functional testing. The total circuit fill volume with water was first estimated using the calculated pipe volumes. A functional test method was developed to manually fill the CIP circuit with water and directly measure the water in the circuit. This was repeated to ensure consistency. The time duration required for acid and alkaline reagent addition was then calculated. Functional testing runs verified that the target concentration was consistently met. Data was accumulated in a spreadsheet. The biggest source of variability in the makeup process was determined to be the acid pump flow rate. Occasional drifts would occur. When concentrations started to drift, the acid pump flow rate was tested and verified. Adjustments were made if necessary. Initial calculated parameters determined by engineering principles never required adjustment.

Test Data for Consistency

Between Project A and B the method of filling the circuit was modified and a direct measurement of circuit volume was more difficult to perform. For Project B, the initial reagent addition time was estimated for the first trial. The addition time was adjusted based on the resulting concentration. Data was not accumulated and tracked. No fill volumes were determined. As pumps drifted or other variability occurred throughout the functional testing, reagent times were simply tweaked. At one point late in the functional testing, make-up failures started to occur in some circuits. With little or no accumulated data, the problems were difficult to troubleshoot.

A request was made for help to troubleshoot the problem. Historian data was first accumulated in a spreadsheet for all previous functional testing (and reagent pump testing) runs for both alkaline and acid. Average fill volumes were calculated for circuits based on the alkaline concentrations achieved and the alkaline addition times. Calculations were then done to determine what the acid reagent addition times should be. Inconsistencies were found. It was possible to correlate and attribute the inconsistencies to adjustments made in the acid pumps. It was also possible to determine the acid conductivity probe for one of the CIP skids was failing.

Fix Hardware Mistakes

It is very tempting during functional testing to compensate for hardware deficiencies by developing software workarounds. Trying to outsmart the hardware can even be very satisfying. This may not cause great difficulties for very simple systems. However, for complex systems, especially where multiple workarounds are performed, these patches make the software far more complex and may ultimately prevent you from having the system operate the way you want.

Early in Project A, a problem was discovered with the type of flowmeter used on the CIP skid. If the flow was ever blocked while the pump was running, the flowmeter did not read "zero." Instead it actually spiked up to a value close to its upper calibrated range value. The flowmeter could have been replaced with a more suitable flowmeter; however, workarounds were implemented instead. These workarounds got very complex during Project A. It has not been quite as complicated for Project B, but the following problems remain:

1. Sequences have to be designed without ever blocking flow, otherwise total volumes will be way off.
2. The control valve output was limited to a 15% minimum to avoid problems with the flow control loop.
3. Low flow alarms are not useful for detecting closed manual valves or other setup problem.

CONCLUSION

Complex CIP systems can be the most challenging commissioning task in you project. Planning the commissioning can be made easier by understanding the design and design basis, identifying all necessary activities and planning them in the appropriate order. Using sound scientific principles in your execution will pay great dividends in reduced commissioning time and robustness of CIP operations.

19 Cleaning Validation Strategies

Charles Lankford
PharmaSys, Inc., Cary, North Carolina, U.S.A.

OVERVIEW OF CLEANING VALIDATION STRATEGIES AND METHODS

This chapter is an overview of cleaning validation strategies and methods. Principles and concepts employed in the design and implementation of an effective cleaning validation program will include cleaning validation study design and implementation, analytical techniques, establishment of acceptance criteria, and single versus multi-product issues.

It is impossible to provide a thorough in-depth dissertation of the cleaning validation process within the confines of this chapter as it is a complex subject and quite specific to the product or products that are being manufactured. Volumes have been written on the subject. There are, however, common principles and industry accepted techniques involved and this chapter will focus on the basic principles of validating the cleaning process for clean-in-place (CIP) systems which translates to validation of manual cleaning procedures as well.

Cleaning validation is uncomplicated in concept. Simply put, it is proving that a specific cleaning process adequately cleans the contact surfaces of a specific equipment train configuration of a specific substance to a predetermined acceptable limit; and further that the cleaning agent is adequately removed from the contact surfaces of the equipment train to a predetermined acceptable limit. To be valid, a cleaning process must also be consistent and reproducible.

Many factors are taken into consideration when designing a cleaning validation study program. These factors include physical product characteristics, the toxicity of the product and the cleaning agent, the dirty/clean hold times for the equipment, the sample/recovery techniques and locations, and the analytical techniques used to evaluate the effectiveness of the cleaning process. All of these factors are highly specific to a given substance and therein constitutes the problem with a single source reference on the subject.

CLEANING VS. SANITIZATION VS. STERILIZATION

To start with, let's contrast the following terms to clarify what we are talking about:

- Cleaning is the process of removing particulates and residue from a surface. A clean surface is free of dirt, grime, or other residue. It is easy to clean a surface but cleaning doesn't mean sanitizing.
- Sanitization is generally thought of as killing the more sensitive vegetative cells but not heat-resistant spores. Sanitization does not necessarily include sterilization, although some processes of sanitization accomplish sterilization.

- Sterilization refers to the removal or destruction of all microorganisms, including pathogenic and other bacteria, vegetative forms, and spores. Techniques of sterilization include heat, gases (such as formaldehyde or ethylene oxide), solutions of various chemicals, ultraviolet irradiation, or gamma irradiation.

CLEANING VALIDATION

Cleaning validation is documented evidence that a specific cleaning procedure adequately cleans specific contaminates from an equipment train to a specified level. Cleaning validation is defined by the United States Food and Drug Administration as follows:

> The purpose of cleaning validation is to demonstrate that a particular cleaning process will consistently clean the equipment to a predetermined limit; the sampling and analytical test methods should be scientifically sound and provide adequate scientific rationale to support the validation (1).

This chapter will focus only on validation of the cleaning process rather than sanitization or sterilization processes.

Cleaning Validation Study Design and Implementation

To discuss cleaning validation study design and implementation, commentary on the hierarchy of the cleaning validation program is in order.

An effective cleaning validation program will have two or three tiers of documentation as in any robust quality documentation system. At the top level, a cleaning validation policy and/or a cleaning validation master plan will establish the why, what, where, when, how and by whom of the cleaning validation program in general terms. Under that document, a cleaning validation master plan will establish the same information for a given project. Individual cleaning validation protocols will establish the procedures, acceptance criteria, sample methods and other pertinent design information for each cleaning validation study. This section will highlight the general expectations of content for each of these documents.

The FDA expects:

- Written procedures on how cleaning processes will be validated.
- The general validation procedures to address who is responsible for performing and approving the validation study, the acceptance criteria, and when revalidation will be required.
- Written validation protocols approved in advance for the specific studies to be performed on each system and/or equipment train. Each protocol should specify sampling procedures and analytical methods.
- Validation studies to be conducted in accordance with the pre-approved protocols and the results documented.
- Final validation reports presenting study results. Reports should include a declaration whether or not the cleaning process is valid based upon results indicating that residues have been reduced to an "acceptable level" (2).

Cleaning Validation Policy

An effective cleaning validation program starts with a cleaning validation policy. The cleaning validation policy should serve as a general guideline and direction for company personnel, regulatory authorities, and customers as to how the company conducts a cleaning validation study. The policy should include:

- Definition of terms employed during validation.
- A statement specifying what company policy is on validation of cleaning procedures related to equipment and processes.
- Company policy regarding dedication of equipment in different areas.
- Analytical method validation policy.
- Rationale for the methods by which acceptance criteria is determined.
- Cleaning Revalidation policy (3).

The cleaning validation policy may be a standalone document or a section of a validation policy that includes other validation disciplines (e.g., equipment, process, cleaning, and computer validation).

Cleaning Validation Master Plan

Cleaning validation master plans define all cleaning validation activities, documentation requirements, overall approach, and schedule for a specific project. They are typically utilized for new facilities, facility expansions and new process scale-ups. A cleaning validation master plan may be a stand alone document or it may be included in an overall validation master plan as a relevant section.

For ongoing cleaning validation and/or revalidation activities, oftentimes a master plan is not necessary as it typically duplicates or references the cleaning validation policy and procedure set.

When a cleaning validation master plan is utilized it typically has the following sections:

- Overview of the project
- Approach to the project in respect to cleaning validation
- Definitions
- Responsibility of personnel for the execution of cleaning and cleaning validation
- Plant/process/product descriptions
- References to cleaning policies/procedures, analytical methods, and cleaning validation policies/procedures
- Specific process considerations such as selection of worst case cleaning scenarios
- List of products, processes or equipment trains to be validated
- Schedule(s)

Cleaning Validation Studies

A cleaning validation study should confirm the effectiveness of a cleaning procedure. Acceptance criteria should consider residual limits of active drug substances, cleaning agents, and microbial contamination with corresponding rationale for their selection. All sampling procedures utilized during the study must be approved and validated to determine their recovery factors (the percentage of total residue recovered during the sampling process for a specific contaminant).

All analytical methods utilized during a study should have the sensitivity to detect the target contaminant at levels consistent with the acceptable residue limits and must be validated.

Typically, a cleaning validation protocol includes or references approved procedures for the following information:

- Study objective
- Cleaning procedure(s) to be validated
- Study scope
- The process parameters to be verified
- Sampling/inspection procedure to be utilized including rationale for selecting the specific sampling/inspection procedure and references to recovery studies for the sampling/inspection procedure
- Sampling diagrams
- Personnel responsibilities for the study
- Test methods to be used including rationale for their selection and references to the method validation studies
- Acceptance criteria including rationale for acceptance criteria along with calculations for residue limits. (If worst case products are being grouped, for the study, rationale for selection of the products should be included.)

Validation studies should be conducted in accordance with the pre-approved protocols and the results documented. Personnel cleaning the equipment, whether via manual procedure(s) or CIP, should be trained in the cleaning procedure(s). In addition, the personnel conducting the validation study should be trained on and have a thorough understanding of sampling techniques.

Prerequisites for conducting the cleaning validation study include all required equipment be installed and operational, cleaning procedures/CIP cycles developed, critical parameters identified, residue limits established, recovery studies performed for the sampling/inspection procedures and all analytical methods validated.

Typically, installed and operational implies all commissioning, installation qualification and operational qualification has been performed for the equipment train, the CIP skid, ancillary equipment, and controls systems. For CIP systems, one operation qualification test that is necessary to perform before actual cleaning validation is a spray ball coverage test.

A validation report is necessary to present the results and conclusions of the study. In cases where it is unlikely that further batches of the product or product group will be manufactured for a period of time it is prudent to produce interim reports on a batch by batch basis until the cleaning validation study has been completed.

A cleaning validation study report should include:

- A summary of or reference to procedures used to clean, sample, and test.
- Test results and any applicable observations.
- Conclusions on the validity of the process based on the supporting test results.
- Recommendations based on the results or other relevant observations obtained during the study.
- Review and documentation of any deviations that occurred during testing including the resolution and the rationale of why the resolution is acceptable.

Establishment of Acceptance Criteria

An acceptance criterion is a condition that must be met before the cleaning process is deemed successful. Each should be specific and verifiable. Before a cleaning validation study can be executed, acceptance criteria must be established. Typically, acceptance criteria are specified for three critical features. The equipment should be visually clean, the contaminant/cleaning agent residue limits should be specified and the number of consecutive iterations for which the results meet specifications must be specified. Other acceptance criteria can be selected specific to a particular processes and analytical techniques; however, these are the "big three."

Obliviously, establishment of acceptance criteria for dedicated equipment is much simplified because cross-contamination issues are eliminated. Unless there are issues which exist with the cleaning agent reacting with the target contaminant, acceptance criteria is often limited to visually clean and detection of the cleaning residue to levels less than the LD_{50} value of the agent times a safety factor.

Visually Clean

An acceptance criterion of visually clean is exactly what it implies. Contact surface is visually free of residue when inspected by the test executor.

Visual examination is one of the most popular methods to determine the cleanliness of a visible process contact surface and is one of the simplest methods to use as most residues are visible at 100 pg for a 4 in.^{2} area (4). It gives the operator an immediate indication of the cleanliness or non-cleanliness particularly if visual residues are present (residues can be seen and therefore the surface is not clean). Visual examination alone is not necessarily adequate to confirm that residue limits for a product have been met. For some residues, it may be possible to establish the approximate level of concentration that is visible but in most cases; an acceptance criterion of visually clean is used in conjunction with other sampling techniques such as rinse sampling or swab sampling.

When using an acceptance criterion of visually clean in conjunction with swabbing, care must be taken to clarify the relationship between the two techniques because due to the concentrated residue on the swab, it could result in a visual indication (5).

Contaminant/Cleaning Agent Residue Limits

The most defendable acceptance criterion for cleaning validation is setting contaminant and cleaning agent residue limits for the worst case batch that could be processed after the cleaning process. They are then converted for use in a specific analytical technique. The calculations are relatively easy to complete and must be performed whether rinse sampling, swab sampling, or visual identification of residue is utilized. This worst case limit should be established for the smallest batch/largest dosage combination following the cleaning exercise. (The largest possible concentration of contaminants per dose should be considered.)

To establish this acceptance criteria one must consider the materials to be cleaned and their acceptable daily intakes (ADI). The active compound/ biologic that is being cleaned, any additives and cleaning agents must be considered. Additionally, sample recovery factors and possible interference between multiple contaminants must be taken into account. Acceptance criteria should be based upon scientific rationale, be practical, be achievable, and always be verifiable (6).

There are several ways to determine the ADI, with one example being the largest therapeutic dose divided by 1000. Another is based on the acute toxicity of the contaminant. The second method will be presented in this text.

Using the acute toxicity method of calculating the acceptance criterion, the acceptable limit of contamination is based on the toxicity (i.e., LD_{50}) of each contaminant and the cleaning agent. Calculations should always be based upon the smallest expected batch yield and its maximum daily dosage of the next process going into the equipment. Calculations may be performed in a single equation for rinse sampling or swab sampling with slight changes to the equation. In this text, for the sake of clarity, the calculations will be broken into several steps.

The maximum allowable carryover (MACO) of a substance from one batch to the next is calculated by:

Equation (1)—MACO Calculation

$$\mathrm{MACO} = \mathrm{ADI} \times \frac{Y_{\mathrm{min}}}{D_{\mathrm{max}}} \tag{1}$$

where MACO, maximum allowable carryover; ADI, acceptable daily intake; Y_{min}, smallest expected yield for the next process using the equipment; D_{max}, largest normal dosage rate of the drug product produced during the next process which uses the active pharmaceutical ingredient (API).

Note: the object is to obtain the smallest number of doses for the next batch (smallest $(Y_{\mathrm{min}}/D_{\mathrm{max}})$ value). It could be possible that a larger batch of large doses will yield less doses than that of a smaller batch of smaller doses. Consider the following example: the smallest batch of Product A contains 20 kg of active and its maximum daily dosage is 10 mg. The smallest batch of Product B contains 30 kg of active and its maximum daily dosage is 100 mg. Product B would be used for the calculation because the number of possible doses per batch is less than that for Product A.

The ADI may be calculated using the acute toxicity value of the contaminant in rats (LD_{50}), a conversion factor and the worst case average human body weight for the system. A conversion factor of 1/10,000 or 1/1000 is normally accepted in the industry depending on the LD_{50} value. The larger conversion factor is used for smaller LD_{50} values (7). An additional safety factor is sometimes utilized to assure that the ADI is well within range. A typical ADI calculation is as follows:

Equation (2)—ADI Calculation

$$\mathrm{ADI} = \mathrm{HBW}_{\mathrm{avg}} \times \frac{\mathrm{LD}_{50}}{\mathrm{CF} \times \mathrm{SF}} \tag{2}$$

where $\mathrm{HBW}_{\mathrm{avg}}$, average human body weight of the target group (~70 kg for Adults); LD_{50}, acute toxicity of the contaminant in rats; CF, conversion factor for converting LD_{50} to the no observable effect limit (NOEL). This is typically 10,000 or 1000 depending on the LD_{50} value; SF, safety factor. This is an additional conversion factor used to assure that the levels of contaminants are within the NOEL under most circumstances. This value is typically 100.

The MACO may then be converted into an acceptance criterion by one of the two formulas below depending whether a rinse sampling method or swab sampling method is used:

Equation (3)—Limit Calculation for Swab Sampling

$$C_{\mathrm{s\_max}} = \mathrm{RF_s} \times \frac{\mathrm{MACO}}{A_\mathrm{t}} \tag{3}$$

where $C_{\mathrm{s\_max}}$, maximum allowable concentration for swabbing expressed in weight/surface area; $\mathrm{RF_s}$, recover factor for the contaminant using the swab method as established by swab recovery studies; MACO, maximum allowable carryover of the contaminant; A_t, total contact surface area of the process train.

Rinse Sampling Method

Equation (4)—Limit Calculation for Rinse Sampling

$$C_{\mathrm{r\_max}} = \mathrm{RF_r} \times \frac{\mathrm{MACO}}{V_\mathrm{r}} \tag{4}$$

where $C_{\mathrm{r\_max}}$, maximum allowable concentration of rinse sampling expressed in weight/volume; $\mathrm{RF_r}$, recover factor for the contaminant using the rinse method as established by rinse recovery studies; MACO, maximum allowable carryover of the contaminant; V_r, total volume of rinse.

The above calculations should be utilized for each active ingredient, potentially harmful excipient or other additive and the cleaning agent itself.

Taking an example of cleaning aspirin from a process train with the following process's smallest yield being 50 kg and largest dosage form being 10 mg. ADI is calculated using equation (2) above.

$$\mathrm{ADI} = \mathrm{HBW_{avg}} \times \frac{\mathrm{LD_{50}}}{\mathrm{CF} \times \mathrm{SF}} = 70\ \mathrm{kg} \times \frac{200\ \mathrm{mg/kg}}{10000 \times 100} = 0.014\ \mathrm{mg}$$

Plugging these results into equation (1), the following is obtained:

$$\mathrm{MACO} = \mathrm{ADI} \times \frac{Y_{\mathrm{min}}}{D_{\mathrm{max}}} = 0.014\ \mathrm{mg} \times \frac{50\ \mathrm{kg} \times 10^6\ \mathrm{mg/kg}}{100\ \mathrm{mg}} = 7\ \mathrm{g}$$

Both the FDA and Pharmaceutical Inspection Corporation Scheme (PIC/S) imply in guidance documents that a default MACO limit of 10 ppm is appropriate for cleaning. The calculated MACO in the above equation is 7 g per 70 kg or 14 ppm. However, when using the above default limit of 10 ppm, the residue limit acceptance criteria should be 10 ppm or 5 g.

Consecutive Iterations for Validation

It has been a long standing custom in the biopharma industry that any process including cleaning must produce the output within specifications for three consecutive sequences. The old adage applies that the chance of obtaining a given result in a process one time is happenstance, two times is coincidence, and three times is a truly consistent pattern.

It's important to explicitly state what constitutes three consecutive sequences. The phrase "three consecutive sequences" typically means three consecutive cleaning sequences of the same product. Where multiple products are produced in a process train, it is acceptable to process the other product but once execution of a cleaning validation protocol has been initiated for a specific product, the next two batches of the same product should be used for the process. By skipping a batch, it will call into question the validity of the study.

Analytical Techniques

The analytical methods used for cleaning validation vary according to the characteristics of the potential contaminants thus making it impossible to specify one preferred analytical method in this dissertation. A brief survey of methods currently utilized and factors that should be considered when selecting cleaning validation methods will be discussed.

The basic considerations for use of analytical methods for cleaning validation are the ability of the method to detect the target substance(s) at levels consistent with the acceptance criteria and the ability to detect the target substance(s) in the presence of other materials that may also be present in the sample (8,9).

Regardless of the method(s) utilized, they should be validated for the specific application. Method validation is beyond the scope of this dissertation; however, each method should be validated for specificity, linearity, precision, accuracy, limit of detection (LOD), and limit of quantitation (LOQ) (10). Validated analytical techniques utilized in evaluating residues in cleaning validation are critical to the success of the program and each substance may require different methods. Method validation is beyond the scope of this chapter.

Analytical methods can be grouped into two categories, specific and non-specific. Each category has advantages and disadvantages which will be covered later in the text. Regardless of the analytical method utilized for cleaning validation, it must be scientifically applicable and validated for its intended application.

Care should be taken when using analytical methods to test for acceptable residual amounts of active material. Several opportunities exist to introduce unknown material into the sample. A common occurrence is the introduction of leached material into the sample by rinsing or swabbing agents. Plastic and rubber material from tubing, seals, and components are often the source of these unknowns. A sound investigation and reporting procedure covering unknown material in samples should be developed. A typical investigation path would be to check the unknowns against the two previous active materials and the detergent(s) used for cleaning.

Specific or Selective Methods

An analytical method that is specific will provide a quantitative assessment of a particular target contaminant. The primary advantage of using a specific method is the fact that the target contaminant can be detected and measured.

The use of specific methods is traditionally considered the conservative approach for testing for active contaminants and cleaning agents during cleaning validation. Health Canada and PIC/S recommend that specific methods be used in cleaning validation (6,2). However, the FDA, the European Community Working Party on Control of Medicines and Inspections, and Active Pharmaceutical Ingredients Committee (APIC) do not require a specific method to be used for cleaning validation analysis (8,11,12). Indeed APIC directly addresses that non-specific methods are acceptable for cleaning validation.

In most cases, companies will already have analytical methods to detect active products as part of their stability and/or product release program. It should be relatively easy to adapt and validate those methods for cleaning validation analysis. Some examples of specific methods include high presure liquid chromatography (HPLC), mass spectrometry (MS), gas chromatography (GC), atomic absorption (AA), ion mobility spectrometry (IMS), and thin layer chromatography (TLC).

HPLC methods are some of the most popular specific methods utilized for cleaning validation because they are sensitive, specific and quantitative (13). Additionally, because almost every biopharma company has multitude of the instruments and personnel who operate them on hand, they are easily adopted for cleaning validation analysis. It should be noted that analytical methods will invariably result in increased laboratory investigation time for determining the source of unknown compounds eluting during the analyses. HPLC methods are relatively straight forward to develop.

Studies have indicated the IMS can be applied to cleaning validation for some contaminants with reliable results (14). Since IMS analysis is less expensive and less time consuming than HPLC analysis, it may gain popularity in the future.

Nonspecific Methods

A nonspecific method is a method that will measure the presence of certain physical properties to which a number of compounds and/or contaminants can contribute. Some examples of nonspecific methods are total organic carbon (TOC) analysis, pH testing, conductivity measurements, particulate counts, and titration.

Whenever nonspecific methods are utilized, care should be taken to understand the positive or negative contribution of coexisting contaminates such as the active residue and the cleaning agent. As such, these methods should be validated for the specific active and cleaning agent to be analyzed.

TOC analysis is a widely used method for the detection of a wide range of active contaminants and cleaning agents for pharmaceuticals and biotech.

When using TOC analysis, the worst case acceptance limits for a contaminant must be characterized by its carbon content. The active contaminant, the cleaning agent, excipient and/or other additives can all contribute to TOC. Since the resulting residue is cumulative, this scenario could result in the method failing the acceptance criteria while a specific method to detect and measure each contaminant would pass.

Example: a process train is attempting to clean Active A (PA), Active B (PB) with Cleaning Agent 1 (CA1). PA has a residue limit of 10 ppm, PB has a residue limit of 8 ppm. CA1 has a residue limit of 10 ppm. Using specific methods for each of the compounds, the residue is found to be 6 ppm for PA, 4 ppm for PB and 2 ppm for CA. The acceptance criteria pass. If using TOC or any nonspecific method for that matter the TOC for the residue is determined to be the sum of that for all three contaminants, 12 ppm. Since the worst case acceptance limit is 8 ppm, the potential acceptance criterion fails.

Studies have demonstrated that TOC analysis is particularly suitable for measuring organic residues on stainless steel surfaces and has the ability to detect trace amounts of organic contaminates. Most cleaning processes have the ability to clean well below the dose limits of a product, and TOC analysis is highly sensitive, accurate and easy to perform. It can also be used across a number of analytes. Finally, the FDA recognizes it as an acceptable method for cleaning validation so long as studies have been conducted on the potential contaminants for carbon content (15).

In summary, TOC analysis is a reliable method for cleaning validation. It has excellent linearity and enough precision to detect and measure extremely low levels of carbon (as low as 5 ppb C). Studies have demonstrated that it may be used to detect a variety of active contaminants and cleaning agents (16).

Sampling Techniques

In general, three sampling procedures are routinely utilized for cleaning validation. They include visual inspections, rinse sampling, and swab sampling. Other less utilized sampling procedures include coupon sampling, solvent sampling, placebo sampling, production sampling and in-situ monitoring.

A single sampling procedure is rarely sufficient for all cleaning validation within a facility so a brief discussion of advantages and disadvantages of each will be discussed. The most effective cleaning validation study will utilize a combination of several of the above sampling methods.

Visual Inspection

When visual inspection is used, it should be performed with appropriate lighting and viewed at multiple angles. The dry surface should be visually free of residue when inspected. Wiping the surface with a cloth is a technique that is also used for visual inspections. For this technique, the cloth should be lint free and of a contrasting color of the target residue. After wiping the surface, the cloth should be free of residue. In both of the above cases, detailed procedures of where, how, and under what conditions to perform the wipe and/or inspection should be specified.

Swab Sampling

Swabbing is a widely used sampling method that directly measures the amount of residue left on the surface of equipment. Swab sampling is applicable for the detection of active, microbial and cleaning agent deposits. Swabs can be performed dry or with solvent.

The primary advantages of using the swab sampling method is that it is a direct measurement of residue on the surface and can be utilized to verify the ability of the cleaning process to clean the worst case areas.

One of the disadvantages of using the swab sampling method is that its accuracy is dependent upon the location that the sample is taken and that it is an invasive technique which makes it problematic to fully evaluate complex process trains. (Just a note here: one requirement for the equipment train should be a swab site rationale matrix for each piece of equipment. Swab samples should always be taken from the worst case area from a cleanability standpoint. Additionally, samples should be taken from each contact surface.)

Calculations for determining the total amount of contaminate in the system are as follows:

Equation (5)—Calculation of Total Residue for Swab Sampling

$$R_t = \sum_{i=\text{material type}}^{n} \frac{1}{\text{RF}_i} \times \frac{\text{R}_{\text{S}_i}}{A_{\text{S}_i}} \times A_i \tag{5}$$

where R_t, total residue on contact surface of equipment train; RF_i, recovery factor for material i (the percentage of total residue recovered from material i using the specific sampling technique); R_{S_i}, amount of residue recovered from swabbing for material i; A_{S_i}, sample area swabbed for material i; A_i, total contact surface area of material i in equipment train; i, each contact surface material in equipment train; n, total number of contact surface materials in equipment train.

Typically, swabbing involves using a wipe or swab that is moistened with high purity water (water for injection) or other solvent and wiped over a defined area in a systematic multi-pass way always going from clean to dirty areas to avoid recontamination—i.e., 10 side by side strokes vertically, 10 horizontally and 10 each with the flip side of the swab in each diagonal direction (7).

See the following example of a swabbing procedure:

- Pretreat the swab(s) in the sample solvent, and squeeze the excess solvent from the swab(s) using a clean glove.
- Swab the surface of the tested metal firmly and evenly with one side of the swab(s) in a horizontal direction, and with the other side in a vertical direction back and forth (one stroke back and one stroke forward) to cover the entire area.
- Cut off the handle of the swab into a clean container.
- Use a specific amount of sample solvent (also called recovery solvent or extractable solvent) to extract the drug residue.
- Filter the extracted sample and analyze the sample by the specified analytical method.

Rinse Sampling

Perhaps the most commonly used method of measuring the effectiveness of the cleaning is the rinse solution sampling method (O'Brien and Voss 1999). For this method the product for which one wishes to test must be soluble in the solvent used for the rinse. Also, a residue recovery factor must be established for the product/rinse solution combination.

The primary advantages of sampling the rinse solution are that it is relatively fast and easy to collect samples and may be used to access the residue in areas of the equipment and piping that are comparatively inaccessible.

One of the disadvantages of using rinse solution sampling is the conception that rinse solution sampling has not traditionally been accepted as the sole method of determining cleaning effectiveness by regulatory agencies. The FDA stated in the December 1998 Human Drug CGMP Note: "... for the purposes of cleaning validation, rinse samples alone would not be acceptable unless a direct measurement of the residue or contaminant has been made." Another disadvantage of this method is the fact that rinse solution sampling does not directly measure the cleanliness of the equipment, but rather provides an indication of the amount of extractable dissolvable residue present in the system (if the material is baked on surfaces or located behind obstructions, it may not completely dissolve in a rinse cycle).

The total amount of active contaminate residue is calculated by an equation derived from equation (6) below, which has been expanded to more easily present the concept.

Equation (6)—Calculation of Total Residue for Rinse Sampling

$$R_{\mathrm{t}} = \frac{1}{\mathrm{RF}} \times \frac{R_{\mathrm{s}}}{V_{\mathrm{s}}} \times V_{\mathrm{t}} \tag{6}$$

where R_{t}, total residue; RF, recovery factor (the percentage of total residue recovered during a rinse); R_{s}, residue recovered from rinse sample; V_{s}, volume of rinse sample; V_{t}, total volume of rinse.

In-Situ

In-situ measurement techniques have shown promise of late due to the fact of their ease of use. They are quick, require no sample preparation, are non invasive and are in line with the FDA's position on process analytical technology (PAT).

Two in-situ measurement techniques that are being researched for use in cleaning validation are spectroscopy and TOC measurements.

Mid-infrared (IR) grazing-angle spectroscopy is one of the most sensitive in-situ techniques for measuring low contaminant concentrations on reflective surfaces. One study used a mid-IR spectroscopy method and a grazing-angle sampling fiber optic probe to detect and quantify small amounts (a few $\mu g/cm^2$) of organic material on metal surfaces. Results suggest that it provides a performance advantage over traditional HPLC-swab methods (18).

Another in-situ measurement technique that is applicable but is an indirect measurement of residual containments is the use of in-process TOC to measure the rinse water at the end of the rinse cycle.

Considerations for Cleaning Validation of Single vs. Multi-Products

For multi-product equipment, cleaning and cleaning validation is quite expensive. For a small campaign, it doubles the cost of a product. A one week campaign can result in another week of cleaning, cleaning verification and cleaning validation (17). Some companies have developed fairly innovative ways to manage cleaning new products by not only grouping products by their physical recovery properties, but also grouping equipment trains into logical assembles of use. It is then possible to perform three consecutive batches of a particular product group instead of a specific product.

The degree or level of cleaning and validation required depends largely on whether or not the equipment is dedicated to a single product, the stage of manufacture, and the nature of the potential contaminants (toxicity, solubility, etc.). In general, the higher the potential for finished drug product contamination, the more important it is to validate cleaning procedures to assure product safety.

For cleaning validation programs in multiple product facilities, it is important to create a matrix of the solubility coefficients for the actives to determine the worst case from a cleanability standpoint. Since most cleaning procedures are independent of the material, the worst case active material can be the indicator material for

TABLE 1 Levels of Cleaning Validation

Cleaning level	Situation	Validation
LEVEL 2	Product changeover of equipment used in final step Intermediates of one batch to final step of another	Validation is essential
LEVEL 1	Intermediates or final step of one product to intermediate of another Early step to intermediates in a product sequence	Progression between level between 0 and 2 depending on process and nature of contaminant based on scientific rational
LEVEL 0	In-campaign, batch to batch changeover	No validation required

the validation and the other actives that have higher solubility can be considered qualified. Where there exists separate cleaning procedures for water soluble and water insoluble materials, a worst case active should be chosen for each.

It is desirable to categorize different levels of cleaning requirements based upon the above parameters. Table 1 is taken from the APIC Guide to Cleaning Validation in API plants and suggests three levels of cleaning validation (8).

Using the above model, dedicated equipment for different batches would fall under LEVEL 1 cleaning validation and could utilize visually clean acceptance criterion while multiple products would fall into the LEVEL 2 category and require full cleaning validation.

REFERENCES

1. Human Drug cGMP Notes. (Volume 6, Number 4). U.S. FDA, Center for Drug Evaluation and Research, Office of Compliance, December, 1998. (Accessed July 2007 at www.fda.gov/cder/hdn/cnotesd8.htm)
2. Guide to Inspections Validation of Cleaning Processes. U.S. FDA, Office of Regulatory Affairs. (Accessed July 2007 at www.fda.gov/ora/inspect_ref/igs/valid.html)
3. APIC Cleaning Validation in Active Pharmaceutical Ingredient Manufacturing Plants. APIC "The Active Pharmaceutical Ingredients Committee," a Sector Group within the European Chemical Industry Council, 1999:7. (Accessed July 2007 at www.apic.cefic.org/pub/4CleaningVal9909.pdf)
4. Brunkow R, Delucia D, Haft S, et al. Cleaning and Cleaning Validation: A Biotechnology Perspective. Bethesda, MD: PDA, 1996:108.
5. LeBlanc DA. Is a dirty swab a visually clean failure. Cleaning Memos: Volume 3 January 2003 to December 2003. San Antonio: Cleaning Validation Technologies, 2003.
6. Recommendations on Validation Master Plan, Installation and Operational Qualification, Non-Sterile Process Validation, Cleaning Validation, PI 006-2, Pharmaceutical Inspection Convention, Pharmaceutical Inspection Co-Operation Scheme (PIC/S). Geneva, Switzerland, 2001:21. (Accessed July 2007 at www.picscheme.org/publis/recommandations/PI%200062%20Recommendat_121CF0.pdf)
7. Pharmaceutical Cleaning Validation Method References for Alconox, Inc. Detergents. White Plains, NY: Alconox, Inc. (Accessed July 2007 at www.alconox.com/static/section_top/gen_cleanval.asp)
8. APIC Cleaning Validation in Active Pharmaceutical Ingredient Manufacturing Plants. APIC "The Active Pharmaceutical Ingredients Committee," a Sector Group within the European Chemical Industry Council, 1999:9. (Accessed July 2007 at www.apic.cefic.org/pub/4CleaningVal9909.pdf)
9. LeBlanc DA. Analytical methods for cleaning validation. Validated Cleaning Technologies for Pharmaceutical Manufacturing. Boca Raton, FL: Interpharm/CRC, 2000.
10. Good Manufacturing Practices—Cleaning Validation Guidelines. Health Canada, Health Products and Food Branch Inspectorate, 2001:7. (Accessed July 2007 at www.hc-sc.gc.ca/dhp-mps/alt_formats/hpfb-dgpsa/pdf/compliconform/cleaning-nettoyage_e.pdf)
11. Annex 15 to the EU Guide to Good Manufacturing Practice, Qualification and Validation. European Commission, Enterprise Directorate–General, Working Party on Control of Medicines and Inspections. Brussels, 2001:8–9. (Accessed July 2007 at www.picscheme.org/publis/recommandations/PI%20006-2%20Recommendat_121CF0.pdf)
12. Bismuth G, Neumann S. Cleaning Validation: A Practical Approach. Boca Raton, FL: CRC Press, 2000:54–5.
13. Davis M, Stefanou S, Walia G. Ion mobility spectrometry speeds cleaning validation. Pharmaceutical Processing. Rockaway, NJ: Advantage Business Media, 2003.
14. Melling P, Thomson M, Mehta N, Goenaga-Polo J, Hernández-Rivera S, Hernández D. Development of an in-situ spectroscopic method for cleaning. Spectroscopy 2003; 18(4):13–9. Advanstar Communications, Iselin, NJ.

15. Karen A. Clark, Product Manager, Anatel Corporation. Total Organic Carbon Analysis for Cleaning Validation in Pharmaceutical Manufacturing. Newsletter. Technical Archive. ISPE Boston Chapter, October 2000 (www.ispe.org/boston/articles/arch10-1000.htm)
16. Goenaga-Polo J, et al. Development of an in-situ spectroscopic method for cleaning validation using mid-IR fiber-optics. BioPharm 2002.
17. Mettler HP. Personal correspondence with author. Visp, Switzerland, 2005.
18. O'Brien RW, Voss JM. In: Kenneth A, Carmen W, Vincent W, eds. Cleaning and Validation of Cleaning in Pharmaceutical Processing: A Survey. Boca Raton, FL: Interpharm/CRC Press, 1999:221–50.

BIBLIOGRAPHY

Validation of analytical procedures, ICH-Q2A. In: International Conference on Harmonization (ICH) of Technical Requirements for the Registration of Pharmaceuticals for Human Use. Geneva, 1995.

Recommendations on Validation Master Plan Installation and Operational Qualification Non-Sterile Process Validation 1 July 2004 PI 006-2 Pharmaceutical Inspection Convention Pharmaceutical Inspection Co-Operation Scheme PIC/S, Geneva. (www.picscheme.org/BAK/docs/pdf/PI%20006-2%20Recommendation%20on%20Validation%20Master%20Plan.pdf)

20 International Regulations

Albrecht Killinger
Uhde GmbH, Biotechnology Division, Leipzig, Germany

Joachim Höller
Boehringer Ingelheim Austria GmbH, Vienna, Austria

INTRODUCTION

This chapter outlines the regulative requirements concerning cleaning in the pharmaceutical industry for non-U.S. countries including the European Union (EU) (France, Germany and the U.K.), and Japan.

In general, the regulations for the life cycle of a pharmaceutical drug are based on a body of rules known as Good Manufacturing Practices (GMPs). These rules describe the manufacturing requirements to produce quality pharmaceutical products.

TO UNDERSTAND GMP

The GMP concept is based on the idea that all drug manufacturing activities should lead to a product, meeting its pre-determined specifications and quality attributes. As the pharmaceutical entrepreneur is responsible for specifying his product and as he is the process owner, he is the only one who knows, based on a risk assessment, what is necessary for the intended use. The GMPs therefore contain only performance criteria and rely on the entrepreneur to document in which way and through which technical specifications he and his process achieve the required performance.

Deviations from the pre-determined specifications can occur through contamination of the product from two sources:

- Contamination from the environment due to insufficient cleaning of the rooms and
- Cross-contamination from the process equipment by residues from the previous batch due to insufficient cleaning of the equipment.

Cleanability and cleaning procedures are therefore a central and major concern of the GMP.

HISTORY

The Canadian Specifications Board of the Supply and Services Department issued the first modern code that could be considered as GMPs in 1957. Soothe success of the regulations stimulated the launch of GMP by regulatory agencies at a rapid pace and by the beginning of the 1980s, more than 20 countries had issued their own regulations. These GMPs show an almost uniform world-wide content and style except in Japan where they reworked their rules in 2002. (Prior to that time, the Japanese regulations contained more detailed job descriptions with duties and responsibilities of the staff in a pharmaceutical company.)

Actually, most countries revise and update their GMPs about twice a decade to keep up with changes to manufacturing and testing technology and changes in the industry. Nevertheless, the regulations of the individual countries show localized variations.

REGULATIONS IN THE UNITED STATES

Responsible for the execution of the laws regulation the drug manufacturing is the Food and Drug Administration (FDA), which performs inspections of the production facilities. The homepage (1) gives an overwhelming amount of information about laws, guidelines, interpretations, and finding on inspections.

The legal requirements for the production of a drug are provided by Code of Federal Regulation (CFR) 21 Part 210—Current GMPs in the Manufacturing, Processing, Packing, or Holding of Drugs; General (2) and CFR 21 Part 211—Current GMPs for Finished Pharmaceuticals (Figs. 1–5) (3). These GMPs are the legal basis for manufacturing equipment including washing and cleaning equipment.

Subpart D (Figs. 1 and 2) deals with equipment and furthermore defines requirements for cleaning the equipment used in the production process. It's interesting that the term "clean-in-place (CIP)" isn't mentioned in special and nevertheless, a cleaning in general has to be taken into consideration.

§ 211.63 Equipment design, size, and location.
Equipment used in the manufacture, processing, packing, or holding of a drug product shall be of appropriate design, adequate size, and suitably located to facilitate operations for its intended use and for its cleaning and maintenance.

§ 211.65 Equipment construction.
(a) Equipment shall be constructed so that surfaces that contact components, in-process materials, or drug products shall not be reactive, additive,or absorptive so as to alter the safety, identity, strength, quality, or purity of the drug product beyond the official or other established requirements.
...

§ 211.67 Equipment cleaning and maintenance.
(a) Equipment and utensils shall be cleaned, maintained, and sanitized at appropriate intervals to prevent malfunctions or contamination that would alter the safety, identity, strength, quality, or purity of the drug product beyond the official or other established requirements.
(b) Written procedures shall be established and followed for cleaning and maintenance of equipment, including utensils, used in the manufacture, processing, packing, or holding of a drug product. These procedures shall include, but are not necessarily limited to, the following:
(1) Assignment of responsibility for cleaning and maintaining equipment;
(2) Maintenance and cleaning schedules, including, where appropriate, sanitizing schedules;
(3) A description in sufficient detail of the methods, equipment, and materials used in cleaning and maintenance operations, and the methods of disassembling and reassembling equipment as necessary to assure proper cleaning and maintenance;
...
(c) Records shall be kept of maintenance, cleaning, sanitizing, and inspection ...

FIGURE 1 Excerpt from CFR 21 Part 211, Subpart D—Equipment. *Source*: From Ref. 3.

§ 211.68 Automatic, mechanical, and electronic equipment.

(a) Automatic, mechanical, or electronic equipment ... may be used in the manufacture, processing, packing, and holding of a drug product. If such equipment is so used, it shall be routinely calibrated, inspected, or checked according to a written program designed to assure proper performance. ...

(b) Appropriate controls shall be exercised over computer or related systems to assure that changes in master production and control records or other records are instituted only by authorized personnel....

FIGURE 2 Excerpt from CFR 21 Part 211, Subpart D—Equipment. *Source*: From Ref. 3.

The performance of the CIP process has to be documented to prove a proper cleaning of the equipment used. This documentation represents a part of the whole production process not to be underestimated and the importance of this is underlined by the extensive regulations in Subpart J—Records and Reports (Figs. 4 and 5).

REGULATIONS IN THE EU

European Union (EU) Guideline 2001/83 provides a guideline for the pharmaceutical industry by regulating registration and production of drugs. In Article 41 and 46 (Fig. 6), the first hints are given as to how the manufacturer can obtain a manufacturing license by using suitable and sufficient technical equipment. These legal requirements for equipment are summarized under one topic as guidelines for GMP for medicinal product. This GMP is the legal basis for manufacturing equipment including washing and cleaning equipment.

The EU has revised its GMP regulations in a more detail in the form of the Commission Directive 91/356/EC of 13 June 1991 (5) to cover GMP of investigational medicinal products. Volume 4, Commission Directive 2003/94/EC, of 8th of October 2003, "Principles and guidelines of GMP in respect of medicinal products for human use and investigational medicinal products for human use" and its 18 annexes outline a complete guideline and provide a periodic review to ensure technical and scientific progress. In fact these rules are not real laws in the sense, that a specific punishment will follow any infringement but the EU member countries are required to adopt these rules in their own legal framework.

The GMP covers all aspects of producing drugs including quality control, labeling, etc., but chapter 3 of the 4th Annex to the Commission Directive (Fig. 7) deals with equipment and therefore this chapter includes the fundamental GMP requirement that "The equipment has to fit the intended purpose which has to be phrased as design requirements."

Regulations in Germany

In general, the handling of drugs is regulated by the Arzneimittelgesetz (AMG) (7) which states in section 13.1 that authorization by the responsible authorities is needed to produce drugs to be marketed and delivered to end users which can be denied or revoked as stated in section 14.1 if appropriate facilities are missing or the producer can not ensure production and testing of the drug according to scientific and technological state-of-the-art.

§ 211.100 Written procedures; deviations.

(a) There shall be written procedures for production and process control designed to assure that the drug products have the identity, strength, quality, and purity they purport or are represented to possess. Such procedures shall include all requirements in this subpart. These written procedures, including any changes, shall be drafted, reviewed, and approved by the appropriate organizational units and reviewed and approved by the quality control unit.

(b) Written production and process control procedures shall be followed in the execution of the various production and process control functions and shall be documented at the time of performance. Any deviation from the written procedures shall be recorded and justified.

...

§ 211.105 Equipment identification

(a) All compounding and storage containers, processing lines, and major equipment used during the production of a batch of a drug product shall be properly identified at all times to indicate their contents and, when necessary, the phase of processing of the batch.

(b) Major equipment shall be identified by a distinctive identification number or code that shall be recorded in the batch production record to show the specific equipment used in the manufacture of each batch of a drug product. In cases where only one of a particular type of equipment exists in a manufacturing facility, the name of the equipment may be used in lieu of a distinctive identification number or code.

...

FIGURE 3 Excerpt from CFR 21 Part 211, Subpart F—Production and Process Controls. *Source*: From Ref. 3.

Definitive requirements for cleaning are missing in the AMG as this regulation only provides reasons to deny a production license as the regulations originate from a time, when the focus was on quality control and not on quality management and GMP.

In addition to the AMG the Pharma-Betriebsverordnung (Fig. 8) provides more specific requirements for GMP.

As Germany is an EU member, regulations, the AMG, and the Pharma-Betriebsverordnung are being harmonized with the EU regulations as discussed previously concerning drug production and therefore the topics mentioned in both regulations are now nearly the same and have similar intentions.

By defining qualification as a requisite for the authorization to use equipment, the Pharma-Betriebsverordnung exposes an important aspect previously fixed in some, but not all other GMPs. Nevertheless, qualification is a well-established and obligatory process to obtain written and documented evidence that the equipment fits the intended purpose.

REGULATIONS IN CANADA

The Therapeutic Products Directorate regulates pharmaceutical drugs as required by the Food and Drugs Act (9). Comparable to European regulations, government permission is necessary to manufacture and market drugs. The Health Products and Food Branch is responsible for inspections to verify compliance with GMP and to enforce activities related to drug manufacturing. To ensure a uniform application of these requirements and assist the industry with compliance, the Inspectorate has developed the GMP Guidelines as well as a series of guides and other related documents. To reduce the number of audits to be performed, Canada participates to several Mutual Recognition Agreements (MRAs) of PIC/S GMPs Compliance Programs where collaboration between different countries permits an acceptance of another country's audits.

Compared to other country's regulations, the Canadian GMPs Guidelines (Figs. 9–12) contains more details on equipment and cleaning. It is interesting that in addition to the very short regulation, rationale and interpretations are included. It is explicitly written, that validated CIP equipment should be dismantled for periodic verification of the cleaning success.

Explicitly mentioned are the requirements for documentation of equipment qualification and periodic maintenance and calibration of all sensors and automated control systems involved in the cleaning processes.

The Canadian GMP regulations demand a sanitation program, which includes not only the cleaning procedures but also the cleaning intervals as cleanliness is not forever and has its own life cycle. Moisture, particles and handling of the equipment may contaminate the surfaces. Furthermore, it has to be shown and validated that the cleaning agents to be used are removed by the last steps of the cleaning procedures. As contaminants they are as critical as all the other potential cross contaminations.

At the same time, Canada issued Guidelines to be used for biological drugs (Fig. 13) requiring that equipment has to be cleanable and wherever possible CIP procedures should be used indicating that CIP is recognized by the authorities as a "high quality" cleaning process.

They stress the fact that facilities producing biological drugs by means of recombinant micro-organisms are very often multi-product facility shaving a particular propensity for cross-contamination.

§ 211.180 General requirements.

(a) Any production, control, or distribution record that is required to be maintained in compliance with this part and is specifically associated with a batch of a drug product shall be retained for at least 1 year after the expiration date of the batch or, in the case of certain OTC drug products lacking expiration dating because they meet the criteria … 3 years after distribution of the batch.

…

§ 211.182 Equipment cleaning and use log.

A written record of major equipment cleaning, maintenance (except routine maintenance such as lubrication and adjustments), and use shall be included in individual equipment logs that show the date, time, product, and lot number of each batch processed. If equipment is dedicated to manufacture of one product, then individual equipment logs are not required, provided that lots or batches of such product follow in numerical order and are manufactured in numerical sequence. In cases where dedicated equipment is employed, the records of cleaning, maintenance, and use shall be part of the batch record. The persons performing a double-checking the cleaning and maintenance shall date and sign or initial the log indicating that the work as performed. Entries in the log shall be in chronological order.

…

FIGURE 4 Excerpt from CFR 21 Part 211, Subpart J—Records and Reports. *Source*: From Ref. 3.

§ 211.186 Master production and control records.
(a) To assure uniformity from batch to batch, master production and control records for each drug product, including each batch size thereof, shall be prepared, dated, and signed (full signature, handwritten) by one person and independently checked, dated, and signed by a second person. The preparation of master production and control records shall be described in a written procedure and such written procedures shall be followed.
...

§ 211.188 Batch production and control records.
Batch production and control records shall be prepared for each batch of drug product produced and shall include complete information relating to the production and control of each batch. These records shall include:
(a) An accurate reproduction of the appropriate master production or control record, checked for accuracy, dated, and signed;
(b) Documentation that each significant step in the manufacture, processing, packing, or holding of the batch was accomplished, including:
(1) Dates;
(2) Identity of individual major equipment and lines used;
(3) Specific identification of each batch of component or in-process material used;
(4) Weights and measures of components used int the course of processing;
...
(10) Any sampling performed;
(11) Identification of the persons performing and directly supervising or checking each significant step in the operation;
(12) Any investigation made ...

FIGURE 5 Excerpt from CFR 21 Part 211, Subpart J—Records and Reports. *Source*: From Ref. 3.

REGULATIONS IN JAPAN

Being responsible for the legal aspects of drugs, the Ministry of Health, Labour and Welfare issued GMPs. This ministry is also responsible for granting licenses if the regulations are met. The requirements concerning buildings and facilities of drug manufacturing plants are specified in the "Regulations for Buildings and Facilities for Pharmacies, etc." (12).

Article 41 states:
"In order to obtain the manufacturing authorization, the applicant shall meet at least the following requirements:
...
(b) have at his disposal, for the manufacture or import of the above, suitable and sufficient premises, technical equipment and control facilities complying with the legal requirements ..."
Article 46 states:
"The holder of a manufacturing authorization shall at least be obliged:
...
(f) to comply with the principles and guidelines of good manufacturing practice for medicinal product as laid down by Community law. ..."

FIGURE 6 Excerpt from EU-Guideline 2001/83, Article 41 and 46. *Source*: From Ref. 4.

"...
Premises and equipment must be located, designed, constructed, adapted and maintained to suit the operations to be carried out.
...

3.34 Manufacturing equipment should be designed , located and maintained to suit its intended purpose.
3.35 Repair and maintenance operations should not present any hazard to the quality of the products.
3.36 Manufacturing equipment should be designed so that it can be easily and thoroughly cleaned. It should be cleaned according to detailed and written procedures and stored only in a clean and dry condition.
3.37 Washing and cleaning equipment should be chosen and used in order not to be a source of contamination.
3.38 Equipment should be installed in such a way as to prevent any risk of error or of contamination.
3.39 Production equipment should not present any hazard to the products. The parts of the production equipment that come into contact with the product must not be reactive, additive or absorptive to such an extent that it will affect the quality of the product and thus present any hazard. ..."

FIGURE 7 Excerpt from chapter 3 of Commission Directive 2003/94/EC, 8 October 2003, Volume 4. *Source*: From Ref. 6.

Before reworking their rules in 2002, the Japanese differed significantly from other GMPs by containing detailed job descriptions with duties and responsibilities of the staff in a pharmaceutical company.

The Japanese regulations do not contain requirements, which are more specific or more detailed or more additive than the others discussed above.

OTHERS

In 1967, the World Health Organization (WHO) issued its own set of GMPs. Being a first attempt to provide a basic outline of the minimum steps required to establish

"§ 1 Business and facilities
- have to be according EU GMPs
- have to run a working pharmaceutical quality management system
- have to ensure that the drug has the necessary quality for the intended use

...

§ 3.1 The facility has to
- ensure correct business with proper size, number, location and equipment
- Facility and equipment have to be checked for suitability (Qualification)

...

§ 3.3 The facility and equipment have to be proper cleanable and have to be maintenanced"

FIGURE 8 Excerpt from the Pharma-Betriebsverordnung. *Source*: From Ref. 8.

Equipment

Regulation
C.02.005
The equipment with which a lot or batch of a drug is fabricated,... shall be designed, constructed, maintained,... in a manner that:

(a) permits fective leaning of its surfaces;
(b) prevents the contamination of the drug and the addition of extraneous material to the drug; and
(c) permits it to function in accordance with its intended use.

Rationale
... to prevent the contamination of drugs by other drugs, by dust, and by foreign materials such as rust, lubricant and particles coming form the equipment. ... Equipment arranged in an orderly manner permits cleaning of adjacent areas ...

FIGURE 9 Excerpt from Good Manufacturing Practices Guidelines, 2002 Edition, Version 2 concerning Equipment (Regulation and Rationale). *Source*: From Ref. 10.

acceptable standards for pharmaceutical facilities throughout the world. Nevertheless there is still a global quality in-equilibrium as the level of quality is a function of the economic development of the country.

GUIDELINES OF NON-GOVERNMENTAL ORGANIZATIONS

International Society for Pharmaceutical Engineering

The International Society for Pharmaceutical Engineering (ISPE) (13) is a global, non-governmental organization (NGO) that provides education, training, and

INTERPRETATION

1. The design, construction and locaton of equipment permit cleaning, sanitizing, and inspection of the equipment.
1.1 Equipment parts that come in contact with raw materials, in-process drugs or drugs are accessible to cleaning or are removable.
1.2 Tanks used in processing liquids and ointments are equipped with fittings that con be dismantled and cleaned. Validated Clean-In-Place (CIP) equipment can be dismantled for periodic verification. ...
2. Equipment does not add extraneous material to the drug. ...
3. Equipment is operated in a manner that prevents contamination. ...
4. Equipment is maintained in a good state of repair when in use. ...
5. Production equipment is designed, located, and maintained to serve its intended purpose. ...
5.3 ... Equipment qualification is documented.
5.4. Automatic, mechanical, electronic, or other types of equipment including computerized systems that are used in the fabrication, packaging/labeling, and storing of a drug is routinely calibrated, inspected or checked according to a written program designed to assure proper performance. Written records of these calibration checks and inspections are maintained. ...

FIGURE 10 Excerpt from Good Manufacturing Practices Guidelines, 2002 Edition, Version 2 concerning Equipment (Interpretation). *Source*: From Ref. 10.

Sanitation

Regulation
C.02.007
...
(2) The sanitation program ... shall include:
(a) cleaning procedures ... for the equipment used in the fabrication... of the drug; ...

Rationale
... The quality requirement for drug products demand that such products be fabricated and packaged in areas that are free from environmental contamination and free from contamination by another drug. ...

FIGURE 11 Excerpt from Good Manufacturing Practices Guidelines, 2002 Edition, Version 2 concerning Sanitation (Regulation and Rationale). *Source*: From Ref. 10.

technical publications to pharmaceutical manufacturing professionals. In working parties, members of the pharmaceutical industry discuss their experience and exchange their knowledge, thus establishing the latest technological and regulatory trends. Although not a governmental organization and not authorized to issue official regulations and interpretations, ISPE offers a means to reduce discussions by establishing a baseline of standardization. ISPE Baseline© Pharmaceutical Engineering Guides (14) includes a discussion of CIP design and procedures as well as qualification of production plants. CIP Cleaning is specifically mentioned in the following volumes:

- Volume 1—Bulk Pharmaceutical Chemicals
- Volume 3—Sterile Manufacturing Facilities
- Volume 5—Commissioning and Qualification
- Volume 6—Biopharmaceuticals

Pharmaceutical Inspection Convention

In the late 1980s, some northern European countries agreed to accept inspections performed by the authorities of other countries to save time and reduce their own

Interpretation
...
2. The sanitation program contains procedures that outline the following: ...
2.2 cleaning requirements applicable to processing equipment;
2.3 cleaning intervals;
2.4 products for cleaning and disinfection, along with their dilution and the equipment or be used; ...
3. The sanitation program is implemented and is effective in preventing unsanitary conditions.
3.1 Cleaning procedures for manufacturing equipment are validated based on the Cleaning Validation Guidelines.
3.2 Residues from the cleaning process itself (e.g., detergents, solvents, etc.) are also removed from equipment; ...

FIGURE 12 Excerpt from Good Manufacturing Practices Guidelines, 2002 Edition, Version 2 concerning Sanitation (Interpretation). *Source*: From Ref. 10.

...
C.02.004
...
4. Pipework systems, including valves, pumps, vent filters and housings, that come into contact with final product, or with material used in final product, or with material that contacts surfaces which contact final product, must be designed to facilitate cleaning ... Where possible, 'clean-in-place' (CIP)...systems should be used.
...
C.02.007 and C.02.008
...
3. Pipework systems, valves and vent filters are designed to facilitate cleaning ... Where possible CIP and SIP systems are used.
4. Equipment cleaning processes are designed to remove endotoxins, bacteria, toxic elements and residual contaminating proteins and/or other identified contaminants.
5. Cleaning validation of equipment is critical of a multi-product facility involved in campaign or concurrent production of biological drugs...

FIGURE 13 Excerpt from the Guideline GMP for Schedule D Drugs, Part 1, Biological Drugs. *Source*: From Ref. 11.

inspection efforts and to eliminate costly re-inspections. Acceptance of inspections performed by foreign authorities needs the same intensity and inspection philosophy in both or all countries. The countries of the European Free Trade Association founded "The Convention for the Mutual Recognition of Inspections in Respect of the Manufacture of Pharmaceutical Products" shortened to the Pharmaceutical Inspection Convention (PIC) in 1970. The original goals were uniform inspection systems, harmonization of the GMP requirements, mutual recognition of inspections, training of inspectors, mutual confidence and exchange of information. Later, the formal inter-governmental convention was changed into an association of national health authorities, the Pharmaceutical Inspection Cooperation Scheme (PIC/S). Due to the success in achieving these goals, the PIC has expanded and the present member countries include the EU members, Australia, Canada, Malaysia, and Singapore.

There have also been discussions between the U.S. and the EU on enacting a "MRA." The Mutual Inspection Agreement (MRA) approach is an effective way to enhance international regulatory cooperation and maintain high standards of product safety and quality while reducing the regulatory burden on manufacturers. An FDA study recognized the GMP inspection standards of most of the EU countries as acceptable except for those of Southern Europe. Unfortunately, the EU considers its political unity more important than common quality standards and took the position that the FDA should accept either everyone or no one. Despite the effort already invested, the U.S. authorities have chosen not to accept the MRA on the basis of all or none.

Despite this present disagreement, the FDA and WHO maintain observer status in the PIC/S leaving the option of a future agreement open, because the MRA is the largest and most universally.

International Conference on Harmonization

International Conference on Harmonization (ICH) is another project organized to achieve greater harmonization in the interpretation and application of technical guidelines and requirements for product registration. The ICH was founded in 1990. The "ICH of Technical Requirements for Registration of Pharmaceuticals for Human Use" brings together experts from the pharmaceutical industry and regulatory authorities of Europe, Japan, and the U.S. Using a defined procedure the topics of general interest are discussed by experts and finally adopted by the health authorities of the EU, Japan, and the U.S.

"ICH Q7A GMP Guide for Active Pharmaceutical Ingredients (APIs)" (Figs. 14 and 15) representing GMPs for the production of APIs has been agreed and published, and has achieved some degree of status in the member countries. When designing and a facility for manufacturing APIs this GMP is relevant and gets executed.

Like other GMPs, these guideline mentions topics discussed before including designing for intended use, cleaning, maintenance, etc., and therefore will not be discussed again. But the ICH guidelines contain some interesting additions:

- An operating range of the equipment should be defined and qualified. To ensure proper function, the limits of the operating range should be written down in user requirements and tested in the entire operating range. Of course, the operating range has to be within the process range of the equipment.
- Processing lines, equipment should be identified to prevent mix up. This can be achieved by labeling.
- To visualize the present state of the equipment and installations a set of current drawings should be available. This is important for maintenance, service and repair.
- Furthermore, the equipment should be identified as to its cleanliness. This can be achieved by removable labels or by electronic status management. The aim is to prevent using unclean contaminated equipment for manufacturing.
- Qualification is mentioned once more as obligatory (Figs. 14–15).

...

5.10 Equipment used in the manufacture of intermediates and APIs should be of appropriate design and adequate size, and suitably located for its intended use, cleaning, sanitization (where appropriate), and maintenance.

5.11 Equipment should be constructed so that surfaces that contact raw materials, intermediates, or APIs do not alter the quality of the intermediates and APIs beyond the official or other established specifications.

5.12 Production equipment should only be used within its qualified operating range.

5.13 Major equipment (e.g., reactors, storage containers) and permanently installed processing lines used during the production of an intermediate or API should be appropriately identified.

...

5.15 Closed or contained equipment should be used whenever appropriate. Where open equipment is used, or equipment is opened, appropriate precautions should be taken to minimize the risk of contamination.

5.16 A set of current drawings should be maintained for equipment and critical installations (e.g., instrumentation and utility systems). ...

FIGURE 14 Excerpt from ICH Q7A Good Manufacturing Practice Guide for Active Pharmaceutical Ingredients. *Source*: From Ref. 15.

5.20 Schedules and procedures (including assignment of responsibility) should be established for the preventative maintenance of equipment.

5.21 Written procedures should be established for cleaning of equipment and its subsequent release for use in the manufacture of intermediates and APIs. Cleaning procedures should contain sufficient details to enable operators to clean each type of equipment in a reproducible and effective manner.
These procedures should include:
- Assignment of responsibility for cleaning of equipment;
- Cleaning schedules, including, where appropriate, sanitizing schedules
- A complete description of the methods and materials, including dilution of cleaning agents used to clean equipment;...

5.25 Acceptance criteria for residues and the choice of cleaning procedures and cleaning agents should be defined and justified.

5.26 Equipment should be identified as to its contents and its cleanliness status by appropriate means....

5.30 Control, weighing, measuring, monitoring and test equipment that is critical for assuring the quality of intermediates or APIs should be calibrated according to written procedures and an established schedule. ...

12.3 Before starting process validation activities, appropriate qualification of equipment and ancillary systems should be completed. Qualification is usually carried out by conducting the following activities, individually or combined: ...

FIGURE 15 Excerpt from ICH Q7A Good Manufacturing Practice Guide for Active Pharmaceutical Ingredients. *Source*: From Ref. 15.

HOW TO COMPLY WITH THE REGULATIONS

Europe, Japan, Canada, and the U.S. represent the highest current standard of pharmaceutical regulations. As we can see, all the regulations are more or less similar. Nevertheless, the performance criteria are comprehensively described but do not contain specific technical specifications to ensure the entrepreneur flexibility in complying with the rules.

The fact is that the way to comply with these regulations is open, if the process owner can defend his approach. Competent consulting and engineering companies, non-governmental auditors, official audits, and NGOs like ISPE together provide worldwide think tank where ideas, solutions and arguments circulate and get discussed. Together they represent the technical and scientific state-of-the-art. Nevertheless, the opinions and solutions to a specific problem differ and the discussions may get philosophical and abstract.

At the end of the day, the owner has to decide what to do, how to do it and to prove, why he did it in that way. If he has another solution for a specific problem that leads to the same results and he has a defendable rationale, it's his decision to go that way.

PUTTING THE REGULATIONS INTO PRACTICE

Finally we arrive at a summary of regulations and criteria to enable a detailed discussion of how these regulations can specifically being applied during design and operation of CIP equipment. This summary is useful because, due to the ongoing efforts in harmonization and due to the tendency for mutual agreement of foreign inspections, the current different regulations become more and more similar.

The aim is to achieve a reproducible cleaning process by:

- Providing written procedures and definition of all the steps, parameters, limits, and responsibilities.
- Defining the responsibility for cleaning procedures. This is an organizational action to place this important step in the hands of trained and responsible staff.
- Wherever possible, Cleaning in Place procedures and CIP systems should be used. CIP is the preferable cleaning process, because it minimizes manual handling and it can be highly automated. Furthermore, they can facilitate documentation of cleaning using electronic batch recording (CFR 21 Part 11 compliant in the U.S.).
- The equipment has to be maintained to preserve the qualified status. Planning this maintenance process including replacing gaskets and membranes, which may have deteriorated and show surface cracks. Recalibrating measuring instruments, etc., ensures the proper control of the system within the defined and qualified limits.
- A set of current drawings should be available to visualize the present state of the equipment and installations. This is important, for example, for maintenance, service, repair, and for conception of new production processes.
- Automatic and computerized systems have to be routinely calibrated and maintained. Remember that these systems require a high degree of maintenance to safeguard proper operation. CIP systems normally are automated and operated by computerized process control systems. The routine calibration of measuring instruments should be managed and documented.

Qualification (often called equipment validation) of the system should be performed. The FDA defines validation as following: "Establishing documented evidence, which provides a high degree of assurance that a specific process will consistently produce a product meeting its pre-determined specifications and quality attributes" (16). Qualification is an established process consisting of four steps (Design qualification, Installation qualification, Operation qualification, and Performance qualification). These steps prove in a documented and formal manner that the equipment is designed to suit its pre-defined intended use as discussed before. Attention should be laid on the wording "a high degree." To claim an absolute assurance would make qualification not only extremely expensive but sometimes impossible without getting better results when manufacturing the product. A pragmatic risk analysis shows the critical parameters and items concerning the manufacturing process and gives a good guide to rank the critically and to select which systems have to be qualified. Qualification itself is a very comprehensive topic of filling books on its own and therefore will not be discussed here in detail.

The process equipment, whether it is cleaned manually or by CIP-procedures, has to be cleanable.

- Equipment, pipework systems, valves ... should be designed to facilitate cleaning. Hygienic design is the magic word to achieve a proper CIP of equipment. Hygienic design is an extensive topic consisting of some basic rules. Level surfaces, no dead ends, completely drainable equipment and slopes to draining points in piping, no crevices, minimizing installations in vessels, etc., makes effective cleaning possible. Furthermore CIP equipment has to be designed with return lines on the low points of the equipment. Pipes should be designed to achieve turbulence during the CIP step. Bigger equipment like

TABLE 1 Directives of the European Union

Directive 90/219 EEC on the contained use of genetically modified micro-organisms
Directive 90/220 on the deliberate release of genetically modified micro-organisms
Directive 90/679 on the protection of workers from risks related to exposure to biological agents at work

vessels, centrifuges should be equipped with spray balls to rinse the entire inner surface. A lot more design criteria could be mentioned but other chapters of this book deals with this topic in detail.

- The surface of the production equipment must not affect the quality of the product. The material has to be a chemical resistant against process and cleaning agents. Stainless steels are first choice. If not applicable plastics are an alternative. Authorities give good advice that a material doesn't affect the product quality.
- Cleanability of process equipment will be verified through cleaning validation.

In case different equipment can be connected with the same CIP circuit, the equipment has to be appropriately identified. Preventing mix up of equipment can be achieved by manually or automatically labeling the equipment. Because most facilities have to manage a lot of equipment, a system to create an individual number for each item should be established.

The cleanliness status of the equipment has to be identified. The aim is to prevent using contaminated equipment for manufacturing. This can be achieved by removable labels or by electronic status management if using automated control systems.

Closed or contained equipment should be used for cleaned-in place process as well as for CIP equipment. This is important to prevent contamination from the environment. When designing cleaned-in place process equipment fix installed spray balls, closed CIP supply and return loops are a few interesting possibilities.

The above discussion includes the most important points to be considered for design and maintenance of a CIP system to fulfill the GMP requirements and to comply with the performance criteria as set in the GMP.

EUROPEAN STANDARDS FOR BIOTECHNOLOGY

The European Commission mandated the European Committee for Standardization Comité Européen de Normalisation (CEN) to develop standards, which support the existing and future legislation and specifically those directives, which are concerned directly with Biotechnology (17) (Table 1). Furthermore, they should facilitate and guarantee compliance with the legal requirements at technical level. The standards should define in concrete terms the technical specifications, codes, methods of analysis, etc., and such establish fundamentals for the basic requirements of the directives for safety, environmental and workers protection (18).

TABLE 2 Working Groups of European Institute for Normation (CEN)

WG 1	Research, development and microbial analysis in laboratories
WG 2	Large scale process and production
WG 3	Modified organisms for application in the environment
WG 4	Guidance on testing procedures
	Performace criteria for equipment

TABLE 3 Standards for Equipment in Biotechnology

EN 12296	Biotechnology—Equipment—Guidance on testing procedures for cleanability
EN 12297	Biotechnology—Equipment—Guidance on testing procedures for sterilizability
EN 12298	Biotechnology—Equipment—Guidance on testing procedures for leak tightness
EN 12347	Biotechnology—Performance criteria for steam sterilizers and autoclaves
EN 12462	Biotechnology—Performance criteria for pumps
EN 12469	Biotechnology—Performance criteria for microbiological safety cabinets
EN 12690	Biotechnology—Performance criteria for shaft seals
EN 12884	Biotechnology—Performance criteria for centrifuges
EN 12885	Biotechnology—Performance criteria for cell disrupters
EN 13091	Biotechnology—Performance criteria for filter elements and filtration assemblies
EN 13092	Biotechnology—Guidance on sampling and inoculation procedures
EN 13095	Biotechnology—Performance criteria for off-gas systems
prEN 13311-1	Biotechnology—Performance criteria for vessels—Part 1: General performance criteria (Draft)
prEN 13311-4	Biotechnology—Performance criteria for vessels—Part 4: Bioreactors (Draft)
prEN 13312-1	Biotechnology—Performance criteria for piping and instrumentation—Part 1: General performance criteria (Draft)
prEN 13312-3	Biotechnology—Performance criteria for piping and instrumentation—Part 1: Sampling and inoculation devices (Draft)
prEN 13312-4	Biotechnology—Performance criteria for piping and instrumentation—Part 1: Tubes and pipes (Draft)

Four major areas have been foreseen by the Commission to be followed (Table 2). As they serve the safe use of micro-organisms with different hazardous potential, they contain criteria and recommendations for the safe handling of micro-organisms and requirements for the equipment. In the context of this book, the results of working party four are of specific interest, to be described in more detail (Table 3).

The technical committee CEN/TC 233 Biotechnology identified three performance criteria (Table 4) to secure the safety of equipment used with hazardous micro-organisms. Leak tightness during operation and sterilization prior to opening should prevent any contact of operators and the environment with hazardous micro-organisms. Cleanability is considered as a prerequisite for sterilization as deposits or soil in the equipment could jeopardize the sterilization procedure. It is further stated that cleaning procedures are intended to remove and inactivate micro-organisms to make the equipment safe for handling without using any other sterilization or inactivation procedure (17).

TABLE 4 Performance Classes

- Leak tightness
- Cleanability
- Sterilisability

TABLE 5 Cleanability Performance

Cl-A	Visible soil or cleanliness not defined
Cl-B	Cleanablility tested and quantified under defined conditions or designed with regard to specified technical criteria
Cl-C	Cleanablility tested and quantified under defined conditions and soil below detection limit or threshold value

TABLE 6 Selection of Performance Classes for a Pump

To be classified as Type I, a pump should comply with the performance criteria Cl-A, SI-C, and LI-B. This type may be used together with organisms of Group 2 according Article 2 of EU directive 90/679/EWG (20)
To be classified as Type II, a pump should comply with the performance criteria Cl-B, SI-C, and LI-C, and has to be cleaned in place and sterilized in place. This type may be used together with organisms of Group 3 and 4 according Article 2 of EU directive 90/679/EEC

Level B is specifically remarkable, because a mixture of performance criteria, e.g., cleanability tested and quantified under defined conditions or designed with regard to specific technical conditions (Table 5). Alternatively to any test, the equipment can be classified Cl-B, if it is designed according to EN 1672-2 (19) and the surface roughness of the metallic material applied in the equipment is less than or equal to Ra = 1.6 μm (in other standards later reduced to Ra = 0.8 μm). An appropriate documentation should be delivered with the equipment, containing drawings, how well the design complies with the cited standards.

Since this roughness value has been published in 1991 (20), it is under controversial discussion, whether one roughness value is enough and significant to describe the cleanability of a surface or whether a "mixed value" could give more information and better classification, including roughness, topographic parameters, material parameters, etc. (21).

After all performance criteria have been defined, the equipment can be classified and in some cases the use of the performance classes to choose the appropriate machinery for the use together with hazardous organisms is described. An example is given below for a pump, which can be classified into two categories (Table 6) (22).

For type II, the use of CIP cleaning methods is mandatory to protect workers and the environment.

REFERENCES

1. www.fda.gov
2. CFR 21 Part 210—Current Good Manufacturing Practice in the Manufacturing, Processing, Packing, or Holding of Drugs; General. Washington, DC: Governmental Printing Office (current issue).
3. CFR 21 Part 211—Current Good Manufacturing Practice for Finished Pharmaceuticals. Washington, DC: Governmental Printing Office (current issue).
4. Directive 2001/83/EC of the European Parliament and of the Council of 6 November 2001 on the Community Code Relating to Medicinal Products for Human Use. Brussels: European Parliament and the Council of the European Union, 2001.
5. Commission Directive 91/356/EEC of June 1991 Laying Down the Principles and Guidelines of Good Manufacturing Practice for Medicinal Products for Human Use. Brussels: Commission of the European Communities, 1991.

6. Commission Directive 2003/94/EC of 8 October 2003 Laying down the Principles and Guidelines of Good Manufacturing Practice in Respect of Medicinal Products for Human Use and Investigational Medicinal Products for Human Use. Vol. 4. Brussels: Commission of the European Communities, 2003.
7. Gesetz über den Verkehr mit Arzneimitteln (Arzneimittelgesetz AMG); BGBl I 1976, 2445, 2448; Stand: Neugefasst durch Bek. v. 11.12.1998 I 3586; zuletzt geändert durch Art. 17 G v. 21. 6.2005 I 1818, Bonn, 1998.
8. Betriebsverordnung für pharmazeutische Unternehmer (PharmBetrV).
9. Food and Drugs Act. Ottawa: Department of Justice Canada, 2004.
10. Manufacturing Practices Guidelines. 2002nd ed. Version 2. Canadian Health Products and Food Branch.
11. GMP for Schedule D Drugs, Part 1, Biological Drugs; Canadian Health Products and Food Branch.
12. Regulations for Buildings and Facilities for Pharmacies etc. (MHW Ministerial Ordinance No. 2 dated February 1 1961; Amended: MHLW Ministerial Ordinance No. 92 dated May 20, 2003); Tokyo, 2003.
13. Accessed August, 2005 at www.ispe.org
14. ISPE, ISPE Baseline (Pharmaceutical Engineering Guides for New and Renovated Facilities. Vols. 1, 3, 5, 6. ISPE (current edition).
15. ICH Q7A Good Manufacturing Practice for Active Pharmaceutical Ingredients, CPMP/ICH/4106/00. London: EMEA, 2000.
16. Guideline on General Principles of Process Validation. Rockville, MD: FDA, Center for Drug Evaluation and Research, dated May 1987 (reprinted February 1993).
17. Council Directive 90/679/EEC of 26 November 1990 on the protection of workers from risks related to exposure to biological agents at work. No. L374. OJEC 31.12.1990:1.
18. Berg B. DIN-Taschenbuch Bd. 308 Biotechnik, 2001.
19. EN 1672-2: Food processing machinery—Safety and hygiene requirements, Part 2; Hygiene requirements.
20. DECHEMA. Standardisierungs- und Ausrüstungsempfehlungen für Bioreaktoren und periphere Einrichtungen. DECHEMA: Frankfurt/Main, 1991.
21. Kohler W. Materialoberfläche und Reinigbarkeit. In: Kohler W, Schmidt R, eds. Materialoberflächen in der Reinstraum- und Steriltechnik. Frankfurt/Main: VDMA Verlag, 2005:25–34.
22. EN ISO 12462 Biotechnology—Performance criteria for pumps.

Index